Process Control, Intensification, and Digitalisation in Continuous Biomanufacturing

Process Control, Intensification, and Digitalisation in Continuous Biomanufacturing

Edited by Ganapathy Subramanian

Editor

Dr. Ganapathy Subramanian
44 Oaken Grove
SL6 6HH Maidenhead, Berkshire
United Kingdom

Cover Image: © icestylecg/Getty Images

All books published by **WILEY-VCH** are carefully produced. Nevertheless, authors, editors, and publisher do not warrant the information contained in these books, including this book, to be free of errors. Readers are advised to keep in mind that statements, data, illustrations, procedural details or other items may inadvertently be inaccurate.

Library of Congress Card No.: applied for

British Library Cataloguing-in-Publication Data
A catalogue record for this book is available from the British Library.

Bibliographic information published by the Deutsche Nationalbibliothek
The Deutsche Nationalbibliothek lists this publication in the Deutsche Nationalbibliografie; detailed bibliographic data are available on the Internet at <http://dnb.d-nb.de>.

© 2022 WILEY-VCH GmbH, Boschstr. 12, 69469 Weinheim, Germany

All rights reserved (including those of translation into other languages). No part of this book may be reproduced in any form – by photoprinting, microfilm, or any other means – nor transmitted or translated into a machine language without written permission from the publishers. Registered names, trademarks, etc. used in this book, even when not specifically marked as such, are not to be considered unprotected by law.

Print ISBN: 978-3-527-34769-8
ePDF ISBN: 978-3-527-82732-9
ePub ISBN: 978-3-527-82733-6
oBook ISBN: 978-3-527-82734-3

Typesetting   Straive, Chennai, India

10 9 8 7 6 5 4 3 2 1

Contents

Preface *xiii*

Part I Continuous Biomanufacturing *1*

1 Strategies for Continuous Processing in Microbial Systems *3*
Julian Kopp, Christoph Slouka, Frank Delvigne, and Christoph Herwig
1.1 Introduction *3*
1.1.1 Microbial Hosts and Their Applications in Biotechnology *3*
1.1.2 Regulatory Demands for Their Applied Cultivation Mode *5*
1.2 Overview of Applied Cultivation Methods in Industrial Biotechnology *6*
1.2.1 Batch and Fed-Batch Cultivations *7*
1.2.1.1 Conventional Approaches and Their Technical Limitations *7*
1.2.1.2 Feeding and Control Strategies Using *E. coli* as a Model Organism *8*
1.2.2 Introduction into Microbial Continuous Biomanufacturing (CBM) *9*
1.2.2.1 General Considerations *9*
1.2.2.2 Mass Balancing and the Macroscopic Effects in Chemostat Cultures *11*
1.2.3 Microbial CBM vs. Mammalian CBM *13*
1.2.3.1 Differences in Upstream of Microbial CBM Compared with Cell Culture *13*
1.2.3.2 Downstream in Microbial CBM *14*
1.3 Monitoring and Control Strategies to Enable CBM with Microbials *16*
1.3.1 Subpopulation Monitoring and Possible PAT Tools Applicable for Microbial CBM *16*
1.3.2 Modeling and Control Strategies to Enable CBM with Microbials *19*
1.4 Chances and Drawbacks in Continuous Biomanufacturing with *E. coli* *21*
1.4.1 Optimization of Plant Usage Using CBM with *E. coli* *21*
1.4.2 Reasons Why CBM with *E. coli* Is Not State of the Art (Yet) *23*
1.4.2.1 Formation of Subpopulation Following Genotypic Diversification *23*
1.4.2.2 Formation of Subpopulation Following Phenotypic Diversification *25*

1.4.2.3	Is Genomic Integration of the Target Protein an Enabler for CBM with *E. coli*? *26*
1.4.3	Solutions to Overcome the Formation of Subpopulations and How to Realize CBM with *E. coli* in the Future *27*
1.5	Conclusion and Outlook *29*
	References *30*

2 Control of Continuous Manufacturing Processes for Production of Monoclonal Antibodies *39*
Anurag S. Rathore, Garima Thakur, Saxena Nikita, and Shantanu Banerjee

2.1	Introduction *39*
2.2	Control of Upstream Mammalian Bioreactor for Continuous Production of mAbs *40*
2.3	Integration Between Upstream and Downstream in Continuous Production of mAbs *46*
2.3.1	Continuous Clarification as a Bridge Between Continuous Upstream and Downstream *46*
2.3.2	Considerations for Process Integration *48*
2.4	Control of Continuous Downstream Unit Operations in mAb Manufacturing *49*
2.4.1	Control of Continuous Dead-End Filtration *49*
2.4.2	Control of Continuous Chromatography *50*
2.4.3	Control of Continuous Viral Inactivation *53*
2.4.4	Control of Continuous Precipitation *54*
2.4.5	Control of Continuous Formulation *56*
2.5	Integration Between Adjacent Unit Operations Using Surge Tanks *57*
2.6	Emerging Approaches for High-Level Monitoring and Control of Continuous Bioprocesses *59*
2.6.1	Artificial Intelligence (AI) and Machine Learning (ML) Control *60*
2.6.2	Statistical Process Control *61*
2.6.3	Process Digitalization *62*
2.7	Conclusions *63*
	References *63*

3 Artificial Intelligence and the Control of Continuous Manufacturing *75*
Steven S. Kuwahara

3.1	Introduction *75*
3.2	Continuous Monitoring and Validation *84*
3.3	Choosing Other Control Charts *84*
3.4	Information Awareness *85*
3.5	Management and Personnel *86*
	References *90*

Part II Intensified Biomanufacturing 93

4   Bioprocess Intensification: Technologies and Goals   95
William G. Whitford
4.1   Introduction   95
4.2   Bioprocess Intensification   98
4.2.1   Definition   98
4.2.2   New Directions   100
4.2.3   Sustainability Synergy   102
4.3   Intensification Techniques   103
4.3.1   Enterprise Resource Management   103
4.3.2   Synthetic Biology and Genetic Engineering   104
4.3.3   New Expression Systems   105
4.3.4   Bioprocess Optimization   106
4.3.5   Bioprocess Simplification   107
4.3.6   Continuous Bioprocessing   108
4.4   Materials   109
4.4.1   Media Optimization   109
4.4.2   Variability   110
4.5   Digital Biomanufacturing   110
4.5.1   Data   111
4.5.2   Bioprocess Control   112
4.5.3   Digital Twins   113
4.5.4   Artificial Intelligence   114
4.5.5   Cloud/Edge Computing   114
4.6   Bioprocess Modeling   114
4.7   Automation and Autonomation   115
4.8   Bioprocess Monitoring   117
4.9   Improved Process and Product Development   118
4.9.1   Design of Experiments   118
4.9.2   QbD and PAT   119
4.9.3   High-Throughput Systems   119
4.9.4   Methods   120
4.9.5   Commercialized Systems   120
4.10   Advanced Process Control   121
4.11   Bioreactor Design   121
4.12   Single-Use Systems   122
4.13   Facilities   123
4.14   Conclusion   126
         Abbreviations and Acronyms   126
         Acknowledgment   129
         References   129

5		**Process Intensification Based on Disposable Solutions as First Step Toward Continuous Processing** *137*
		Stefan R. Schmidt
5.1		Introduction *137*
5.1.1		Theory and Practice of Process Intensification *137*
5.1.2		Current Bioprocessing *140*
5.1.3		General Aspects of Disposables *140*
5.2		Technical Solutions *141*
5.2.1		Process Development *141*
5.2.2		Upstream Processing Unit Operations *142*
5.2.2.1		High-Density, Large-Volume Cell Banking in Bags *143*
5.2.2.2		Seed Train Intensification *144*
5.2.2.3		Cell Retention and Harvest *145*
5.2.3		Downstream Processing Unit Operations *149*
5.2.3.1		Depth Filtration *149*
5.2.3.2		In-line Virus Inactivation *151*
5.2.3.3		In-line Buffer Blending and Dilution *152*
5.2.3.4		Chromatography *153*
5.2.3.5		Tangential Flow Filtration *159*
5.2.3.6		Drug Substance Freezing *161*
5.3		Process Analytical Technology and Sensors *162*
5.3.1		Sensors for USP Applications *163*
5.3.2		Sensors for DSP Applications *164*
5.4		Conclusions *165*
5.4.1		Transition from Traditional to Intensified Processes *165*
5.4.2		Impact on Cost *169*
5.4.3		Influence on Time *170*
		References *171*
6		**Single-Use Continuous Manufacturing and Process Intensification for Production of Affordable Biological Drugs** *179*
		Ashish K. Joshi and Sanjeev K. Gupta
6.1		Background *179*
6.2		State of Upstream and Downstream Processes *180*
6.2.1		Sizing Upstream Process *181*
6.2.2		Sizing Downstream Process *182*
6.2.3		Continuous Process Retrofit into the Existing Facility *184*
6.2.3.1		Upstream Process *184*
6.2.3.2		Downstream Process *184*
6.2.4		Learning from Chemical Industry *185*
6.3		Cell Line Development and Manufacturing Role *186*
6.3.1		Speeding Up Upstream and Downstream Development *188*
6.3.2		The State of Manufacturing *189*
6.4		Process Integration and Intensification *190*

6.4.1	Intensification of a Multiproduct Perfusion Platform	*190*
6.4.2	Upstream Process Intensification Using Perfusion Process	*192*
6.5	Process Intensification and Integration in Continuous Manufacturing	*192*
6.6	Single-Use Manufacturing to Maximize Efficiency	*194*
6.6.1	The Benefits of SUT in the New Era of Biomanufacturing	*195*
6.6.2	Managing an SUT Cost Profile	*195*
6.6.3	In-Line Conditioning (ILC)	*196*
6.6.4	Impact of Single-Use Strategy on Manufacturing Cost of Goods	*197*
6.6.5	Limitations of SUT	*198*
6.7	Process Economy	*199*
6.7.1	Biopharma Market Dynamics	*200*
6.7.2	Management of the Key Risks of a Budding Market	*201*
6.8	Future Perspective	*202*
	References	*203*

Part III Digital Biomanufacturing *209*

7 Process Intensification and Industry 4.0: Mutually Enabling Trends *211*
Marc Bisschops and Loe Cameron

7.1	Introduction	*211*
7.2	Enabling Technologies for Process Intensification	*213*
7.2.1	Process Intensification in Biomanufacturing	*213*
7.2.2	Process Intensification in Cell Culture	*214*
7.2.3	Process Intensification in Downstream Processing	*214*
7.2.4	Process Integration: Manufacturing Platforms	*216*
7.2.5	The Two Elephants in the (Clean) Room	*217*
7.3	Digital Opportunities in Process Development	*220*
7.4	Digital Opportunities in Manufacturing	*222*
7.5	Digital Opportunities in Quality Assurance	*223*
7.6	Considerations	*224*
7.6.1	Challenges	*224*
7.6.2	Gene Therapy	*226*
7.7	Conclusions	*227*
	References	*227*

8 Consistent Value Creation from Bioprocess Data with Customized Algorithms: Opportunities Beyond Multivariate Analysis *231*
Harini Narayanan, Moritz von Stosch, Martin F. Luna, M.N. Cruz Bournazou, Alessandro Buttè, and Michael Sokolov

8.1	Motivation	*231*
8.2	Modeling of Process Dynamics	*232*

8.2.1	Hybrid Models	*234*
8.2.2	Conclusion	*238*
8.3	Predictive Models for Critical Quality Attributes	*238*
8.3.1	Historical Product Quality Prediction	*238*
8.3.2	Synergistic Prediction of Process and Product Quality	*242*
8.4	Extrapolation and Process Optimization	*242*
8.5	Bioprocess Monitoring Using Soft Sensors	*247*
8.5.1	Static Soft Sensor	*248*
8.5.2	Dynamic Soft Sensors	*250*
8.5.3	Concluding Remarks	*251*
8.6	Scale-Up and Scale-Down	*251*
8.6.1	Differences Between Lab and Manufacturing Scales	*252*
8.6.2	Scale-Up	*253*
8.6.3	Scale-Down	*254*
8.6.4	Conclusions	*255*
8.7	Digitalization as an Enabler for Continuous Manufacturing	*255*
	References	*257*

9 Digital Twins for Continuous Biologics Manufacturing *265*
Axel Schmidt, Steffen Zobel-Roos, Heribert Helgers, Lara Lohmann, Florian Vetter, Christoph Jensch, Alex Juckers, and Jochen Strube

9.1	Introduction	*265*
9.2	Digital Twins in Continuous Biomanufacturing	*269*
9.2.1	USP Fed Batch and Perfusion	*273*
9.2.2	Capture, LLE, Cell Separation, and Clarification	*273*
9.2.2.1	Fluid Dynamics (Red)	*277*
9.2.2.2	Phase Equilibrium (Blue)	*277*
9.2.2.3	Kinetics (Green)	*277*
9.2.3	UF/DF, SPTFF for Concentration, and Buffer Exchange	*278*
9.2.4	Precipitation/Crystallization	*282*
9.2.5	Chromatography and Membrane Adsorption	*282*
9.2.5.1	General Rate Model Chromatography	*282*
9.2.5.2	SEC	*284*
9.2.5.3	Adsorption Mechanism	*284*
9.2.5.4	IEX-SMA	*284*
9.2.5.5	HIC-SMA	*285*
9.2.5.6	Modified Mixed-Mode SMA	*285*
9.2.5.7	Modified HIC-SMA Process Model Exemplification by mab Purification	*287*
9.2.5.8	Model Parameter Determination	*289*
9.2.5.9	Phase Equilibrium Isotherms	*290*
9.2.5.10	Mass Transfer Kinetics	*292*
9.2.6	Lyophilization	*293*
9.2.6.1	Thermal Conductivity of the Vial	*293*
9.2.6.2	Product Resistance	*293*

9.2.6.3	Product Temperature	*295*
9.2.6.4	Water Properties	*295*
9.3	Process Integration and Demonstration	*295*
9.3.1	USP Fed Batch and Perfusion	*301*
9.3.2	Capture, LLE, Cell Separation, and Clarification	*306*
9.3.3	UF/DF, SPTFF for Concentration, and Buffer Exchange	*309*
9.3.4	Precipitation/Crystallization	*311*
9.3.5	Chromatography and Membrane Adsorption	*314*
9.3.6	Lyophilization	*314*
9.3.7	Comparison Between Conceptual Process Design and Experimental Data	*319*
9.4	PAT in Continuous Biomanufacturing	*320*
9.4.1	State-of-the-Art PAT	*321*
9.4.2	QbD-based PAT Control Strategy	*322*
9.4.3	Process Simulation Toward APC-Based Autonomous Operation	*323*
9.4.4	Applicability of Spectroscopic Methods in Continuous Biomanufacturing	*328*
9.4.5	Proposed Control Strategy Including PAT	*332*
9.4.6	Evaluation and Summary of PAT	*337*
9.5	Conclusion	*338*
	Acknowledgments	*339*
	References	*339*

10 Regulatory and Quality Considerations of Continuous Bioprocessing *351*
Britta Manser and Martin Glenz

10.1	Introduction	*351*
10.2	Integrated Processing	*352*
10.3	Process Traceability	*353*
10.3.1	Batch and Lot Definition	*353*
10.3.2	Lot Traceability and Deviation Management	*354*
10.4	Process Consistency	*355*
10.4.1	Process Control	*356*
10.4.1.1	Automation	*356*
10.4.1.2	Process Analytical Technologies (PAT)	*357*
10.4.1.3	Data Analysis	*359*
10.4.1.4	Real-Time Release Testing	*360*
10.4.2	Quality by Design	*360*
10.4.2.1	Multicolumn Protein A Chromatography	*361*
10.4.2.2	Continuous Virus Inactivation	*362*
10.4.2.3	Bind/Elute Cation Exchange Chromatography	*362*
10.4.2.4	Flow-Through Anion Exchange Chromatography	*363*
10.4.2.5	Ultrafiltration and Diafiltration	*363*
10.4.2.6	Sterile Filtration	*363*
10.4.2.7	Virus Reduction Filtration	*363*

10.4.2.8	Connection of Unit Operations	*364*
10.5	Patient Safety	*365*
10.5.1	Contamination Control	*365*
10.5.2	Virus Safety	*366*
10.5.2.1	Virus Reduction in Chromatography	*367*
10.5.2.2	Low-pH Virus Inactivation	*367*
10.5.2.3	Virus Reduction Filtration	*368*
10.6	Equipment Design	*369*
10.7	Conclusion	*370*
	References	*371*

Index *377*

Preface

During the past two decades, we have seen great trends in the advancements of technologies especially in accelerating the bioprocessing sector for manufacturing. The current Covid-19 pandemic and the currently existing bioprocessing technologies have been accelerated to meet the global needs of vaccines. One of the several issues that the current pandemic has highlighted is the need to be in the forefront to quickly meet the global market demand. The requirements to bring vaccines to the market and scale up the production at a much higher speed have accelerated the research and development. Clinical trials and supply chain strategies were carried out with their established processes. The rapid development of vaccines has set a positive precedent and the ongoing expectation that biological products will continue to reach the market much faster than in the past. The pandemic has certainly disturbed the equilibrium, and hence, Pharma 4.0 is underway.

Over the past ten years, pharmaceutical industries have adopted continuous processing and have invested to manufacture the products economically, to ease the method of operation, and to minimize the operational cost.

Process intensification is an ongoing trend in the bioprocessing sector that focuses on continuous bioprocessing. Over the years, industries have been regularly adapting process intensification methods.

This book presents the current advances in the intensified bioprocessing process and its application in biomanufacturing. Each chapter brings out detailed information and its values in the bioprocessing sector.

We hope that this volume will stimulate great appreciation of the usefulness, efficiency, and its potential in continuous processing of biological products and propel further progress in advancing continuous processing to meet the ever-increasing challenges and demands in the manufacturing of therapeutic products.

This book has been completed with the help and support of my friends and colleagues. It is a great pleasure for me to acknowledge the authors with deep gratitude for their contribution toward the chapters and for spending their valuable time. During this Pandemic period, one of the contributors and his family has unfortunately been the victim, and I would like to sincerely thank him and his family for still completing the chapter.

Finally, I would like to thank Felix Bloeck, Sakeena Qurashi of Wiley, and their team for their great encouragement and support throughout the preparation of this book.

Maidenhead
19 April 2021

Ganapathy Subramanian

Part I

Continuous Biomanufacturing

1

Strategies for Continuous Processing in Microbial Systems

Julian Kopp[1], Christoph Slouka[2], Frank Delvigne[3], and Christoph Herwig[1,2]

[1] Vienna University of Technology, Institute of Chemical, Environmental and Biological Engineering, Christian Doppler Laboratory for Mechanistic and Physiological Methods for Improved Bioprocesses, 1060, Vienna, Austria
[2] Vienna University of Technology, Institute of Chemical Environmental and Bioscience Engineering, Research Division Biochemical Engineering, Department of Chemical, Environmental and Biological Engineering, Gumpendorferstr. 1a, 1060, Vienna, Austria
[3] University of Liège, Terra Research and Teaching Center, Microbial Processes and Interactions (MiPI), Gembloux Agro-Bio Tech, Département GxABT, Bât. ABT09 G140 - Microbial, food and biobased technologies, Avenue de la Faculté d'Agronomie 2B, 5030, Gembloux, Belgium

1.1 Introduction

1.1.1 Microbial Hosts and Their Applications in Biotechnology

With regard to microbial cultivation technology, first associations might be drawn between classical food technological applications like ethanol fermentation in beer and wine and production of dry yeast for baking dough. Nevertheless, microbial systems play a fundamental role in all parts of biotechnology in a multitude of industrially used processes. Table 1.1 gives a – certainly not complete – list for possible application of microbes in today's industrial biotechnology.

There is a high variety of possible applications for a high number of different microorganisms (MOs) as shown in Table 1.1. There are classical working horses like *Escherichia coli*, *Saccharomyces cerevisiae*, and *Bacillus* spp. that can be cultivated easily to high cell densities and produce high amounts of the desired product. Other applications and microorganism suffer from inhibitory effects (e.g. inhibition from contaminants in waste water) and low biomass and product yields. Continuous cultivations are referred to increase the time–space yield (TSY) of many processes and provide optimal usage of installed assets. Still, most these processes are established for biomass generation or detoxification. Only very few continuously operated processes involve the production of recombinant compounds. The benefits and drawbacks of continuous cultivation will be discussed throughout this book chapter, focusing especially on microbial hosts. Hence, the ideal cultivation mode must be chosen wisely.

Table 1.1 Applications of microbial biotechnology.

Microbes	Benefit	Application in biotechnology	Cultivation mode	Source
Aspergillus niger, *Enterobacteria*	Overproduction of raw chemical by MOs, e.g. citric acid, lactic acid, vitamins	Bulk chemicals	Batch, fed-batch, and continuous cultivations	[1–3]
Thermophilic microbes – genera *Picrophilus*, *Thermoplasma*, *Sulfolobus*	High-temperature stable enzymes	Food, feed, textile, chemical, pharmaceutical, and other industrial sectors	Continuous cultivation	[4, 5]
Thiobacillus/ Leptospirillum	Noble metal recovery	Bio-oxidation	Bioleaching	[6]
High diverse group, e.g. *R. eutropha*	Conversion of toxic organic compounds, surface binding of heavy metals	Bioremediation	Batch and continuous processing	[7, 8]
E. coli, Bacillus, S. cerevisiae, P. pastoris	Drug production, antibiotics, etc.	Biopharmaceutical industry, enzyme industry, agricultural industry	Fed-batch technology	[9, 10]
Lactobacillus and *Bifidobacterium*	Functional food	Probiotics	Batch cultivation	[11]
S. cerevisiae, Zymomonas mobilis, Klebsiella oxytoca, Streptococcus fragilis	Biomass fuels based on waste streams	Biofuels	Batch and continuous cultivations	[12, 13]
Wild type: *Ralstonia eutropha, Alcaligenes latus*; Recombinant: *Aeromonas hydrophila, E. coli* Photosynthetic: *Synechocystis* sp.	Environmentally friendly non-petrochemical-based plastics	Bioplastics (polyhydroxyalka-noates)	Batch and fed-batch cultivation	[14, 15]
Haloferax mediterranei, other halophiles	Tolerate high salt concentrations	Detoxification in chemical waste streams	Continuous cultivation	[16, 17]
High diverse groups – depending on application	Waste to value	PHA production; enzymes/organic acids	Batch and fed-batch cultivations	[18, 19]
Mixed cultures, e.g. *Proteus vulgaris, Rhodoferax ferrireducens, Geobacter sulfurreducens*	Energy generation from waste	Microbial fuel cells	Batch and continuous cultivations	[20]

1.1.2 Regulatory Demands for Their Applied Cultivation Mode

The batch definitions in continuous manufacturing, preciously defined for mammalian cultivations, apply for microbial processes as well: "A Batch means a specific quantity of a drug or other material that is intended to have uniform character and quality, within specified limits, and is produced according to a single manufacturing order during the same cycle of manufacture. In the case of a drug product manufactured by a continuous process, it is a specific identified amount produced in a unit of time or quantity in a manner that assures its having uniform character and quality within specified limits" 21 CFR 210.3 2, or "a batch may correspond to a defined fraction of the production. The batch size can be defined either by a fixed quantity or by the amount produced in a fixed time interval" EU GMP Guide, Part II (ICH Q7).

More important than batch definition is the application of the quality-by-design (QbD) context to continuous processing. Generally, QbD mainly urges to relate critical quality attributes (CQAs) to critical process parameters (CPPs) and raw material attributes (RMA) to form a design space [21]: "A multidimensional combination and interaction of input variables and process parameters that have been demonstrated to provide assurance of quality of the product" for demonstrating process understanding. As proposed by current validation guidelines [22], stage 1 validation includes the execution of process characterization studies (PCS), which is the "collection and evaluation of data, from the process design stage throughout production. This establishes scientific evidence that a process is capable of consistently delivering quality product." PCS finally leads to the awareness of the mutual interplay of CPPs on CQAs. This demonstrates process robustness within multivariate normal operating ranges (NOR) and therefore finally proposes the control strategy including process and analytical controls. Currently, this is achieved by fusing development and manufacturing data.

Using an enhanced PCS approach, the determination of appropriate material specifications and process parameter ranges could follow a sequence such as the following [23]:

(i) Identify potential sources of process variability.
(ii) Identify the material attributes and process parameters likely to have the greatest impact on drug substance quality.
(iii) Design and conduct studies (e.g. mechanistic and/or kinetic evaluations, multivariate design of experiments, simulations, modeling) to identify and confirm the links and relationships of material attributes and process parameters to drug substance CQAs.
(iv) Analyze and assess the data to establish appropriate ranges, including the establishment of a design space.

Even more, continuous processes require a different level of process understanding: as an example, classical recombinant protein production (RPP) using *E. coli* as a host pools the product solution after four days of processing. The time-variant dependency of CPPs and CQAs is finally integrated in one analytical result, and the process is also registered as such. Hence, batch processes are characterized by operating subsequent steps on the *integral* outcome of the current process step. Implementing continuous operations, we must *understand the time dependency* between CPPs and CQAs with the goal to have a time-invariant CQA process result. Hence, as microbial processes are more dynamic in terms of kinetics and stoichiometry, proper understanding of a dynamic design space and establishment of a robust control strategy are more relevant from a regulatory point of view ("A planned set of controls derived from product and process understanding that assure process performance and product quality") [21].

For continuous processing, time-variant interrelations between CPPs and CQAs must be transformed into a control concept. This calls for the enhanced use of metabolic and kinetic models integrated in experimental designs for elucidating the design space. Of course, initially, we relate CPPs and CQAs in a classical QbD manner. However, we need to enhance this context: for example, classical design of experiment (DoE) approaches can only capture the response of the system to time-invariant factors of the integral experiment. For continuous operation, and in contrast to conventional development strategies, we aim to operate with time-invariant process variables and CQAs. Therefore, we need to understand the time-variant dependencies on their CPPs for control. Thus, we do not change the CPPs to analyze the integral outcome of the CQAs, but we change the CQAs to analyze the integral outcome of the CPPs, which can compensate for their variability of time. Hence, the development strategy is enhanced. We may need dynamic model-based experimental designs to develop a control strategy able to cope with process variability over time; those experimental setups need to be established in the R&D environments [24].

Hence, continuous processing requires a *much earlier definition of the process control strategy directly during the process development and characterization phase*. We need this control strategy earlier as prerequisite for process design. NORs must be defined earlier and turned into a real process control strategy based on PAT, models, and controls. Hence, the tasks of PCS need to be done already during the development. Those elements may include data mining, risk assessments, characterization of process performance, screening studies, criticality assessment [23], and integrated process modeling [25], as shown in the workflow in Figure 1.1. On the other hand, scale-down model qualification tasks may not be necessary, as the development scale may already be the production scale, since productivity is scaled by processing time or scale-out techniques.

Thus, R&D labs need therefore higher data management and data science orientation, as well as advanced PAT and process control environments as skill set, which will be addressed in Chapter 3.

1.2 Overview of Applied Cultivation Methods in Industrial Biotechnology

As an easy rule, the cultivation mode resulting in the highest TSY should be pursued. TSY could be defined in pharmaceutical applications as the highest throughput from pre-culture inoculation until purified drug substance, in gram product per operating

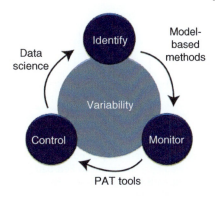

Figure 1.1 Tackling process variability in the process development phase for continuous biomanufacturing through early establishment of identification, monitoring, and control for the identification of a dynamic control strategy. Source: Dream et al. [24].

liter per day [26], or for waste streams as the process enabling the highest catalytic capacity to degrade toxic compounds.

1.2.1 Batch and Fed-Batch Cultivations

1.2.1.1 Conventional Approaches and Their Technical Limitations

The golden standard in RPP is batch and fed-batch cultivations. With regard to batch, all ingredients are added to the reactor, and microorganism react until limiting component inhibits further growth. Common limitation elements in industrial biotechnology are carbon, nitrogen, or phosphor. Problems in batch cultivation are that MOs grow at maximal specific growth rate and causing problems in aeration and heat transfer, discussed in more detail later. This limits the maximal limiting component concentration and results in low overall biomass concentrations. For this purpose, fed-batch technology is currently applied. Additional feeding is conducted (e.g. high concentrated sugar feeds), which results in higher biomass concentrations. Controlled addition of limiting substrate can also overcome several problems like carbon catabolite repression and substrate inhibition [27]. High cell density cultivations are referred to increase the overall titer. However depending on the media and reactor setup employed, biomass concentrations should not exceed physiological levels [28]. This is because in high cell density fermentations, (i) non-controlled nutrient limitation might occur, (ii) $K_L a$ levels might not cope for the demands of high cell densities, and (iii) reactor cooling capacity might be exceeded [29]. To cope for demands of limited oxygen transfer, additional oxygen could be supplemented, but at industrial scale, additional oxygen supply might lead to unfeasible cultivation costs. pO_2-limited cultivations tend to increase secondary metabolite production to synthetize their needed reduction equivalents. Furthermore, amino acid mis-incorporation in recombinant produced proteins has been found as a side effect of oxygen-limited cultivations [30, 31]. Moreover, biomass concentrations must be kept within the reactor cooling capacity. High growth rates monitored for many microorganism can cause high heat formation, being especially a problem in yeast fermentation: as methanol is commonly used for the induction in *Pichia pastoris* systems containing alcohol oxidase (AOX) promoters, high heat is generated by methanol on its own [32]. To stay within reactor cooling

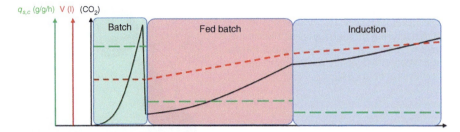

Figure 1.2 General procedure for microbial Fed-batch cultivations: A Batch phase is followed by a non-induced Fed-batch phase and an Induction phase, with $q_{s,c}$ being the specific carbon uptake rate, with V being the reactor volume and CO_2 coding for residual carbon dioxide.

capacities, the Mut^S strain was invented, showing decreased methanol uptake rates in *P. pastoris*. Hence a compromise between maximum biomass concentration and reactor cooling capacity must be made at an industrial level.

1.2.1.2 Feeding and Control Strategies Using *E. coli* as a Model Organism

A sketch for industrial fed batch used for *E. coli* cultivation in red biotechnology is shown in Figure 1.2. Maximum specific feeding rates ($q_{s,max}$) are generally applied through batch. Fed batches are operated at specific feeding rate values far below the $q_{s,max}$.

The batch phase is followed by an exponential fed batch for biomass production according to

$$F(t) = \frac{q_{s,(C)} * X(t) * \rho_f}{c_f} \text{ with } X(t) = X(t=0) * e^{\mu * t} \tag{1.1}$$

where F is the feeding rate (g/h), $q_{s,(C)}$ is the specific uptake rate (g/g/h), $X(t)$ is the absolute biomass at the time point t (hours), ρ_F is the feed density (g/l), c_F is the feed concentration (g/l), $X(t=0)$ is the biomass before start of the fed batch in (g), and μ is the specific growth rate (1/h). After the first exponential fed-batch phase, cells are induced for RPP and fed until harvest. Besides the classic exponential fed batch, different feeding profiles can be employed, which is often done throughout induction phase [33, 34]. Cells are mainly grown carbon limited after batch phase, as a desired specific growth rate (μ) can be adjusted easily, with a set μ beneath $\mu_{max}/2$ to reduce acetate formation and reduce stress onto host cells. Common control strategies for carbon-limited growth are either basic feed-forward protocols (see Eq. (1.1)) or soft-sensor approaches [35]. Throughout feed-forward control strategies, a constant q_s value is set for a fixed timeframe to achieve a targeted biomass within a certain time. The amount of fed carbon is calculated into biomass, assuming a constant biomass yield. As overall biomass is increasing, feed rate is thus increased via higher pump set points, which are adjusted using a PID controller (proportional, integral, and derivative control terms). However, in this strategy, no feedback control is implied. In soft-sensor approaches, a feedback loop to off-gas signals by mass balancing is implemented in the feeding strategy. Hence, feeding rate can be adjusted to unexpected process deviations. The usage of a noncontrolled feeding strategies might lead to substrate accumulation in carbon-limited feeding approaches. Off-gas signals are used to predict biomass formations due to the stoichiometric balances, hence adjusting pump set points [26, 36].

1.2.2 Introduction into Microbial Continuous Biomanufacturing (CBM)

1.2.2.1 General Considerations

In some branches of biotechnology, continuous processing is already established (i.e. bioleaching and oxidation, using several stirred tank reactors serially connected) [6]. Also in the field of biofuels, the trend leads to a continuous production platform [37, 38]. Moreover, continuous processing is well suited for the degradation of toxic compounds. As cell growth-inhibiting compounds are fed, growth rates can be very low, and thus retentostat setups (Figure 1.3b) can increase the detoxification efficiency as shown for the halophile *Haloferax mediterranei* [16]. Large-scale detoxification can be found in wastewater treatment plants, also using retentostat principles. Retentostat cultivation used a retention device (i.e. 0.2 µm pore size membrane) to maintain a controlled number of cells in the cultivation device. Hence, a feed/bleed system can be maintained at feasible cell densities compared with common chemostat cultivation, especially advantageous for slow-growing organisms. Problems such as changing media composition and changing yields and inhibitory substances often make continuous cultivations challenging in diverse branches. For recombinant protein expression, using microbial hosts, continuous biomanufacturing (CBM) is still far from its industrial application. Despite the several benefits coming with fed-batch cultivation, product quality is highly time dependent. Furthermore, high batch-to-batch variations may result in severe problems in the subsequent downstream process for red biotechnology. The following benefits could be expected from the establishment of CBM for microbial cultivations:

- Small reactor systems reduce investment costs and enable efficient and highly flexible production even for small companies ("**small footprint facilities**").
- Cleaning in place (CIP) and steam in place (SIP) can be reduced to a minimum, as cultivation times are increased from some days to several weeks, making this cultivation mode **sustainable**.
- Quality of the product is not batch performance dependent but can be expressed at **constant quality**.
- Continuous waste streams may be used for certain applications (whey from milk industry, molasses from sugar industry, etc.). This would decrease the costs for the product drastically, leading to **circular economy** approaches.
- Continuous upstream enables continuous downstream, leading to an **integrated process**, and enables robust downstream processing, e.g. usage of "simulated moving bed chromatography."

In this chapter, three different cultivation modes that are often implemented in the upstream processing (USP) of microbial continuous systems will be discussed. Figure 1.3a) shows the classic chemostat process for microbial systems. A feed is

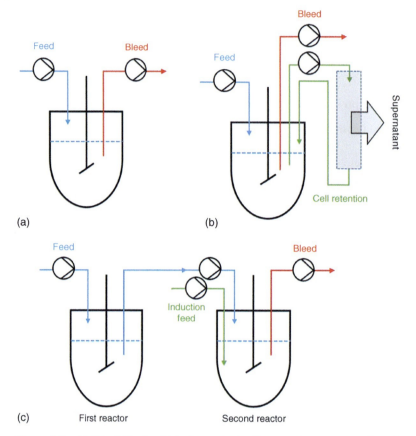

Figure 1.3 (a) Classical microbial chemostat for fast-growing organism. Feed is pumped at a fixed dilution rate and bled out at a certain volume including the product. (b) Cell retention system for slow-growing microorganism. Cell concentrations are increased until the theoretical biomass to substrate yield is reached. Product is usually concentrated after (c) Cascade systems for sequential/serial addition of bioreactors. First reactor is used for biomass production only, biomass is transferred from reactor 1 to reactor 2 indicated by the blue line. Only reactor 2 is fed with an induction feed and bleed out of reactor 2 is containing target product. This cultivation system could be used for different MOs digesting the same feedstock or for recombinant protein production. Source: Refs. [39–41].

added at a constant rate to the reactor. The bleed is removed using pneumatic valves, connected to peristaltic pumps, enabling constant volume throughout cultivation. This system is preferably used for fast-growing MOs. The main benefit is easy process control as generally only monitoring is necessary and no control circuits need to be used, like PI or PID controllers.

Retentostats, also called perfusion systems, shown in Figure 1.3b), contain a common chemostat setup with an additional hollow fiber membrane to retain cells in the reactor. Pumps (feed, bleed, cell retention) must be adjusted accordingly to guarantee a stable process performance. Retentostats are common for slow-growing cultures and are therefore often used in cell culture. The second advantage is that extracellular product can easily be harvested using cell retention modules and waste-to-value approaches can clear contaminants effectively through higher biomass concentrations inside the reactor. The third cultivation system is shown in Figure 1.3c, which is regarded as a serial combination of chemostats or retentostat systems. Cascaded cultivation systems can be successfully applied for red biotechnology approaches in decoupling biomass production from induction of the cells in a spatially resolved manner [39, 40] (see Section 1.4.3). Hereby cells in reactor one is grown "burden-free," whereas the second reactor is operated in an induced stage. Continuous application is given as two feed/bleed systems are serially connected with each other: feed, free of inducer, is supplemented to the burden-free stage (first reactor), and non-induced biomass is transferred to the induced reactor (second reactor). Further ongoing, the second reactor is supplied with an inducer-containing feed to initiate RPP [40]. Using this system, the benefits of time-dependent cultivations can be included in a continuous system as (i) burden-free cell growth, equal to non-induced biomass growth, can be maintained in the first stage and (ii) adequate induction times can be set via the residence time in the induced stage. Cascaded or serial combinations can also be used in waste-to-value approaches and circular economy thoughts combining aerobic cultures producing CO_2 that may be recycled in the second reactor using autotroph/chemolithotroph MOs [39], implementing a neutral carbon footprint.

The cultivation method of choice has of course always to be adapted to the current aim. A rough overview about the desired aim can be gained via proper mass balances.

1.2.2.2 Mass Balancing and the Macroscopic Effects in Chemostat Cultures

Mass balancing can be perfectly used to highlight benefits of a continuous system in favor of the classical fed-batch approach. The general macroscopic mass balance for an ideal stirred tank reactor is given in Eq. (1.2):

$$\dot{V}_{in} * c_{i,in} + \dot{V}_{out} * c_{i,out} + V_R * r_i = V_R * \frac{\partial c_i}{\partial t} + c_i * \frac{\partial V_R}{\partial t} \quad (1.2)$$

where $\dot{V}_{in}$ is the volume flux in the reactor, $\dot{V}_{out}$ is the flux of the bleed, $c_{i,in}$ is the concentration of component i in the influx, $c_{i,out}$ is the concentration of component in the bleed, V_R is the reactor volume, r_i is the reaction rate for component i, and t is the time. As one of the strong benefits of continuous reactor systems is the time independence of the reactor upon tuning, the balance reduces to

$$\dot{V}_{in} * c_{i,in} + \dot{V}_{out} * c_{i,out} = -V_R * r_i \quad (1.3)$$

As flux in and flux out are constant in a classic chemostat and solving for the reaction rate and substituting $\frac{\dot{V}}{V_R} = D$, with D being the respective dilution rate in 1/h,

$$r_i = \Delta c_i * D \quad (1.4)$$

It is clearly visible that every volumetric rate r_i is dependent upon the applied dilution rate of the bioreactor and on the concentration of components in the media. TSY, being the volumetric productivity, is directly dependent on these two

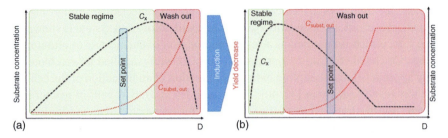

Figure 1.4 (a) A set point for stable biomass production was chosen, all fed substrate is consumed, and biomass is formed based on the yield coefficient. Upon induction, the $Y_{X/S}$ changes, and less substrate can be metabolized. (b) shows that the stable set point before might now suffer from a decrease in yield. Hence sugar biomass formation might be reduced at the given set point, and substrate is accumulating.

factors. Consequently, high dilution rates and high concentrations of the limiting component in the feed should positively influence the TSY. Connections between the different rates can be easily drawn in using yield coefficients in Eq. (1.5):

$$Y_{\frac{a}{b}} = \frac{r_a}{r_b} \tag{1.5}$$

Postulating constant yield coefficients, the continuous reactor can be set up and operated at optimal conditions. In general, this hypothesis holds true for chemostats with defined media of constant quality and sole biomass production. However, in RPP using, for example pET plasmids, induction with isopropyl-β-D-thiogalactopyranosid (IPTG), lactose or related inducer is necessary. The effects of induction onto the cell itself will be discussed in a later chapter. We regard the reactor as a black box for now and just look at the effects of a changing substrate to biomass yield ($Y_{X/S}$). A visualization is given in Figure 1.4.

Starting with induction feeds at identical carbon concentrations, the yield coefficient changes within some hours as several stress responses affect cell growth. With decreasing yield, the same set point of dilution rate might possibly lead to sugar accumulation. As the stable set point moves toward the washout regime and consequently biomass concentrations are reduced, substrate is washed out of the reactor. This is no stable process and brings again a time dependence of the yield coefficient into consideration. These effects act also upon productivity and make single-vessel chemostat cultures very unstable at fast-growing MOs.

Similar problems are observed upon changes in the feed substrate quality. These changes may be based on fluctuations in substrate concentration but could also be fluctuating in inhibitory substances. Simple Michaelis–Menten kinetic considerations show effects upon the process in Eq. (1.6). We assume competitive inhibition as cells are directly affected by the inhibitory substance in the reactor:

$$\mu = \mu_{max} * \frac{[S]}{[S] * K_S * \left(1 + K_I * [I]\right)} \tag{1.6}$$

where μ is the specific growth rate, which is identical to the dilution rate, $[S]$ is the substrate concentration, μ_{max} is the maximal possible growth rate, K_S is the reaction constant for substrate uptake, [I] is the inhibitor concentration, and K_I is the reaction constant for the inhibitory reaction. Therefore, changing inhibitory concentrations [I], as well as substrate concentrations [S], has effects on the specific growth rate and may shift the critical specific growth rate. Close to μ_{max}, $\mu = D$ is not valid anymore, as washout starts and hence results in an unstable process.

So even simple macroscopic mass balance and kinetic considerations, considering the biomass in the reactor, show the complexity of the system. Further cell physiological effects might occur in RPP. However, the high expression of recombinant protein and the extremely high doubling rates may make microbial continuous cultivation a promising alternative to state-of-the-art fed-batch approaches.

1.2.3 Microbial CBM vs. Mammalian CBM

1.2.3.1 Differences in Upstream of Microbial CBM Compared with Cell Culture

The first remarkable difference between microbial and cell culture-based expression systems are the differences in cell doubling times. While cell culture-based cultivations take up several hours for a cell division, the maximal doubling time in E. coli can be 20 minutes. Table 1.2 compares the three most important organisms regarding their growth rates upon the production of recombinant proteins. Absolute values may differ from strain to strain and expressed recombinant protein but give a certain lead to compare different continuous approaches.

Cell cultures (Chinese hamster ovary [CHO] cells) exhibit exceptionally low doubling rates (13.8–85 hours per cell doubling). This also results in long preparation times for pre-cultures (seed flasks) up to four weeks and the starting batch phase before enabling continuous feeding. For E. coli preparation, pre-culture and batch phase take approximately 30 hours, depending on applied sugar concentrations [35, 46]. It was already stated that microbial systems show very high dynamics in metabolism and recombinant protein expression, based on the high number of cell divisions during a continuous process [44]. Taking mean dilution rates in Table 1.1 and comparing generation times to one week of cultivation, which corresponds to 168 hours, CHO cells doubled in mean 4.3 times, E. coli cells doubled 50 times, and P. pastoris, as frequently used expression host for yeast-based expression, doubled 18.5 times.

While CHO and yeast cells have a eukaryotic translation and posttranslational modification (PTM) mechanism (through golgi apparatus), prokaryotic

Table 1.2 Growth rates and approximated generations for cultivation times found in the literature.

Organism	Growth rate (1/h)	General process duration (h)	Generations (–)	Source
CHO cells	0.0008–0.05	650–2160	2–47	[42, 43]
E. coli	0.1–0.49	up to 300 h	43–212	[39, 44]
P. pastoris	0.009–0.2	up to 1000 h	13–290	[44, 45]

Source: Refs. [23, 27, 28, 33].

microorganism lack these systems [47, 48]. As the product is generally secreted into cultivation supernatant, cell culture processes rely mainly on retentostat/perfusion technology, where product can be harvested in the broth, without dealing with the intact host cell. Yeasts also have the possibility to translocate the product to the broth while having a sufficient high growth rate. Recombinant proteins produced in yeast, however, are highly mannose glycosylated, and no human like N-glycosylation can be performed. Hence, products need cost- and time-intensive treatment prior to clinical application [49]. Recombinant proteins in *E. coli* are located primary inside the cell. Most of these products are expressed in the cytoplasm and kept in this place, where no signal sequences for transport into the periplasm are attached to the protein. The reducing milieu in the cytoplasm does not allow disulfide bond creation and makes correct folding of complex proteins difficult. The result is often the expression of inclusion bodies (IBs), misfolded proteins with hydrophobic character. Hence, continuous purification in microbial systems might be leading to challenging technical applications, owing to different product loci.

1.2.3.2 Downstream in Microbial CBM

An integration of the process from up- to downstream would be the desired future perspective in a modular design. This would ease the way for "small-footprint facilities" as high modular elements can be easily exchanged and stuck together for a new product. Furthermore, costs can be strongly reduced especially in the downstream, heading toward smaller columns [50–52]. Continuous purification methods for extracellular proteins have been established [53]. Filtration steps, followed by continuous chromatography systems (making use of simulated moving bed principles), have been established for the purification of products derived from mammalian cells [51].

However, other downstream unit operations, especially such operations for intracellular proteins, are still considered problematic. Figure 1.5 shows the schematic downstream chain for intracellular proteins and highlights the additional steps needed for misfolded protein aggregates derived by *E. coli*, which are known as IBs.

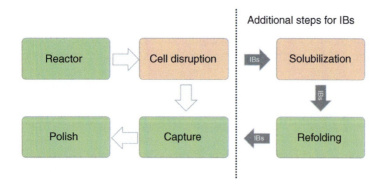

Figure 1.5 Simplified process chain for production of a recombinant product in *E. coli*. Green unit operations can be accessed in a continuous mode; red operations are hard to realize. IBs need at least two additional steps during downstream.

Cell disruption is the first bottleneck to be dealt with in integrated CBM as link between up- and downstream for intracellular products. Generally, cell disruption techniques can be separated between mechanical and non-mechanical approaches [54, 55]. The most frequently used techniques for small-scale cell disruption are performed via bead mill or french press technology and suitable for a small volume of broth in a batch approach. Large-scale cell disruption is performed with high-pressure homogenization, as high-pressure homogenization is the only scalable form of cell disruption [56, 57]. These homogenizers can be operated at high velocities, realizing cell disruption within one passage, and the implementation of a continuous cell disruption mode is rather easy [58]. Besides high-pressure homogenization, other methods are currently under investigation for continuous cell disruption. Ultrasonic devices may be used for quick energy-saving cell disruption using flow cells. However, problems with the abrasion of titanium elements responsible for energy transduction must be considered. Strong pulsed electrical fields may also be used for cell lysis, as the cells are effectively opened, but intracellular components are not affected. However, bubble generation and joule heating issue cause problems, and first tests were only performed in microscale [54].

A major bottleneck of misfolded proteins (IBs) is the solubilization and refolding steps. Even though, IBs exhibit some disadvantages, there are many benefits, especially in continuous cultivation, shown in Table 1.3.

Solubilization of IBs is performed using high concentrations of chaotropic detergents, like urea or guanidine hydrochloride [59]. However, mild solubilization has also shown to result in high yields of biological active protein [60]. Continuous approaches for the refolding step revealed positive effects in *in silico* studies (especially buffer consumption) and were found to yield promising results in experimental studies [50, 61]. A fully integrated continuous downstream protocol for the purification of inclusion bodies has been described recently [62]. Even though no integration between upstream and downstream is performed up until now, results for continuous inclusion body treatment in downstream seem promising.

Table 1.3 Pros and cons of IB expression in *E. coli*.

Positive aspects	Negative aspects
Nano-particulate matter, which can be highly concentrated upon cell disruption	Necessary cell disruption for capture of the protein
High initial purities before capture step (up to 90%)	No posttranslational modifications (also true for soluble proteins)
Active (so-called nonclassical IBs [ncIBs]), which require no refolding	Refolding needs to be performed for classical IBs, which reduces the yield drastically
High volumetric titers of up to 15 g/l in fed batch	Time- and cost-intensive downstream
Expression of inactive toxic proteins	—

Once continuous cell disruption can be performed and promising capture steps are established, the production of intracellular proteins could be fully integrated in a continuous mode. As continuous chromatography principles are established for the purification of extracellular proteins, these methods are transferable. By adapting separation techniques and resins to the needed downstream step (i.e. capture or polishing [Figure 1.5]), continuous purification might be possible for *E. coli* – derived products. Detailed information for continuous purification of intracellular products has been given in recent reviews [33, 63].

1.3 Monitoring and Control Strategies to Enable CBM with Microbials

1.3.1 Subpopulation Monitoring and Possible PAT Tools Applicable for Microbial CBM

Besides genetic instabilities mentioned earlier (Section 1.2.3.1), microbial cell population can also exhibit phenotypic differentiation. This differentiation arises from biological noise taking place in the intracellular environment, leading cell-to-cell differences in the amount of key intracellular components, such as regulatory proteins or metabolites [64]. Ultimately, this biological noise has been recognized to confer functionality to microbial populations through the appearance of subpopulations with distinct metabolic functions [65]. In natural ecosystems, the occurrence of these subpopulations is generally recognized as a beneficial factor known for increasing the global fitness of the whole population when facing challenging environmental conditions [66]. However, in the context of bioprocesses, such subpopulations are generally unwanted, leading to the simultaneous occurrence of producing and non-producing cells requiring the use of advanced process analytical technology (PAT) tools for the proper characterization of assessing the real productivity of the biological system [67].

Even though some continuous cultivation techniques (i.e. cascaded continuous cultivation) can overcome the issues in microbial continuous bioprocessing, monitoring of occurring dynamics needs to be performed to establish a knowledge platform. PAT applications for *E. coli* continuous manufacturing have been summarized in recent reviews [33, 68]. Classic measurements (such as pH, pO_2, and oxygen transfer rate [OTR] measurements) are implemented in the process chain, but also new applications have been described. Label-free technologies such as Raman microspectroscopy [69] and nano-SIMS [70] begin to emerge as promising alternatives to fluorescence-dependent approaches. Especially, flow cytometry has been found to be a powerful tool to distinguish between productive and nonproductive subpopulations [71]. As flow cytometry is a single-cell analytic, highly accurate information of the subpopulation state can be given [29]. Online flow cytometry was successfully established to describe the metabolic state of *E. coli* and *P. putida* cultivations [30]. The usage of the fluorescent dye propidium iodide (PI) in flow cytometry analysis can distinguish the so called red-but-not-dead phenotype

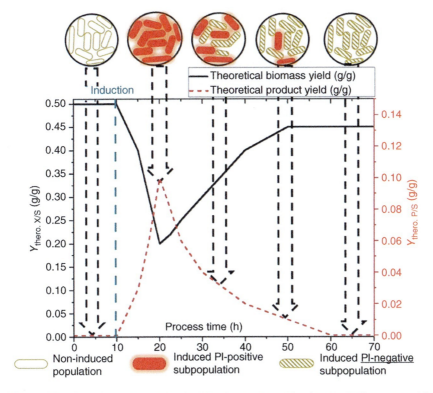

Figure 1.6 Time-dependent behavior of a chemostat cultivation producing a recombinant protein (Figure 1.10a) on a subpopulation level: it can be seen that the un-induced population starts to enhance its PI potential upon induction. However, the formation of PI-negative subpopulation takes over at elongated cultivation times, and no more fluorescence can be monitored.

[65, 66]. These are referred to show enhanced membrane potential, thus allowing PI to penetrate cell wall. Implementation of PI as a marker stain helped to determine and monitor the membrane potential throughout chemostat cultivation with online flow cytometry [30]. The possibility of online flow cytometry in combination with online propidium iodide staining for subpopulation monitoring is given and shown as a hypothetical example for a common chemostat behavior in Figure 1.6.

In large-scale *E. coli* fermentations, a high percentage of the cultivation broth was found to have increased membrane activity, detected via PI staining. Local glucose limitations might have caused enhanced membrane potential, whereas the effects were not monitored in well-mixed scale-down experiments [31]. In this study, green fluorescent protein (GFP), fused to a target protein, was transcribed as a fluorescent marker protein, an interesting approach to gain gene-specific information [29]. Autofluorescent reporter genes and frequent measurement of such with flow cytometry throughout cultivations might help to get more insights in population dynamics. Hence, this may help to distinguish subpopulation noise occurring throughout microbial fermentations [32]. However, reporter genes can be also be used to derive a real-time metabolic state conditions [72]. Using GFP as a marker protein to distinguish the intracellular burden, in combination with high-frequent flow cytometry measurements, might be a powerful tool to distinguish important gene clusters and test optimal cultivation conditions in real time [33]. Rapid screening and information gained can further be used to (i) accelerate strain optimization and (ii) implement data in model-based approaches.

This example shows the benefit and need of PAT in CBM. Key critical quality attributes that could be measured online and/or at-line are as follows:

- Diversification of cell populations.
- Product concentration.
- Host cell protein (HCP).
- Product aggregates.
- Glycosylation profiles (for eukaryotic hosts).
- Metabolites.

The selection of appropriate PAT tools is a crucial step toward setting efficient monitoring and control strategies in continuous processes. They must possess the following characteristics [24]:

- Easy-to-use instrumentation.
- Measuring frequencies.
- Ability to monitor multiple process parameters.
- Directly measuring CQAs.
- Capturing the real-time process state.
- Eliminating traditional offline techniques and increasing efficiency.

There is quite some advancement in the availability of at-line tools for CQA measurements using NMR or liquid chromatography coupled to mass spectroscopy [73]. However, for continuous processing, those tools must be deployed as real-time PAT tools. Various solutions may bridge this task in the near future such as (i) data-driven model-based approaches using spectroscopic measurements, (ii) robust online sampling solutions allowing to link gold standard analytics in online mode, and (iii) model-based approaches linking CQAs to easier real-time measurable components, as models, applied with observability analysis, offer a clear advantageous means to measure less. Hence, we need to diversify the PAT program to do the following:

- Minimize incoming material variation.
- Reduce CQA and CPP variations.
- Perform timely in-process measurements.
- Define representative sampling.
- Develop chemometric models and set appropriate acceptance criteria.
- Characterize the propagation of changes and disturbances through the system.

Data gained from adequate PAT measurements can be added to a knowledge platform, establishing model-dependent control, such as digital twins in the future. The usage of model predication can be a powerful tool to facilitate process control strategies, as discussed in Section 1.3.2.

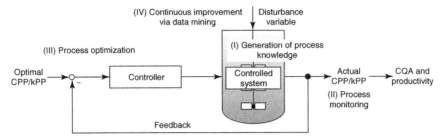

Figure 1.7 Multivariate control loop for the establishment of a robust process. Source: Kroll et al. [74].

1.3.2 Modeling and Control Strategies to Enable CBM with Microbials

As derived from above sections 1.1.2 and 1.3.1, we need the following prerequisites for a robust process control strategy:

- Analysis of the output process variables of the unit operations (key performance indicators [KPIs] and CQA) using PAT (see Section above).
- A controller varying the CPPs to achieve a robust process to achieve time-invariant process performance and CQAs (Figure 1.7).

The development of the multivariate controller is the more difficult challenge, which allows varying multiple CPPs as function of a multitude of CQAs. Those multiple-input multiple-output (MIMO) controllers are available in other market segments; however, they are hardly applied in R&D bioprocessing labs up until now.

The main trigger for this transition will be the integrated real-time architecture combining data management, Namur Open Architecture (NOA) data architectures, real-time execution of models, and advanced control algorithms and workflows for model generation using good modeling practice [75] and maintenance using SaaS tools [75, 76], as recently reviewed [74]. A possible implementation is given in Figure 1.7:

In addition to the understanding of single unit operations, the interplay between the unit operations needs to be efficiently elaborated for robust continuous processing, because we have to consider the variation of the output of the preceding unit operation (intermediate CQAs or also defined by the FDA as RMA). Integrated models have been used in other market segments successfully (ASPEN, G-Proms) and need to be applied to integrated bioprocesses, irrespective from the mode of cultivation. The integrated process model (IPM) should quantitatively display the process understanding and include the elaborated NORs of individual unit operations [23]. Subsequently, this allows linking the individual unit operations and assessing error propagation within the variation in the NOR using sensitivity studies. Acceptance criteria for any process step can be established and fused together in the total process chain via Monte-Carlo simulations, for example, as shown in Figure 1.8. As a result, integrated process modeling identifies PPs that are holistically critical for the entire process chain and therefore allows identifying necessary control strategies along the entire process to meet acceptance criteria of regulatory authorities [25].

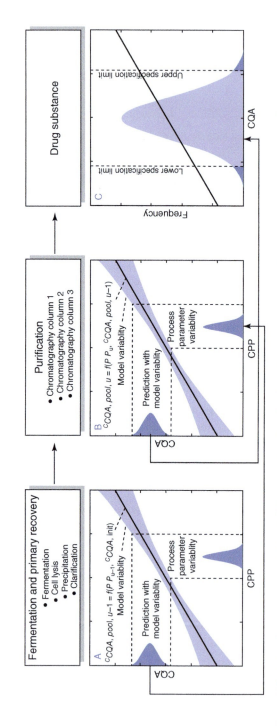

Figure 1.8 Integrated process modeling, defining the criticality of each unit operation and the effect on the process outcome. Source: Zahel et al. [25].

1.4 Chances and Drawbacks in Continuous Biomanufacturing with *E. coli*

The choice of a suitable expression host is highly dependent on the produced protein and the final application. For example, the production of monoclonal antibodies (mABs) cannot be performed in prokaryotic hosts due to required PTMs. In contrast, fragment antigen binding (fABs) can be successfully expressed in *E. coli*, due to the lack of PTMs and the oxidizing environment of the periplasm [77]. Expression of recombinant proteins with *E. coli* exhibits certain benefits, such as high cell specific productivity, throughout continuous upstream. TSY might be boosted in microbial continuous cultivation systems, as overall dilution rates and protein expression rates are higher in microbial hosts than in mammalian cells. Hence, we wanted to demonstrate the chances and drawbacks of microbial continuous applications using the host *E. coli* as a model organism.

1.4.1 Optimization of Plant Usage Using CBM with *E. coli*

As drug manufacturing is operated in large scales, setup and cleaning are time and energy intensive to meet guidelines of authorities in the industry [78]. Even though energy-effective sterilization and cleaning procedures have been established, a reduction of such "downtimes" would facilitate the overall TSY. Continuous processes meet these demands and in addition can reduce the amounts of chemicals needed for cleaning, thus lowering overall costs.

RPP in *E. coli* is usually employed with inducible promoters [79], where cells are grown to high cell densities prior to induction, therefore resulting in a higher amount of catalyst for ongoing reactions [65–67]. Consequently, time-dependent cultivations (such as batch and fed-batch) profit from non-induced cultivation times as cells grow burden-free [64]. Timeframe for batch cultivation is highly dependent on the achieved cell density throughout pre-culture, the amount of pre-culture added, and the glucose concentration in the batch medium [68]. Generally, sugar concentrations supplied throughout batch phase should be below 30 g/L, as long growth at μ-max results in high acetate formation [69, 70], and dO_2 limitations might occur (see Section 1.2.1). Therefore, we assumed an average batch phase to be conducted within six hours. Due to the given reasons, it is beneficial for the overall product yield to conduct a non-induced fed-batch phase. Timeframes for non-induced fed-batch are highly dependent on (i) the set μ and (ii) the targeted biomass concentration before induction. Process conditions throughout induction might vary between target products, and thus it is hard to generalize this approach. Theoretical biomass per substrate yields prior to induction should be $Y_{X/S} = 0.5$ g/g. Therefore, set growth rates in the range of 0.08–0.15 h^{-1} for *E. coli* cultivations should result in a linear correlation of biomass growth and substrate uptake allowing burden-free doubling of batch biomass within 4.6–8.6 hours [71]. Hence, we assumed fed-batch duration with 8 hours. Induction time is highly dependent on the target product and the achieved biomass prior to induction in combination with the set physiological parameters throughout induction. For the production of fABs in the periplasm

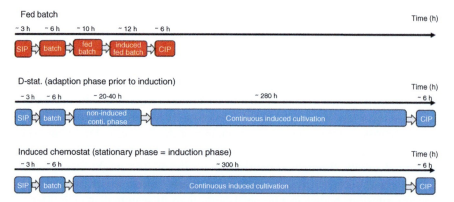

Figure 1.9 Average time needed for each process step comparing a fed-batch cultivation with a continuous cultivation including an adaption phase and a continuous cultivation mode, where induction is performed directly after batch phase for recombinant protein production in *E. coli*.

and intracellular soluble proteins, induction times in the range of 7–10 hours were applied in literature [80, 81]. Time spans in the range of 10–12 hours showed beneficial results for inclusion body production in previous studies [28, 82, 83]. Hence, induction time was calculated with 10 hours. Setup times and cleaning times in the industry are highly dependent on the operating scale. For the comparison of common process strategies as shown in Figure 1.9, we assumed an USP plant in the range of 5–10 m^3, with sterilization and cleaning times of three and six hours, respectively.

A time-dependent biomass growth phase in a batch mode should be conducted to reach a targeted biomass before cultivations are switched to a continuous cultivation mode. In case high biomass concentrations are needed prior to continuous mode, a fed-batch phase can be employed to reach targeted biomass, or a continuous adaption phase can result in the same effect. Furthermore, in case induction is employed throughout continuous phase, either (i) a non-induced adaption phase consisting of four residence times has to be performed to establish an equilibrium state prior to induction (D-stat. cultivation), or (ii) an "induced continuous phase" is performed directly after batch phase (Figure 1.9).

For continuous cultivations with *E. coli*, dilution rates in the range of 0.1–0.5 h^{-1} are commonly employed for the screening of wild-type strains [84]. However, when producing recombinant proteins, biomass yield might decrease, as lower cell capacities for maintenance are given (Figure 1.4) [72]. Hence lower dilution rates in the range of 0.1–0.2 h^{-1} should be employed for induced chemostat cultivation, producing recombinant proteins, to avoid sugar accumulation and minimize the risk of washout.

D-stat cultivation might be an option to adapt cells stepwise to the formation of target molecules. In an induced chemostat, many shifts occur in parallel, as (i) carbon limited growth is started and (ii) a constant washout of cells occurs. As levels of ppGpp and rpoS were shown to alter within minutes when switching to carbon-limited growth, for fed-batch and continuous cultivations [85], "small" shifts might already cause high deviations in the host cell transcriptome. In induced chemostat cultivation, an inducer is supplemented on top, implementing additional shifts such as (i) the establishment of new transport systems and (ii) recombinant molecule growth. Regarding the overall TSY, it would be beneficial to omit the adaption phase and perform induction phase directly after desired biomass concentration is achieved. Still, its shifts must be investigated for any product and host, whether cells tolerate a harsh shift such as an induced chemostat. Cell stress might cause negative side effects, possibly causing a unstable productivity.

In case a continuous cultivation strategy can be found to maintain stable productivity, the downtime can be reduced significantly. Using continuous systems, the percentage of downtime in comparison with total process time can be reduced from 65.7% to 14.3% or 4.7%, comparing fed-batch cultivations with D-stat and induced chemostat cultivations, respectively (Figure 1.9). Therefore, average downtime in continuous processes with *E. coli* is at maximum 22% (14.3% for D-stat vs. 65.7% for fed-batch downtime) of the downtimes required for conventional fed-batch cultivation. Using the timeframe shown in Figure 1.9, a continuous cultivation producing more than 22% of the total throughput achieved in a fed batch would thus be superior. However, in this calculation, no purification procedure is included, and downstream processing is known to be the bottleneck in microbial production of recombinant proteins [86, 87]. Low target protein concentrations provoke highly difficult and expensive downstream applications [88]. To realize the calculation above, (i) continuous systems would have to achieve the same purity (ratio of target protein to impurities) as achieved at fed-batch harvest and (ii) continuous downstream applications would have to achieve the same purification yields as batchwise downstream. In case the same purification yield can be achieved, a continuous cultivation system reducing overall downtime can have a major increase on total product throughput. Taking timeframes depicted in Figure 1.9, continuous cultivation producing constantly 50% of a fed-batch productivity would thus increase total product throughput by more than double. Within this calculation, the saved costs for chemicals and energy are not even implemented; therefore the high potential of increasing TSY via continuous cultivations is most definitely given.

1.4.2 Reasons Why CBM with *E. coli* Is Not State of the Art (Yet)

1.4.2.1 Formation of Subpopulation Following Genotypic Diversification

Biotechnology, unlike many other branches of industry, employs living cells for catalysis. Following Darwin's principles, all living cells always suffer from a certain mutation rate to create a more-fit species [89]. Distinguishing microbial cultivations on a species level would result in a harsh difference; however we can differentiate them into certain subpopulations. Hence, any bacterial cell, bearing a certain mutation, could be the beginning of a new subpopulation formation. Mutation rate probabilities in *E. coli* have been summarized by Rugbjerg and Sommer [90], showing the likelihoods of mutation rates in conventional time-dependent cultivations. Even though mammalian cells bear a higher mutation rate probability than *E. coli*, bacterial cells exhibit much higher growth rates (Table 1.2), and therefore time-dependent effects might occur faster cultivating bacterial cells than in mammalian cells (Table 1.2) [23, 27, 28, 33].

Alterations in the karyotype (chromosome restructuring) were found to increase majorly in mammalian cell cultivations lasting longer than 50 generation. This results in population heterogeneities upon changing genetic material [91]. Furthermore, cloning in CHO cells is an uncontrolled procedure, resulting in a high number of gene copies in the cell with uncontrolled loci of integration [91]. Epigenetic changes based on DNA methylation (decline of recombinant Mab transcript copy number correlated with increased) and methylation of the Mab human cytomegalovirus (CMV) promoter and gene loss are complex interactions making cell culture cultivation hard to predict [92]. Genetic instability for continuous cultivations with *E. coli* is also based on several different effects [44]. Without additional selection pressure, bacterial cells tend to expel plasmids, leading to the reduced plasmid copy numbers per cell, which can be limited by antibiotic resistances or auxotrophic genes [93, 94]. Besides the general belief that single-nucleotide polymorphism (SNP) is the predominant cause for variations in bacterial cultures, major effects in continuous cultures are based on population inhomogeneity. Metabolic burden, also called product burden, decreases fitness and specific growth rates of the producing cultures, which finally results in an overgrowth of the non-productive subpopulation. For several *E. coli* strains, IS elements were identified to be responsible for the inactivation of genes in the production of mevalonic acid [41].

However, differences on subpopulation level can be difficult to monitor, without genome sequencing or transcriptomic analysis. Distinguishing between productive and nonproductive subpopulation using RPP is rather easy to monitor and will be discussed in the following paragraph.

In red biotechnology, plasmid technology is commonly used to produce recombinant target molecules in *E. coli* [68]. Cultivated cells, which have been transformed with plasmids containing gene sequences for target proteins, generally suffer from decreased growth rates and lower yields than wild-type strains [69]. Furthermore, mutation rate probability is increasing upon addition of inducer for recombinant protein producing strains [95]. The so-called transcription-induced mutation has been described to increase mutation rate probability by a factor of 4 compared to the non-induced state [96, 97]. To form an efficient subpopulation, mutations need to decrease burden onto host cell machinery in such an efficient way that the mutated population yields in higher growth rates [41]. In conventional batch and fed-batch cultivations, induction time is relatively short compared with continuous cultivations (Figure 1.9). Thus, a takeover of a non-efficient subpopulation is unlikely to take place. Taking a set μ of $0.1\,h^{-1}$ throughout an induction time of 10 hours [28], cells would only double 1.44 times throughout their induced phase, given there is no observed decrease of growth rate [98]. However, a continuous cultivation at a set $D = \mu = 0.1\,h^{-1}$ running for an induction time of 280 hours (Figure 1.9) would bear 40.4 generations throughout its induction time. By increasing the generation times, the probability of shifting transposable elements, base-pair substitution, and large gene deletion is increased [29, 67]. Furthermore, in conventional fed-batch cultivation, cells are maintained in the fermenter until the time point of harvest, keeping all kinds of subpopulations within one reactor, whereas in chemostat a constant washout occurs. Hence, subpopulations growing "more efficiently" than other subpopulations can be detected more quickly in feed/bleed systems, as they will overgrow the initial population, which will be washed out consequently [99].

1.4.2.2 Formation of Subpopulation Following Phenotypic Diversification

Phenotypic diversification of cell population is driven by the stochasticity of the intracellular biochemical reactions. This phenomenon can be also termed biological noise and comprises two components, i.e. extrinsic noise and intrinsic noise [100]. The intrinsic component of noise results from the low abundance of reacting molecules in the cell (i.e. transcription factors and ribosomes), lowering the probability of collision between reacting species. The extrinsic component of noise is driven by cell-to-cell variation in the number of the reacting species due to the external factors. The cultivation environment thus can have a huge impact onto external factors that is majorly influenced by (i) the design of the cultivation device, (ii) the operation mode of the bioreactor (i.e. batch, fed batch, or continuous), and (ii) the used media taking metabolites or substrate accumulation into account also. The understanding of the dynamic adaptation of microbial populations to environmental perturbations is one of the key missing elements, which is required to allow controllability of the biological system under phenotypic diversification. To analyze phenotypic diversification, the development of appropriate biosensors is required. Furthermore, promoter-based biosensors have been used in many studies to track physiological changes at a single-cell resolution [29, 31, 101]. Models that integrate the stochastic components of biochemical reactions can then be incorporated based on the quality of the acquired single-cell data and can be used to reconstruct transcriptional regulatory networks [102]. Such strategies have allowed to precisely control a microbial population with optogenetics approaches, i.e. by controlling gene expression based on light pulses [103, 104]. However, these stochastic modeling frameworks need to be improved to take account of all factors, leading to the biological noise and affecting the internal state of host cells [32]. Hence, factors such as the noise in gene expression and its consequence on metabolic pathways and cell elongation/division need to be investigated more thoroughly [105–107].

The continuous mode of cultivation (i.e. the chemostat) seems to promote phenotypic diversification since nutrient limitation is a strong driver of such diversification strategies [70, 108]. Switching toward continuous cultivations will have to deal with this kind of phenomena that impair the observability of microbial populations. All the above mentioned methods can be applied to investigate phenotypic diversification of microbial population in CBM systems. However, population control is scarce. A recurrent feature is that population stability cannot be ensured in classical chemostat systems [30, 109]. On the other hand, the very same studies have shown that applying nutrient pulses at given interval during continuous cultivation promoted proteome and subpopulation stabilization. This effect known as periodic forcing has also been shown to be efficient in stabilizing the activity of synthetic gene networks [110]. Whereas the molecular mechanisms behind population stabilization through periodic forcing are still unknown, carbon-limited growth can be avoided, and phenotypic diversification can be lowered by the application of nutrient pulses. Once the puzzle is solved, substrate and inducer pulses at given frequency and amplitude throughout continuous cultivation might be a new paradigm for ensuring cellular stability during CBM.

1.4.2.3 Is Genomic Integration of the Target Protein an Enabler for CBM with *E. coli*?

Genomic integration is believed to solve the stability issues in continuous cultivations as plasmid loss can be avoided and antibiotic-free medium can be employed, being especially important in large scales [111]. As the expression of recombinant protein bears a high burden onto host cells upon induction [112], plasmid loss can be expected as a consequence. Comparing mutation rate probabilities, the probability of base substitution due to the DNA polymerase errors is in the range of 10^{-7} to 10^{-10} per generation per base pair, and disturbance by transposable elements ranges from 10^{-5} to 10^{-8} per generation per base pair, whereas plasmid loss is documented with a higher probability of 10^{-2} to 10^{-6} per generation [67]. Plasmid loss probability can be decreased by diverse selection pressures [113, 114]. Furthermore, the employment of high copy plasmids (obtaining 500–700 plasmid copies per cell) makes plasmid loss rather unlikely of being responsible for a decrease in productivity during continuous cultivation [115, 116]. Constitutive promoters are rarely employed for *E. coli*, whereas inducible promoters present the dominant fraction [59]. Using inducible promoters, no basal expression should be monitored prior to induction, and thus burden onto cells should not be detectable in non-induced growth phases [117]. Genomic integrated strains were found to show a slightly higher μ_{max} than plasmid-based systems [80].

Even though results are promising for genomic integration, the total replication rates of high copy plasmids tend to exceed the numbers achieved by genomic integrated systems. Moreover, the exact interactions in genome-integrated systems have not been understood up until now [16]. Given that the locus of integration is of high relevance for target molecule expression [118], each target sequence would require an efficient screening for the best locus on the genome. As for these reasons, genomic integration technology still needs to be eased, and a toolbox platform needs to be established, favoring plasmid-based systems in industry up to now [119]. However, the expression system and the applied product locus are highly dependent onto the desired goal. Plasmid technology might be the appropriate choice for inclusion body processes, as high copy rates are needed to meet the high titer demand [120, 121]. For the expression of soluble proteins, genomic integration might be feasible as a soft induction of host cells might lead to properly folded protein and lower copy numbers are needed to not overload chaperones and the folding machinery of *E. coli* [122].

As the plasmid loss seems not to be the reason for observed shifts in productivity, it is most likely that genome-based mutations are responsible to cause a "more-fit" subpopulation, having the probability to overtake a cultivation broth. Hence, it would be intriguing to investigate mutation rate likelihood derived (i) by genomic integration vs. plasmid-based systems, (ii) as a function of induction strength, and (iii) as a function of residence time in induced continuous systems.

1.4.3 Solutions to Overcome the Formation of Subpopulations and How to Realize CBM with *E. coli* in the Future

Engineered *E. coli* cells face a certain metabolic burden visible throughout induction time. In batch and fed-batch cultivations, the metabolic burden can be eased during biomass growth due to time dependency, i.e. growing biomass throughout non-induced phases and applying a relatively short induction time. However, recombinant protein formation is a time-dependent process; hence a trade-off between optimal product yield and too high burden has to be done [28, 82]. In continuous cultures, a time-dependent separation is impossible, as cell growth and recombinant protein formation are occuring in parallel [84]. It was clearly shown that biomass and product yields are counteracting throughout a chemostat cultivation [39]. Therefore, separation of growth and recombinant protein formation should be performed in a continuous mode to enhance the product yield, which can be realized in cascaded cultivation mode (Figure 1.3a,c) [41]. In Figure 1.10a), the biomass and product yield trends for chemostat and cascaded continuous cultivation are depicted. The chemostat is producing only for a short duration before a nonproductive subpopulation takes place as shown in the literature [30, 32, 33, 99, 123, 124]. Throughout induction phase, the biomass yield drops as a major part of the energy derived by the substrate is needed to cope for the needs of target molecule production (Figure 1.4). Depending on the set dilution rate and chosen product, the yield decrease might vary; however decreasing biomass yields upon induction cannot be avoided. Throughout the time span of a chemostat cultivation, biomass yield recovery can be observed. However, the yield recovery unfortunately happens at the expense of product formation. As we suggest that a part of the population still tries to produce the target protein, the biomass yield in chemostat cultivation is believed to be lower than the initial biomass yield observed throughout non-induced growth phases.

The two-stage continuous fermentation first described back in 1991 (Figure 1.3c) [125] should enable stable production of recombinant target protein via spatial separation of biomass and product formation. Plasmid concentration was found to stay stable throughout target molecule production using this cultivation mode [125]. For this process mode, the biomass yield in the non-induced reactor should stay stable at the theoretical level of 0.5 g/g, as host cells are not confronted with metabolic burden (Figure 1.10b). Throughout the second stage, there is a short time-dependent behavior, as cells have to adjust their metabolism to induction, and therefore a certain adaption phase can be observed [33]. In case process conditions challenge host cells with a tolerable metabolic burden, a stable production and biomass yield can be achieved, which should stay constant over time. Hence, it has to be kept in mind that the mutation rate probability in the seed reactor increases also with ongoing cultivation time, probably causing time- or generation-based effects in either one of the reactors [41].

Promising results were found using a BL21(DE3) strain transformed with conventional pET-plasmid expressing target proteins under control of the lac promoter. Independently from the target molecule employed, all studies favored higher dilution rates throughout the second stage [39, 40, 125]. We hypothesize that lower

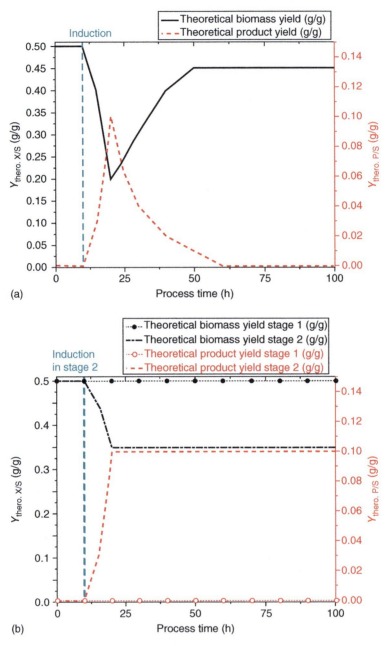

Figure 1.10 Showing theoretical biomass and product yields for (a) chemostat cultivation and (b) cascaded continuous cultivation, showing trends of the seed reactor (stage 1) and the induced producing reactor (stage 2).

residence times are beneficial, as cells do not bear the load of induction for a long duration and are washed out more quickly than in systems where low dilution rates are applied. In fed-batch cultivation, the specific feeding rate was found to have a high dependency on the production of the recombinant protein, thus making the time point of harvest crucial [81]. High feeding rates led to a peaking productivity in the range of 6–8 hours for an inclusion body process [28, 83]. Hence, residence times applied (i) should be set close to the peaking productivity monitored throughout fed-batch cultivation, (ii) should not exceed the peaking productivity to maintain cells in non-stressed metabolic state, and (iii) should not be set too low, as a certain time of induction is necessary to yield in sufficient sugar uptake and RPP. Using high dilution rates, cells might have little time to adapt to cultivation conditions; hence, the evolution of a nonproductive population can be reduced as only little time is given in the cultivation device. However, it must be considered that residence times only calculate the average proportion of cells, which are maintained in the reactor. Therefore, cell populations might also show evolutionary tribes at low residence times. Nevertheless, it has to be kept in mind that very low residence times also implement little time for sugar and inducer uptake. Therefore, the dilution rate has to be adapted to avoid washout upon yield decrease (Figure 1.4), and a trade-off between washout and emerging probability of nonproductive subpopulation must be found. Still, from an industrial point of view, higher productivities at higher dilution rates would be desired as the overall mass flow of product out of the reactor is increased, implementing an increased TSY.

1.5 Conclusion and Outlook

Continuous systems have a huge opportunity to exceed the TSY of time-dependent cultivations (i.e. fed-batch systems). However, due to the high amount in generations achieved by microbial continuous cultivations, the formation of different subpopulations can be observed. Hence, population heterogeneity needs to be dealt with. Flow cytometry measurements of populations via specific staining (red-but-not-dead phenotype) or the use of reporter genes coupled to fluorescent markers might allow to adjust feeding and dilution rates to keep populations at a constant level. Control strategies known as "segregostat" have been recently developed where subpopulations are monitored via online flow cytometry and steered at a certain level via glucose pulses. Furthermore, cascaded continuous cultivation, separating biomass formation and RPP, has been referred to allow feasible productivity throughout continuous processes. Nevertheless, process steps must be investigated thoroughly for their criticality. A yield comparison between time-dependent and time-independent process modes should therefore be carried out for any production process. A continuous process is only sufficient if all unit operations can be combined properly, enabling a fully functioning process chain. Technologies for continuous upstream have been developed, and the continuous purification of inclusion bodies has also been demonstrated. Fusing current state-of-the-art techniques might thus allow to implement a fully integrated microbial continuous process. Once control strategies for microbial continuous systems are sophisticated enough for regulatory authorities, they should be implemented in the corresponding unit operations, and hence continuous processing with microbials should be authorized from a regulatory point of view. So, if companies are questioning whether they should also pursue continuous systems or not, just give it a try - It might increase process efficiency.

References

1 Gavrilescu, M. and Chisti, Y. (2005). Biotechnology – a sustainable alternative for chemical industry. *Biotechnol. Adv.* 23: 471–499.
2 Show, P.L., Oladele, K.O., Siew, Q.Y. et al. (2015). Overview of citric acid production from *Aspergillus niger*. *Front. Life Sci.* 8: 271–283.
3 Vaija, J. and Linko, P. (1986). Continuous citric acid production by immobilized *Aspergillus niger*: reactor performance and fermentation kinetics. *J. Mol. Catal.* 38: 237–253.
4 Satyanarayana, T., Littlechild, J., and Kawarabayasi, Y.J.B.T. (2013). *Thermophilic Microbes in Environmental and Industrial Biotechnology*, vol. 3. Biotechnology of Thermophiles.
5 Quehenberger, J., Pittenauer, E., Allmaier, G., and Spadiut, O. (2020). The influence of the specific growth rate on the lipid composition of *Sulfolobus acidocaldarius*. *Extremophiles* 24: 413–420.
6 Rawlings, D.E. (2013). *Biomining: Theory, Microbes and Industrial Processes*. Springer Science & Business Media.
7 Lovley, D.R. and Lloyd, J.R.J.N. (2000). Microbes with a mettle for bioremediation. *Nat. Biotechnol.* 18: 600–601.
8 Anderson, R.T. and Lovley, D.R. (1997). Ecology and biogeochemistry of in situ groundwater bioremediation. In: *Advances in Microbial Ecology*, 289–350. Springer.
9 Demain, A.L. and Vaishnav, P. (2009). Production of recombinant proteins by microbes and higher organisms. *Biotechnol. Adv.* 27: 297–306.
10 Murooka, Y. (1993). *Recombinant Microbes for Industrial and Agricultural Applications*. CRC Press.
11 Saxelin, M., Tynkkynen, S., Mattila-Sandholm, T., and de Vos, W.M. (2005). Probiotic and other functional microbes: from markets to mechanisms. *Curr. Opin. Biotechnol.* 16: 204–211.
12 Antoni, D., Zverlov, V.V., and Schwarz, W.H. (2007). Biofuels from microbes. *Appl. Microbiol. Biotechnol.* 77: 23–35.
13 Joshi, B., Joshi, J., Bhattarai, T., and Sreerama, L. (2019). Chapter 15: Currently used microbes and advantages of using genetically modified microbes for ethanol production. In: *Bioethanol Production from Food Crops* (eds. R.C. Ray and S. Ramachandran), 293–316. Academic Press.
14 Chen, G.-Q. (2009). A microbial polyhydroxyalkanoates (PHA) based bio-and materials industry. *Chem. Soc. Rev.* 38: 2434–2446.

15 Kamravamanesh, D., Slouka, C., Limbeck, A. et al. (2019). Increased carbohydrate production from carbon dioxide in randomly mutated cells of cyanobacterial strain *Synechocystis* sp. PCC 6714: bioprocess understanding and evaluation of productivities. *Bioresour. Technol.* 273: 277–287.

16 Mainka, T., Mahler, N., Herwig, C., and Pflügl, S. (2019). Soft sensor-based monitoring and efficient control strategies of biomass concentration for continuous cultures of *Haloferax mediterranei* and their application to an industrial production chain. *Microorganisms* 7: 648.

17 Mahler, N., Tschirren, S., Pflügl, S., and Herwig, C. (2018). Optimized bioreactor setup for scale-up studies of extreme halophilic cultures. *Biochem. Eng. J.* 130: 39–46.

18 Nielsen, C., Rahman, A., Rehman, A.U. et al. (2017). Food waste conversion to microbial polyhydroxyalkanoates. *Microb. Biotechnol.* 10: 1338–1352.

19 Panda, S.K., Mishra, S.S., Kayitesi, E., and Ray, R.C. (2016). Microbial-processing of fruit and vegetable wastes for production of vital enzymes and organic acids: biotechnology and scopes. *Environ. Res.* 146: 161–172.

20 Rabaey, K. and Verstraete, W. (2005). Microbial fuel cells: novel biotechnology for energy generation. *Trends Biotechnol.* 23: 291–298.

21 ICH: Q8. (2009). Pharmaceutical development (R2). wwwichorg (accessed 23 June 2021).

22 EMA. (2012). *EMA: Guideline on Process Validation*. EMA/CHMP/CVMP/QWP/70278/2012-Rev1.

23 Zahel, T., Marschall, L., Abad, S. et al. (2017). Workflow for criticality assessment applied in biopharmaceutical process validation stage 1. *Bioengineering (Basel)* 4: 14.

24 Dream, R., Herwig, C., and Pelletier, E. (2018). Continuous biomanufacturing – challenges for implementation. *Pharm. Eng.* 38: 42–51.

25 Zahel, T., Hauer, S., Mueller, E. et al. (2017). Integrated process modeling – a process validation life cycle companion. *Bioengineering* 4: 86.

26 Gustavsson, R. and Mandenius, C.-F. (2013). Soft sensor control of metabolic fluxes in a recombinant *Escherichia coli* fed-batch cultivation producing green fluorescence protein. *Bioprocess. Biosyst. Eng.* 36: 1375–1384.

27 Yamanè, T. and Shimizu, S. (1984). Fed-batch techniques in microbial processes. In: *Bioprocess Parameter Control*, 147–194. Springer.

28 Slouka, C., Kopp, J., Strohmer, D. et al. (2019). Monitoring and control strategies for inclusion body production in *E. coli* based on glycerol consumption. *J. Biotechnol.* 296: 75–82.

29 Baert, J., Kinet, R., Brognaux, A. et al. (2015). Phenotypic variability in bioprocessing conditions can be tracked on the basis of on-line flow cytometry and fits to a scaling law. *Biotechnol. J.* 10: 1316–1325.

30 Sassi, H., Nguyen, T.M., Telek, S. et al. (2019). Segregostat: a novel concept to control phenotypic diversification dynamics on the example of Gram-negative bacteria. *Microb. Biotechnol.* 12 (5): 1064–1075.

31 Delvigne, F., Brognaux, A., Gorret, N. et al. (2011). Characterization of the response of GFP microbial biosensors sensitive to substrate limitation in scale-down bioreactors. *Biochem. Eng. J.* 55: 131–139.

32 Delvigne, F., Baert, J., Sassi, H. et al. (2017). Taking control over microbial populations: current approaches for exploiting biological noise in bioprocesses. *Biotechnol. J.* 12: 1600549.

33 Kopp, J., Slouka, C., Spadiut, O., and Herwig, C. (2019). The rocky road from fed-batch to continuous processing with *E. coli. Front. Bioeng. Biotechnol.* 7: 328.

34 Reichelt, W.N., Brillmann, M., Thurrold, P. et al. (2017). Physiological capacities decline during induced bioprocesses leading to substrate accumulation. *Biotechnol. J.* 12: 1600547.

35 Kopp, J., Slouka, C., Ulonska, S. et al. (2017). Impact of glycerol as carbon source onto specific sugar and inducer uptake rates and inclusion body productivity in *E. coli* BL21(DE3). *Bioengineering (Basel)* 5.

36 Sagmeister, P., Kment, M., Wechselberger, P. et al. (2013). Soft-sensor assisted dynamic investigation of mixed feed bioprocesses. *Process Biochem.* 48: 1839–1847.

37 Ezeji, T.C., Qureshi, N., and Blaschek, H.P. (2013). Microbial production of a biofuel (acetone–butanol–ethanol) in a continuous bioreactor: impact of bleed and simultaneous product removal. *Bioprocess Biosyst. Eng.* 36: 109–116.

38 Ni, Y., Xia, Z., Wang, Y., and Sun, Z. (2013). Continuous butanol fermentation from inexpensive sugar-based feedstocks by *Clostridium saccharobutylicum* DSM 13864. *Bioresour. Technol.* 129: 680–685.

39 Kopp, J., Kolkmann, A.-M., Veleenturf, P.G. et al. (2019). Boosting recombinant inclusion body production-from classical fed-batch approach to continuous cultivation. *Front. Bioeng. Biotechnol.* 7: 297.

40 Schmideder, A. and Weuster-Botz, D. (2017). High-performance recombinant protein production with *Escherichia coli* in continuously operated cascades of stirred-tank reactors. *J. Ind. Microbiol. Biotechnol.* 44: 1021–1029.

41 Buerger, J., Gronenberg, L.S., Genee, H.J., and Sommer, M.O.A. (2019). Wiring cell growth to product formation. *Curr. Opin. Biotechnol.* 59: 85–92.

42 Rasmussen, B., Davis, R., Thomas, J., and Reddy, P. (1998). Isolation, characterization and recombinant protein expression in Veggie-CHO: a serum-free CHO host cell line. In: *Cell Culture Engineering VI*, 31–42. Springer.

43 Gagnon, M., Nagre, S., Wang, W. et al. (2019). Novel, linked bioreactor system for continuous production of biologics. *Biotechnol. Bioeng.* 116: 1946–1958.

44 Peebo, K. and Neubauer, P. (2018). Application of continuous culture methods to recombinant protein production in microorganisms. 6: 56.

45 Curvers, S., Linnemann, J., Klauser, T. et al. (2002). Recombinant protein production with *Pichia pastoris* in continuous fermentation – kinetic analysis of growth and product formation. *Eng. Life Sci.* 2: 229–235.

46 Wurm, D.J., Hausjell, J., Ulonska, S. et al. (2017). Mechanistic platform knowledge of concomitant sugar uptake in *Escherichia coli* BL21(DE3) strains. *Sci. Rep.* 7: 45072.

47 Walsh, G. (2018). Biopharmaceutical benchmarks 2018. *Nat. Biotechnol.* 36: 1136–1145.

48 Walsh, G. (2005). Therapeutic insulins and their large-scale manufacture. *Appl. Microbiol. Biotechnol.* 67: 151–159.

49 Hamilton, S.R. and Gerngross, T.U. (2007). Glycosylation engineering in yeast: the advent of fully humanized yeast. *Curr. Opin. Biotechnol.* 18: 387–392.

50 Wellhoefer, M., Sprinzl, W., Hahn, R., and Jungbauer, A. (2014). Continuous processing of recombinant proteins: integration of refolding and purification using simulated moving bed size-exclusion chromatography with buffer recycling. *J. Chromatogr. A* 1337: 48–56.

51 Ötes, O., Flato, H., Vazquez Ramirez, D. et al. (2018). Scale-up of continuous multicolumn chromatography for the protein a capture step: from bench to clinical manufacturing. *J. Biotechnol.* 281: 168–174.

52 Ötes, O., Flato, H., Winderl, J. et al. (2017). Feasibility of using continuous chromatography in downstream processing: comparison of costs and product quality for a hybrid process vs. a conventional batch process. *J. Biotechnol.* 259: 213–220.

53 Zydney, A.L. (2016). Continuous downstream processing for high value biological products: a review. *Biotechnol. Bioeng.* 113: 465–475.

54 Shehadul Islam, M., Aryasomayajula, A., and Selvaganapathy, P.R. (2017). A review on macroscale and microscale cell lysis methods. *Microorganisms* 8: 83.

55 D'hondt, E., Martin-Juarez, J., Bolado, S. et al. (2017). Cell disruption technologies. In: *Microalgae-based Biofuels and Bioproducts*, 133–154. Elsevier.

56 Lin, Z. and Cai, Z. (2009). Cell lysis methods for high-throughput screening or miniaturized assays. *Biotechnol. J.* 4: 210–215.

57 Balasundaram, B., Harrison, S., and Bracewell, D.G. (2009). Advances in product release strategies and impact on bioprocess design. *Trends Biotechnol.* 27: 477–485.

58 Barazzone, G.C., Carvalho, R., Kraschowetz, S. et al. (2011). Production and purification of recombinant fragment of pneumococcal surface protein A (PspA) in *Escherichia coli*. *Proc. Vaccinol.* 4: 27–35.

59 Slouka, C., Kopp, J., Spadiut, O., and Herwig, C. (2018). Perspectives of inclusion bodies for bio-based products: curse or blessing? *Appl. Microbiol. Biotechnol.*

60 Singh, A., Upadhyay, V., Upadhyay, A.K. et al. (2015). Protein recovery from inclusion bodies of *Escherichia coli* using mild solubilization process. *Microb. Cell Fact.* 14: 41.

61 Walch, N. and Jungbauer, A. (2017). Continuous desalting of refolded protein solution improves capturing in ion exchange chromatography: a seamless process. *Biotechnol. J.* 12.

62 Kateja, N., Agarwal, H., Hebbi, V., and Rathore, A.S. (2017). Integrated continuous processing of proteins expressed as inclusion bodies: GCSF as a case study. *Biotechnol. Progr.* 33: 998–1009.

63 Jungbauer, A. (2013). Continuous downstream processing of biopharmaceuticals. *Trends Biotechnol.* 31: 479–492.

64 Eldar, A. and Elowitz, M.B. (2010). Functional roles for noise in genetic circuits. *Nature* 467: 167–173.

65 Ackermann, M. (2015). A functional perspective on phenotypic heterogeneity in microorganisms. *Nat. Rev. Microbiol.* 13: 497–508.

66 Veening, J.-W., Smits, W.K., and Kuipers, O.P. (2008). Bistability, epigenetics, and bet-hedging in bacteria. *Ann. Rev. Microbiol.* 62: 193–210.

67 Binder, D., Drepper, T., Jaeger, K.-E. et al. (2017). Homogenizing bacterial cell factories: analysis and engineering of phenotypic heterogeneity. *Metab. Eng.* 42: 145–156.

68 Humer, D. and Spadiut, O. (2018). Wanted: more monitoring and control during inclusion body processing. *World J. Microbiol. Biotechnol.* 34: 158–158.

69 Li, M., Xu, J., Romero-Gonzalez, M. et al. (2012). Single cell Raman spectroscopy for cell sorting and imaging. *Curr. Opin. Biotechnol.* 23: 56–63.

70 Schreiber, F., Littmann, S., Lavik, G. et al. (2016). Phenotypic heterogeneity driven by nutrient limitation promotes growth in fluctuating environments. *Nat. Microbiol.* 1: 16055.

71 Fragoso-Jiménez, J.C., Baert, J., Nguyen, T.M. et al. (2019). Growth-dependent recombinant product formation kinetics can be reproduced through engineering of glucose transport and is prone to phenotypic heterogeneity. *Microb. Cell Fact.* 18: 26.

72 Ceroni, F., Algar, R., Stan, G.B., and Ellis, T. (2015). Quantifying cellular capacity identifies gene expression designs with reduced burden. *Nat. Methods* 12: 415–418.

73 Foley, D.A., Wang, J., Maranzano, B. et al. (2013). Online NMR and HPLC as a reaction monitoring platform for pharmaceutical process development. *Anal. Chem.* 85: 8928–8932.

74 Kroll, P., Hofer, A., Ulonska, S. et al. (2017). Model-based methods in the biopharmaceutical process lifecycle. *Pharm. Res.* 34: 2596–2613.

75 Kroll, P., Hofer, A., Stelzer, I.V., and Herwig, C. (2017). Workflow to set up substantial target-oriented mechanistic process models in bioprocess engineering. *Process Biochem.* 62: 24–36.

76 Ulonska, S., Kroll, P., Fricke, J. et al. (2018). Workflow for target-oriented parametrization of an enhanced mechanistic cell culture model. *Biotechnol. J.* 13: e1700395.

77 Rodríguez-Carmona, E., Cano-Garrido, O., Dragosits, M. et al. (2012). Recombinant Fab expression and secretion in *Escherichia coli* continuous culture at medium cell densities: influence of temperature. *Process Biochem.* 47: 446–452.

78 Vaghari H, Anarjan N, Najian Y, Jafarizadeh-Malmiri H: Sterilization process. In *Essentials in Fermentation Technology*. Edited by Berenjian A. *Cham: Springer International Publishing*; 2019: 85-103.

79 Wurm, D.J., Veiter, L., Ulonska, S. et al. (2016). The *E. coli* pET expression system revisited-mechanistic correlation between glucose and lactose uptake. *Appl. Microbiol. Biotechnol.* 100: 8721–8729.

80 Hausjell, J., Kutscha, R., Gesson, D.J. et al. (2020). The effects of lactose induction on a plasmid-free *E. coli* T7 expression system. *Bioengineering*: 7.

81 Hausjell, J., Weissensteiner, J., Molitor, C. et al. *E. coli* HMS174(DE3) is a sustainable alternative to BL21(DE3). *Microb. Cell Fact.* 172018.

82 Slouka, C., Kopp, J., Hutwimmer, S. et al. (2018). Custom made inclusion bodies: impact of classical process parameters and physiological parameters on inclusion body quality attributes. *Microb. Cell Fact.* 17: 148.

83 Kopp, J., Slouka, C., Strohmer, D. et al. (2018). Inclusion body bead size in *E. coli* controlled by physiological feeding. *Microorganisms* 6.

84 Adamberg, K., Valgepea, K., and Vilu, R. (2015). Advanced continuous cultivation methods for systems microbiology. *Microbiology* 161: 1707–1719.

85 Teich, A., Meyer, S., Lin, H.Y. et al. (1999). Growth rate related concentration changes of the starvation response regulators σS and ppGpp in glucose-limited fed-batch and continuous cultures of *Escherichia coli*. *Biotechnol. Progr.* 15: 123–129.

86 Peleg, Y. and Unger, T. (2012). Resolving bottlenecks for recombinant protein expression in *E. coli*. In: *Chemical Genomics and Proteomics: Reviews and Protocols* (ed. E.D. Zanders), 173–186. Totowa, NJ: Humana Press.

87 Gagaoua, M. (2018). Chapter 8: Aqueous methods for extraction/recovery of macromolecules from microorganisms of atypical environments: a focus on three phase partitioning. In: *Methods in Microbiology*, vol. 45 (eds. V. Gurtler and J.T. Trevors), 203–242. Academic Press.

88 Persson, J. and Lester, P. (2004). Purification of antibody and antibody-fragment from *E. coli* homogenate using 6,9-diamino-2-ethoxyacridine lactate as precipitation agent. *Biotechnol. Bioeng.* 87: 424–434.

89 Paul, D.B. (1988). The selection of the "Survival of the Fittest". *J. Hist. Biol.* 21: 411–424.

90 Rugbjerg, P. and Sommer, M.O.A. (2019). Overcoming genetic heterogeneity in industrial fermentations. *Nat. Biotechnol.* 37: 869–876.

91 Wurm, F.M. (2013). CHO quasispecies – implications for manufacturing processes. *Processes* 1: 296–311.

92 Kim, M., O'Callaghan, P.M., Droms, K.A., and James, D.C. (2011). A mechanistic understanding of production instability in CHO cell lines expressing recombinant monoclonal antibodies. *Biotechnol. Bioeng.* 108: 2434–2446.

93 Sieben, M., Steinhorn, G., Müller, C. et al. (2016). Testing plasmid stability of *Escherichia coli* using the continuously operated shaken BIOreactor system. *Biotechnol. Progr.* 32: 1418–1425.

94 Porse, A., Jahn, L.J., Ellabaan, M.M.H., and Sommer, M.O.A. (2020). Dominant resistance and negative epistasis can limit the co-selection of de novo resistance mutations and antibiotic resistance genes. *Nat. Commun.* 11: 1199.

95 Sankar, T.S., Wastuwidyaningtyas, B.D., Dong, Y. et al. (2016). The nature of mutations induced by replication–transcription collisions. *Nature* 535: 178–181.

96 Beletskii, A., Grigoriev, A., Joyce, S., and Bhagwat, A.S. (2000). Mutations induced by bacteriophage T7 RNA polymerase and their effects on the composition of the T7 genome11 Edited by M. Gottesman. *J. Mol. Biol.* 300: 1057–1065.

97 Beletskii, A. and Bhagwat, A.S. (1996). Transcription-induced mutations: increase in C to T mutations in the nontranscribed strand during transcription in *Escherichia coli*. *Proc. Natl. Acad. Sci.* 93: 13919.

98 Scott, M., Gunderson, C.W., Mateescu, E.M. et al. (2010). Interdependence of cell growth and gene expression: origins and consequences. *Science* 330: 1099–1102.

99 Rugbjerg, P., Myling-Petersen, N., Porse, A. et al. (2018). Diverse genetic error modes constrain large-scale bio-based production. *Nat. Commun.* 9: 787.

100 Swain, P.S., Elowitz, M.B., and Siggia, E.D. (2002). Intrinsic and extrinsic contributions to stochasticity in gene expression. *Proc. Natl. Acad. Sci.* 99: 12795.

101 Delvigne, F., Boxus, M., Ingels, S., and Thonart, P. (2009). Bioreactor mixing efficiency modulates the activity of a prpoS::GFP reporter gene in *E. coli*. *Microb. Cell Fact.* 8: 15.

102 Lillacci, G. and Khammash, M. (2013). The signal within the noise: efficient inference of stochastic gene regulation models using fluorescence histograms and stochastic simulations. *Bioinformatics* 29: 2311–2319.

103 Briat, C. and Khammash, M. (2018). Perfect adaptation and optimal equilibrium productivity in a simple microbial biofuel metabolic pathway using dynamic integral control. *ACS Synth. Biol.* 7: 419–431.

104 Milias-Argeitis, A., Rullan, M., Aoki, S.K. et al. (2016). Automated optogenetic feedback control for precise and robust regulation of gene expression and cell growth. *Nat. Commun.* 7: 12546.

105 van Boxtel, C., van Heerden, J.H., Nordholt, N. et al. (2017). Taking chances and making mistakes: non-genetic phenotypic heterogeneity and its consequences for surviving in dynamic environments. *J. R. Soc. Interface* 14: 20170141.

106 van Heerden, J.H., Kempe, H., Doerr, A. et al. (2017). Statistics and simulation of growth of single bacterial cells: illustrations with *B. subtilis* and *E. coli*. *Sci. Rep.* 7: 16094.

107 Nordholt, N., van Heerden, J., Kort, R., and Bruggeman, F.J. (2017). Effects of growth rate and promoter activity on single-cell protein expression. *Sci. Rep.* 7: 6299.

108 Gasperotti, A., Brameyer, S., Fabiani, F., and Jung, K. (2020). Phenotypic heterogeneity of microbial populations under nutrient limitation. *Curr. Opin. Biotechnol.* 62: 160–167.

109 Wright, N.R., Wulff, T., Palmqvist, E.A. et al. (2020). Fluctuations in glucose availability prevent global proteome changes and physiological transition during prolonged chemostat cultivations of *Saccharomyces cerevisiae*. *Biotechnol. Bioeng.* 117: 2074–2088.

110 Lugagne, J.-B., Sosa Carrillo, S., Kirch, M. et al. (2017). Balancing a genetic toggle switch by real-time feedback control and periodic forcing. *Nat. Commun.* 8: 1671.

111 Reisch, C.R. and Prather, K.L.J. (2017). Scarless Cas9 assisted recombineering (no-SCAR) in *Escherichia coli*, an easy-to-use system for genome editing. *Curr. Protoc. Mol. Biol.* 117: 31.38.31–31.38.20.

112 Rozkov, A., Avignone-Rossa, C.A., Ertl, P.F. et al. (2004). Characterization of the metabolic burden on *Escherichia coli* DH1 cells imposed by the presence of a plasmid containing a gene therapy sequence. *Biotechnol. Bioeng.* 88: 909–915.

113 Peubez, I., Chaudet, N., Mignon, C. et al. (2010). Antibiotic-free selection in *E. coli*: new considerations for optimal design and improved production. *Microb. Cell Fact.* 9: 65.

114 Vidal, L., Pinsach, J., Striedner, G. et al. (2008). Development of an antibiotic-free plasmid selection system based on glycine auxotrophy for recombinant protein overproduction in *Escherichia coli*. *J. Biotechnol.* 134: 127–136.

115 Mayer, M.P. (1995). A new set of useful cloning and expression vectors derived from pBlueScript. *Gene* 163: 41–46.

116 Summers, D.K. (1991). The kinetics of plasmid loss. *Trends Biotechnol.* 9: 273–278.

117 Silva, F., Queiroz, J.A., and Domingues, F.C. (2012). Evaluating metabolic stress and plasmid stability in plasmid DNA production by *Escherichia coli*. *Biotechnol. Adv.* 30: 691–708.

118 Englaender, J.A., Jones, J.A., Cress, B.F. et al. (2017). Effect of genomic integration location on heterologous protein expression and metabolic engineering in *E. coli*. *ACS Synth. Biol.* 6: 710–720.

119 Rosano, G.L., Morales, E.S., and Ceccarelli, E.A. (2019). New tools for recombinant protein production in *Escherichia coli*: a 5-year update. *Protein Sci.* 28: 1412–1422.

120 Rosano, G.L. and Ceccarelli, E.A. (2014). Recombinant protein expression in *Escherichia coli*: advances and challenges. *Front. Microbiol.* 5: 172.

121 Spadiut, O., Capone, S., Krainer, F. et al. (2014). Microbials for the production of monoclonal antibodies and antibody fragments. *Trends Biotechnol.* 32: 54–60.

122 Gu, P., Yang, F., Su, T. et al. (2015). A rapid and reliable strategy for chromosomal integration of gene(s) with multiple copies. *Sci. Rep.* 5: 9684.

123 Heins, A.-L., Johanson, T., Han, S. et al. (2019). Quantitative flow cytometry to understand population heterogeneity in response to changes in substrate availability in *Escherichia coli* and *Saccharomyces cerevisiae* chemostats. *Front. Bioeng. Biotechnol.* 7: 187.

124 Leygeber, M., Lindemann, D., Sachs, C.C. et al. (2019). Analyzing microbial population heterogeneity-expanding the toolbox of microfluidic single-cell cultivations. *J. Mol. Biol.* 431 (23): 4569–4588.

125 Kim, J.-Y. and Ryu, D.D.Y. (1991). The effects of plasmid content, transcription efficiency, and translation efficiency on the productivity of a cloned gene protein in *Escherichia coli*. *Biotechnol. Bioeng.* 38: 1271–1279.

2

Control of Continuous Manufacturing Processes for Production of Monoclonal Antibodies

Anurag S. Rathore, Garima Thakur, Saxena Nikita, and Shantanu Banerjee

Indian Institute of Technology Delhi, Department of Chemical Engineering, Hauz Khas, New Delhi, 110016, India

2.1 Introduction

Continuous manufacturing of biotherapeutics is currently of significant interest to both academic and industrial groups in the biotherapeutic industry [1, 2]. In the last decade, rapid growth has been seen in the specialized technologies designed to facilitate continuous manufacturing of a wide range of mammalian and microbial biotherapeutics. On the upstream side, perfusion cell culture systems for continuous production of biomolecules at high titer have seen rapid development and implementation [3]. Various clarification systems such as continuous centrifugation [4, 5], alternating tangential flow filtration [6, 7], and acoustic wave separation [8, 9] have been developed for integration between continuous upstream and downstream processes. Versatile chromatography setups with in-built pumps, valves, and sensors enabling integrated multicolumn operation such as the Cadence™ BioSMB (Sartorius Stedim), Octave™ SMB (Tarpon Systems), ÄKTA PCC (GE Healthcare), Contichrom® CUBE (ChromaCon), and BioSC® (Novasep) have been launched [10]. Customized flow reactors and hold tanks have been developed for achieving continuous low-pH conditions [11–14] and carrying out reactions that were previously done only in batch mode, such as PEGylation [15] and enzymatic cleavage [16]. For continuous formulation, innovative membrane modules have been developed that can achieve high concentration factors and 99.9% buffer exchange in a single pass with continuous flow [17–19].

However, each new innovation enabling conversion of unit operations from batch to continuous mode comes with a set of unique challenges with respect to the implementation and real-time control. In batch mode, the output material of each unit is extensively checked with a variety of offline analytical tools to ensure that the critical quality attributes (CQAs) are within the prescribed range. In the case of any deviation or process nonlinearity, various corrective actions can be easily taken such as repeating or modifying the unit operation or discarding the material, prior to starting the next step. The major challenge of continuous processing is handling deviations in

Process Control, Intensification, and Digitalisation in Continuous Biomanufacturing, First Edition.
Edited by Ganapathy Subramanian.
© 2022 WILEY-VCH GmbH. Published 2022 by WILEY-VCH GmbH.

real time without needing to stop the process, which requires a robust set of real-time online or at-line analytical tools coupled with automated control strategies that can monitor the CQAs and, when required, implement control decisions to modify the process to ensure that they remain in the required range [20–22]. Each unit operation must have a set of real-time sensors coupled with real-time control strategies to automatically handle a range of deviations, including variations in the quality attributes of the upstream material and equipment errors, leading to the changes in the output process stream.

In this chapter, we endeavor to put forth a range of examples for continuous control of the most commonly used unit operations in manufacturing of monoclonal antibodies (mAbs). Today, mAbs are the leading class of biotherapeutic products, comprising more than 50% of all new biotherapeutic approvals in the last five years [23]. The unit operations involved in mAb processing are typically composed of cell culture, clarification, depth filtration, chromatography, viral inactivation, precipitation, and formulation [24]. We also include a section on surge tanks, which are key in-process vessels placed strategically between individual unit operations to allow constant flow throughout the continuous process. The significance of each unit operation involved in downstream processing is given in Figure 2.1. We highlight CQAs that are affected by each unit operation and list a set of guidelines for each that outlines the synergy between novel instrumentation, monitoring tools, and control strategies that are required to implement robust control and ensure smooth continuous operation when faced with process deviations. The importance of process digitalization to achieve these objectives is also considered in detail. Finally, we describe a range of emerging computational approaches including statistical process control, machine learning, and artificial intelligence (AI) that help to transition from an individual unit operation-level view to a global picture of the entire process.

2.2 Control of Upstream Mammalian Bioreactor for Continuous Production of mAbs

The biological cell is like a robust self-replicating intricate machinery with the sole intent to consume all the nutrient in its vicinity for its DNA propagation. Although the intracellular processes in the cell are stochastic in nature, yet these noisy stochastic processes can be averaged to get a deterministic implication, making its behavior predictable and controllable. In this context, the biopharmaceutical industry has been using these cells for manufacturing one of the most complex engineering marvels – the mAb. Over the last few decades involving intensive research in upstream processing, mAb production at industrial scales has pulled off a 100-fold increase of titers to 9–10 g/l for fed-batch processes [25–27] and even up to 25 g/l for improved perfusion processes [28]. With major advances in cell line engineering, leading to the high-titer producing strains, the focus of the industry and regulators has shifted toward the development of the manufacturing process, the latter being responsible for two thirds of biologics drug shortages [29]. With rapidly evolving landscape, product quality will continue to remain the

AWS
- Cell separation efficiency is more than 95% when the cell load density is between 30 and 50 × 10^6 cells per millilitre
- Unaffected quality of HCCF
- Reduction in polishing depth filtration requirement
- Reduction in buffer cost

Chromatography

Capture chromatography
- Increased productivity due to multicolumn method
- Elimination of process hold times
- Higher resin utilization
- Handling high titers and large volumes from enhanced upstream processes Using samll multicolumn setups

Polishing chromatography
- Integration with capture chromatography on a single BioSMB unit
- Real time control of elution pool CQAs using online PAT tools or models to handle upstream variations
- Increased productivity due to multi column method
- Elimination of hold times
- Integration with preceding viral inactivation step by enabling continuous loading

CFIR
- Increased productivity
- Reduced holdup time
- Improved flexibility to handle process volume variation

Ultrafiltration and diafiltration
- Increased productivity due to elimination of hold times of product stream
- Reduced holdup volumes and flush volumes requirements
- Higher recoveries
- Eliminates the need for large recirculation tanks
- Decreases shear exposure and damage
- Reduces aggregation
- Reduced operational cost due to lower flow rates and smaller pumps and tubes
- Higher concentration factor

Surge tank
- Dampens and averages out process gradients
- Emergency safety stop in case of operational failure
- Adds safety feature in process
- Act as short stop for facilitating maintenance without disturbing rest of the process
- Increases process flexibility and robustness to load variations and maintains low operating pressure
- Probes placed in surge tank are useful for in-process monitoring of CPP

Dead end filtration
- Increased productivity and reduction in hold times between steps by incorporating filter switching
- Able to handle filtration after viral inactivation step with potential variation in aggregate content due to low pH

Figure 2.1 Advantage of unit operations involved in downstream processing of monoclonal antibodies.

primary cornerstone of bioproduction, and maintaining high product quality in commercial manufacturing will be an indispensable objective of the biopharma industry. In view of this, FDA had introduced the quality-by-design (QbD) approach that aims to evaluate product and process design and a subsequent robust control strategy to ensure consistent process performance [30]. The PAT initiative acts as a facilitator for QbD implementation by incorporating rigorous process monitoring of substantially large dataset, thereby proliferating traditional offline analytics [21, 31]. While several PAT tools are already in place for measurement of CPPs in the bioreactor [21, 32–34], our study will focus on the process control strategies and its enforcement onto upstream.

The primary focus of upstream is toward high titer, enhanced product quality, and high productivity. For the biopharma companies to continue increasing the volumetric productivity, process intensification with continuous operations is being pursued to boost product yield and quality [35]. The two modes for continuous operation in upstream are chemostat and perfusion. The former is a simpler process where fresh media addition and bioreactor broth removal is done at the same rate, without retaining cells [36]. In the case of perfusion, the process remains similar to chemostat with the notable difference of the cells being retained in the bioreactor. This cell retention opens up different avenues and strategies for process optimization and control.

This chapter will focus on Chinese hamster ovary (CHO) cell expression system, which is the current workhorse of mAb production [23]. The parameters of interest from a control perspective are dissolved oxygen (DO), pH, temperature, metabolites, and substrate concentrations. An introduction to the basic parameters involved in perfusion operation is given as perfusion rate (P) (unit: $V_{med}/V_R/d$), vessel media volume per bioreactor working volume per day; cell-specific perfusion rate (CSPR) (unit: pl/cell/d), amount of media (in picoliter) fed to a single cell per day; and viable cell density (VCD) (unit: $\times 10^6$ cells/ml), concentration of viable cells per unit volume [37, 38].

The control strategies to be implemented on perfusion systems (Figure 2.2) can be broadly classified in four broad ways:

1. *CSPR-based control strategy*: Studies have shown that a lower CSPR improves process performance as the rate of media consumption is lowered. However, CSPR can be lowered only to a certain limit ($CSPR_{min}$), below which the culture cannot be sustained. CSPR-based control links to important perfusion variables such as titer, cellular, and volumetric productivity. The CSPR generally vary between cell lines, and a range between 50 and 500 pl/cell/d is usually selected for operations [39]. This was validated by a study where a key observation revealed that increasing CSPR to 80 pl/cell/d did not have significant impact on cell growth or the maximum VCD. Hence, it was concluded that CSPR range of 50–60 pl/cell/d was optimum for the process [40]. Determining $CSPR_{min}$ can be generally conceived in two ways. In one study, push-to-low approach was used for stepwise reduction of perfusion rate at constant VCD to obtain lower values of CSPR and eventually $CSPR_{min}$ [41]. The drop in the perfusion rate resulted in high antibody titer. The VCD was fixed at 20×10^6 cells/ml as per the cell discard rate (CDR)-based control design. Alternatively, push-to-high approach envisions stepwise increase of

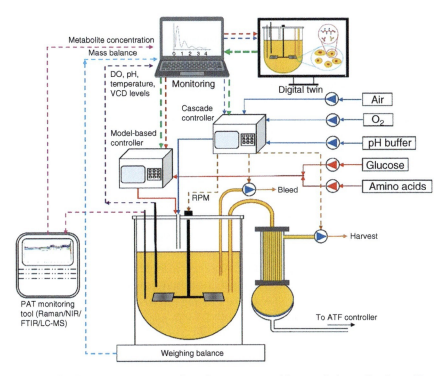

Figure 2.2 Schematic representation of process control in a perfusion cell culture. The bioreactor is connected to various PAT tools and sensors, and all the information is being integrated for real-time monitoring and development of the digital twin. The control can be based on simple cascade control of basic process parameters (such as perfusion rate, DO, pH, temperature, and VCD), a model-based PAT control (such as amino acids, glucose, and lactate), or a consolidated digital twin-based to efficiently control the bioreactor according to the culture and process conditions. The dashed lines represent data flow to or from sensors, and the solid line represents the flow of a physical entity (such as air, metabolites, etc.).

VCD at a constant perfusion rate [42]. The use of automated cell counters (Vi-cell XR, Beckman Coulter) or off-gas analyzers has made it remarkably easier to calculate the oxygen uptake rates and rise in carbon dioxide concentration for online prediction of VCD, which accord better cultivation conditions. Lower CSPR offers lower nutrient availability, thus limiting the production of inhibitors per substrate consumption, which lead to a better metabolic state of the cells [43]. Overall, CSPR-based control strategy ensures better nutrient availability at varying VCDs and results in robust cellular behavior and productivity.

2. *Oxygen uptake rate (OUR)-based control*: OUR provides critical insights on cell metabolic state, especially the oxidative phosphorylation rate in the citric acid cycle (CAC). It is one of the most feasible indicators of VCD and of cellular demand. It can also predict glucose and glutamine consumption rates and thus can provide inputs to designing a feed control strategy using online DO measurement probes [44]. Furthermore, OUR is crucial for design and scale-up of bioreactors in the case of large-scale continuous perfusion systems. One

of the first studies reported the use of online OUR estimation for calculation of the perfusion initiation time and dynamic adjustment of perfusion rate along with achieving high VCDs of almost 10^8 cells/ml [45]. Maintaining DO using PID control, OUR was measured each hour, and accordingly glucose consumption was estimated and maintained at constant level in the bioreactor. The strategy was aimed to use OUR as an indicator of metabolic load and maintain the concentration of measured nutrients at high concentrations while minimizing lactate concentrations. Apart from substrate estimation, OUR was also correlated with the amino acid consumption rate to propose feeding strategy for specific amino acids using a controlled-fed perfusion method [46]. Recently, researchers have calculated OUR using global mass balance that can be applied to long-duration perfusion systems with dynamic $k_L a$ values [47]. This approach has no prerequisites of $k_L a$ determination or any gas phase oxygen measurement; rather the difference between the bioreactor and exit DO levels is measured to calculate OUR. This makes the process robust and simple, simplifying the task of real-time metabolic flux estimation. Another study developed a mathematical model for OUR estimation in CHO cell chemostat ecosystems with enhanced prediction accuracy compared with the existing models [48]. In another recent paper, a computational fluid dynamics (CFD) with population balance modeling (PBM) was used to predict the volumetric kLa in CHO cell systems [49]. The model resulted in significant understanding of oxygen dispersion and can be used for OUR estimation in perfusion cultures operating under various hydrodynamic regimes.

3. *Substrate and metabolite-based control*: Until now, the traditional approach of model and control development have been based on empirical based practice, without integrating sufficient quantitative knowledge of cellular physiological activities. With progress in analytical instrumentation and process understanding, quantitative evaluation of the cell's metabolic activity has become a reality. Lactate and ammonium have been shown to inhibit cell growth, thus mandating the need to control their levels [50–52]. In this context, researchers have used a PAT approach by combining offline Monod model with real-time turbidity and Raman measurements to estimate metabolite and substrate levels [53]. A successful correlation ($R^2 > 0.90$) of glucose and lactate concentration was obtained using the Monod-type model, and a correlation of VCD ($R^2 > 0.95$) with Raman spectroscopy was achieved, making a potential tool for perfusion model-based control. A recent study quantified the intracellular metabolite concentration between different steady states of CHO cell line in perfusion bioreactor using matrix-assisted laser desorption/ionization time-of-flight mass spectrometry (MALDI-TOF-MS) [54]. No significant change was observed in the metabolite profile during the steady-state operation, which exhibited precise and robust control of bioreactor indicating consistent cell metabolism monitoring. In an extension of the same study, transient behavior of cell metabolism and bioreactor hydrodynamics were studied using isotope labeling in perfusion cultures [55]. A better understanding of time scales and dynamic behavior of CHO during its metabolic adaptation to external conditions was developed, which can support

further development of control strategies for achieving metabolic steady states. Adding to the list, dynamic mass balance model, incorporating metabolic flux analysis (MFA), elementary flux analysis (EFM), and extreme pathways (EPs) were used to systematically analyze the central carbon metabolism of glucose and amino acids on hybridoma and GS–CHO cell lines [56]. A macroscopic model was developed by reducing the total pathways to nine linearly independent elementary reactions while evaluating the effects of critical amino acids (glutamine, glutamate, and alanine) to the central carbon metabolism. This dynamic model can be used for further monitoring and control in a perfusion process. Switching over from classical MFA, researchers developed a dynamic MFA based on convex analysis (DMFCA) model to get insights into the metabolic flux changes in hybridoma culture [57]. DMFCA allowed determination of the bounded intervals of intracellular metabolites continuously and allowed better understanding of cell dynamic behavior, reduction of lactate, and ammonia production in perfusion mode and improvement of protein quality.

4. *Digital twin (DT)-based control*: A DT can be defined as an *in silico* or digital replica of an existing physical system composing of multiphysics and probabilistic simulation, which utilizes the maximum data from all available sensors and models to match with its corresponding twin [58]. DTs comprise a physical component (data acquisition sensors and network equipment) and a virtual component (digital representation of the physical component), which are linked to each other via an automated data management system, capable of various data visualization tools [59, 60]. DTs can be an enabler for smart manufacturing; however they require sufficient data and models for robust prediction. The intense development of PAT tools for upstream monitoring and their compelling application for an integral model-based control strategy have laid the foundation for accumulating important data and information for further development of DT. This is a multidisciplinary work and follows nine broad steps as listed by Portela et al. [61]. In context with upstream, the relevant steps include the development and validation of PAT tools and mathematical model, enabling real-time data acquisition and data integrity, constructing the automation and model predictive strategy, and finally integrating the entire strategy into a broader company-level solution. Mechanistic modeling in mammalian cultures can be used to capture the intracellular metabolic pathways, especially the IgG glycosylation in CHO cells [62–66], as well as the bioreactor hydrodynamics [67, 68]. The computationally demanding or application-limited mathematical models can be replaced or coupled with the data driven models to form empirical or hybrid model, respectively. The hybrid models (incorporating artificial neural networks) have lately become more ubiquitous in mammalian processes because of their robustness; being less sophisticated than mechanistic cell models, they require far less computational effort [69, 70]. A key leverage of using machine learning methods is the pattern recognition through volumes of data that would have been missed otherwise. Finally, the most distinctive feature of DT is the data preprocessing and integration into a central data management system and historian. The experimental datasets obtained during real time can be used to train the existing mathematical

models, and dynamic optimization of the process can be performed in parallel to the existing control strategy. Overall, DTs hold an immense advantage over traditional control strategies in upstream and are capable of providing significant technological, economic, and competitive advantage to the manufacturers.

Stringent control is of critical importance in all aspects of production, despite the inclusion of biological complexity. Starting with the control of basic process parameters such as DO, pH to mimicking cellular dynamics in situ using tools such as MFA and FBA, and real-time monitoring by vibration spectroscopy together woven into the threads of AI/ML-based data analytics, modern-day control strategies have made sincere attempts to eliminate the dearth of quality consistency. With lesser variability in upstream, the load on downstream processing thus minimizes to deal with unforeseen hurdles that might increase purification efforts and ultimately cost all while worrying about maintaining the CQA. Therefore, later, consolidation of PAT and data analytics will pave the way of automated and resilient upstream processing, where the product quality and quantity will not deviate from the desired result.

2.3 Integration Between Upstream and Downstream in Continuous Production of mAbs

2.3.1 Continuous Clarification as a Bridge Between Continuous Upstream and Downstream

The cell retention or clarification device is an integral unit of any perfusion system and is used to retain and recirculate cells in the bioreactor. The target is to obtain either a supernatant devoid of cells for an extracellular product or a concentrated slurry in the case of an intracellular product. The cell size and difference in cell density and aqueous medium density decide the separation principle of the clarification device, and generally they are classified in three categories: (i) settling (by means of gravitational or centrifugal field), (ii) filtration (using membranes and filters), and (iii) agglomeration (under the influence of ultrasonic waves). In the case of mammalian cell cultures, the cell retention device should be efficacious in retaining the cells in the bioreactor, providing minimum shear to the latter. From an operational point of view, a long operation device incurs minimum fouling with fail-safe mechanisms and is desirable for perfusion cultures.

The most popular of the cell retention devices has been filtration technologies such as alternating tangential flow (ATF) filtration and tangential flow filtration (TFF) systems. In TFF, the culture broth flow is perpendicular to the permeate direction, whereas ATF moves the broth back and forth by a diaphragm pump. ATF is widely used for achieving and maintaining high cell densities ($>50 \times 10^6$ cells/ml) for long durations [7, 71, 72]. An in-depth mechanistic level modeling of ATF system and its performance comparison with TFF has been published [6]. The ratio "r" of ATF/TFF flow rate to the permeate flow rate was assessed. The flow reversing action in ATF was found accountable for repeated membrane regeneration, relatively lower fouling, and enhanced final clarified volume as compared with TFF. Complete pore blocking was found to be the dominant fouling mechanism in both systems; however, rate of fouling in ATF was found to be lower than TFF. Deep insights in fouling mechanisms provided by the study can be useful for designing optimal filter size and flow rate optimization. Another study compared the performance of ATF and TFF using CHO cell line perfusion culture [7]. Cells undergoing clarification using ATF experienced significantly smaller stress value than TFF. However, the latter was able to retain 50% of the produced mAb, without showing significant difference in the quality. Building on the existing ATF models, Walther et al. meticulously evaluated the effects of various CPPs such as cellular residence time in ATF, hydrodynamic stress, and protein fouling of hollow fiber membranes on high VCD perfusion cultures [73]. Significant process findings indicated a relationship of ATF residence time primarily with ATF exchange rate, suggesting a residence time limit of 75 seconds to avoid decreased cell growth and dead volumes. A positive correlation between exchange rates and hydrodynamic stress was developed, which indicated the increase in stress value at higher exchange rates, leading to the reduction in cell size and productivity. A detailed empirical study was conducted to evaluate the effect of exchange rate, cell density, and filter area on filter fouling rate.

Centrifuge is another device that has been used frequently in biotech industries due to its ability to handle high cell density streams. Cell damage due to the high shear in disc stack centrifuges has been observed in several studies. Shekhawat et al. presented a study for disc stack centrifuge performance optimization and model the correlation between the cell lysis and the turbulent stresses using CFD [74]. Incremental effects of flow rates were observed in reducing cell lysis and increasing productivity while higher rotational speeds resulted in added cell damage, thus highlighting the role of CFD in modeling and optimization of such complex system. Another study investigated the function of a more promising, disposable, and automated centrifuge technology, the kSep400 fluidized bed centrifugation (FBC) [75]. The experimental results were used to find the variables (RPM and flow rate) affecting the bed height for a given volume processed. Two modeling approaches, mathematical and CFD, were used to predict the bed height vs. time relation. CFD model was demonstrated to be better than its counterpart in prediction of optimal conditions for maximizing live cell capture and processing time.

Acoustophoretic separation is another robust technology for cell clarification that has been used for over a decade as a cell retention device. The principle of acoustic cell separation is accomplished due to the formation of standing waves in a chamber, where the cells are agglomerated at cell node/antinodes and eventually settle down as clusters [76]. Several studies have evaluated the effect of CPPs like duty cycle, acoustic power, and flow rate over the cell separation efficiency [77, 78]. Just like all other clarification techniques, the acoustic filters have upgraded from cycle mode to continuous operation, like the Cadence Acoustic Separator (CAS) (Pall Life Sciences, NY). CAS employs three-dimensional standing acoustic wave technology, in contrast to the traditional planar wave, for enhanced cell separation without any significant heat dissipation in the culture broth. The system has been successfully used for clarification of high VCD CHO cell culture (20–60×10^6 cells/ml) and a yield reduction of 7- to 10-fold in secondary filter area [8]. Further modeling and optimization of the technique can result in enhanced cell clarification, with zero cell clogging issue for long-duration perfusion processes.

Such comprehensive studies have augmented our process understanding of cell clarification and have many budding applications in perfusion cell retention. The studies mentioned here have demonstrated how the effect of various CPPs on cell separation efficiency can be quantified using both modeling and experimentation approach, further allowing for the development and implementation of control strategies in perfusion systems. The development of a fully automated large-scale continuous cell clarification will result in processing of large volumes of broth with high separation efficiency and will be an enabler for seamless integration of upstream and downstream processing.

2.3.2 Considerations for Process Integration

Creating a continuous process platform is not simply connecting different unit operations in series. Typically, unit operations require modifications to create a continuous platform, as shown in Figure 2.3. Adjustments in flow rates, loading volume, operating conditions, and management of process interruptions are required. These complications are further amplified by unit integration. Biological processes are sensitive to perturbations; hence, steady-state assumptions need to be handled carefully. Although many automation approaches are available individually for continuous upstream and downstream processing unit operations, these are not integrated to form one complete solution and do not provide the kind of data that can be utilized for real-time control. The first goal in the integration is to establish a flexible and standardized downstream platform composing of multiple unit operations that can be controlled rapidly in real time. This needs to be followed by integration with the control system software designed specifically

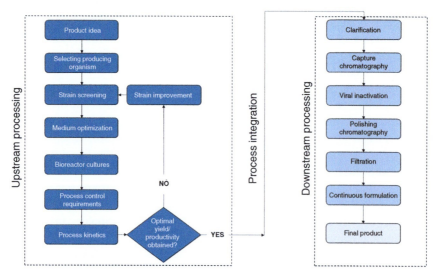

Figure 2.3 Integration of upstream and downstream process.

to address the unique challenges of biopharmaceuticals while automating the system for start-up and shutdown and enabling real time monitoring. The major challenge is hardware–software interfacing and designing a flexible control that can handle various processes and its variability. Proprietary equipment restrictions and limitations of existing software leads to constraints on design and control flexibility, and implementing a control layer of equipment read and write restrictions is a critical design consideration. After integration and placement of analytical tools and automated controllers, the adapted batch process must be run in continuous mode for a minimum of three repeated cycles with detailed assessment of process performance and product quality in the face of a wide range of induced deviations to reach the final goal of continuous process transfer at a clinically relevant scale.

Integration between upstream and downstream is required for developing an end-to-end continuous platform for production of therapeutic proteins. For an end-to-end integrated continuous manufacturing platform, developing an architecture is a challenge [1]. One major concern is the rapid improvements in upstream processing in terms of processability and higher titers, leading to the downstream process becoming the bottleneck as the clarified harvest from the bioreactor should be loaded continuously onto the chromatography columns, requiring high capacity throughout the downstream setup. A recommended approach is the integration of bioreactor with a cell retention device followed by continuous capture chromatography or the use bioreactor in combination with ATF followed by two four-column periodic countercurrent chromatography for capture, viral inactivation, and two polishing steps. It should be compatible in terms of pH, osmolality, and other parameters without any modifications. In addition, equipment changeover time from one unit operation to another and equipment segregation for different classes of therapeutics are also major concerns in downstream processes.

2.4 Control of Continuous Downstream Unit Operations in mAb Manufacturing

2.4.1 Control of Continuous Dead-End Filtration

Clarification of harvest broth is frequently done using dead-end filtration [79, 80]. Batch-mode dead-end filtration has been a common unit operation for decades in mAb purification processes. In batch mode, the turbidity and particle size distribution of the harvest is pre-characterized, and filter membrane area is sized accordingly, along with a safety factor of up to 50% extra membrane area to prevent filter clogging before the entire batch can be processed. The filter capacity is calculated as the volume of filtrate that can be processed by the filter until the pressure buildup or turbidity breakthrough reaches a certain upper threshold, after which the filter must be replaced [81, 82]. This approach is not feasible in continuous processing. This is because incoming feed streams from perfusion bioreactors or multiple sequential fed batches have a range of variations in turbidity, concentration of therapeutic protein, concentration of host cell proteins (HCP) and DNA, and other factors that can affect clarification [83, 84]. The process must be operated continuously over weeks or months without pause to change the filters, ideally under constant flow rate conditions, allowing consistency in clarification volumes and providing a reliable stream to the subsequent downstream unit operations [85].

It is relatively easy to size and deploy filters with an appropriate safety percentage membrane area for well-characterized batches where the turbidity of the entire batch is determined and an appropriate filter is selected prior to the start of the operation. In the case of continuous operation, such calculations are unlikely to be accurate over months of operation with many variations in the upstream material caused by expected variation in the upstream perfusion harvest over time or batch-to-batch variability in the case of an upstream setup consisting of multiple fed batches in series rather than a single long perfusion. It is difficult to accurately predict the variability in feed stream characteristics that can affect the sizing requirements [86–88]. It would be ineffective and cost inefficient to size a single, very large filter for dead-end filtration in continuous operation. Overall, a robust approach for continuous dead-end filtration must incorporate the following key elements:

1. Hardware that allows process streams to be filtered in an uninterrupted manner at constant flow rate over weeks or months of a continuous campaign.
2. Sensors to monitor the CQAs, turbidity, and pressure to ensure that filters are automatically replaced when their capacity is reached.
3. Options for filter switching and replacement that do not affect the filtrate quality, as any turbidity breakthrough or filter damage can have adverse effects on the following unit operations such as precipitation, ultrafiltration, or chromatography.
4. Flexibility and capability of handling unexpected deviations in feed stream attributes such as turbidity, concentration of therapeutic protein, concentration of HCP and DNA, or other factors due to the variability in the conditions of the upstream bioreactor.
5. Freedom from strict sizing calculations prior to filter deployment as in the case of batch mode.

A recent case study showcasing a solution for continuous dead-end filtration involved a filtration skid with pressure and turbidity sensors and multiple small-sized filters. The flow was switched to a fresh filter using an automated valve whenever the filter capacity was reached, by evaluating sensor data against predetermined thresholds of turbidity and pressure in real time [89]. The skid, consisting of real-time sensors and an automated control valve for flow switching, was implemented for a range of feeds and allowed significant cost savings as the sizing safety factor could be eliminated due to the option of automated filter switching. Other approaches for monitoring of continuous dead-end filtration included placing in-line pressure sensors to monitor pressure across dead-end filters and trigger alarms in the case of pressure buildup [90] or using in-line measurements of UV absorbance to measure breakthrough of HCP through depth filters in real time [21].

2.4.2 Control of Continuous Chromatography

Chromatography is used in the downstream stream for capture of the target biotherapeutic molecule and for removal of aggregates and closely related variants that affect its safety and efficacy. Most biotherapeutic processes have two or more chromatography steps. Some of the challenges that arise in converting chromatography from batch to continuous mode are operational complexity arising from having multiple columns, the need for flexibility to handle changes in titer and CQAs of harvest titer over time, and the need to enable processing of high-titer harvest material [91, 92]. Since chromatographic operations consist of multiple steps, including loading, wash, elution, cleaning, and equilibration, it is critical to have optimized scheduling of simultaneous steps in parallel columns. The continuous chromatography equipment setups must contain several columns, inlets, outlets, and valves. Several continuous chromatography systems have been demonstrated and commercialized, such as the Cadence BioSMB (Pall Life Sciences, USA), Akta PCC 75 (GE Life Sciences, Sweden), BioSC Lab (Novasep Inc., France), and multicolumn countercurrent solvent gradient purification (MCSGP) Contichrom system (ChromaCon AG, Switzerland).

Despite the advances in specialized instrumentation for continuous chromatography, there remain a number of challenges. Firstly, due to the very high number of cycles in a continuous campaign of weeks or months, it is critical to have robust tools for flagging/controlling deviated elutions and tracking column health to trigger column replacement or repairs as and when needed, such as due to the resin fouling or column integrity issues [21]. Dynamic scheduling is required to ensure that the process material is received from upstream and directed smoothly downstream with integration on both sides [93]. Furthermore, elution monitoring and pooling are required to collect only the peak(s) of interest and adjust the collection criteria in the case of deviations [94]. These challenges are common to all chromatography operations and must be addressed when designing a control strategy.

There are further challenges specific to different types of chromatography, of which the two major types are capture (for capturing the protein of interest out of a soup of many other different proteins) and polishing (for removing closely related variants and aggregates post-capture). Figure 2.4 summarizes the unique challenges and potential control solutions for each. A challenge specific to capture chromatography of mAbs arises due to the use of Protein A resin, which is very expensive, contributing <60% of downstream consumable costs [95]. The resin also suffers degradation during wash and cleaning cycles [96]. Therefore, it is desirable to have optimal resin utilization, which can be achieved by ensuring that the binding of mAb is done up to the dynamic binding capacity in each cycle [97]. However, additional complexity arises as dynamic binding capacity is a function of multiple variables, including load titer, residence time, and resin health [91], [98].

Various case studies have been published on monitoring and control of continuous capture chromatography using a range of analytical tools. High-pressure liquid chromatography (HPLC) with analytical Protein A columns has been used for determining column breakthroughs using a two-minute analytical method [99], thereby preventing over- and underloading. Other approaches used multi-wavelength UV [100], near-infrared spectroscopy [89], or single-wavelength UV at the entrance and exit of the capture chromatography columns [91, 101], combined with multivariate data analysis to monitor the mAb concentration in the harvest fluid and enable

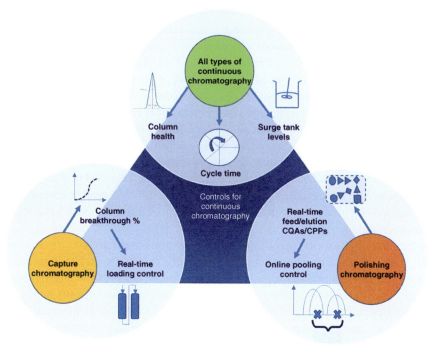

Figure 2.4 Key considerations for effective control of continuous chromatography.

real-time control by switching loading to the next column at a certain breakthrough percentage. In summary, to achieve real-time control of continuous capture chromatography, it is beneficial to incorporate the following elements:

1. Analytical tools or models for monitoring the amount of loading onto the column in near real time.
2. Analytical tools or models for determining the percentage breakthrough in near real time to prevent underloading (enabling utilization of expensive Protein A resin) or overloading (preventing loss of expensive mAb product).
3. Analytical tools or models for assessing column health or resin fouling in near real time.
4. Communication between the sensors, models, and hardware to effect control decisions such as switching the loading to the next column, pausing the loading, switching the flow to a standby safety column, and directing a deviated elution to a waste tank.
5. Communication between the capture chromatography equipment and the rest of the continuous train to carry out the continuous scheduling while preventing other unit operations or surge tanks from running dry or overflowing.

Likewise, there are control challenges specific to polishing chromatography. The major challenge is real-time control of elution pooling. Unlike capture chromatography in which the entire elution peak is typically collected, polishing chromatography requires sensitive decision making to determine the elution collection points to achieve separation of closely related variants, adequate removal of dangerous variants such as acidic and aggregated species, and close matching of the CQA profile of each elution cycle to the desired benchmark level as per the clinical approvals of the mAb biotherapeutic [1, 20, 102, 103].

The pooling decision becomes quite challenging to execute in real time in the case of column degradation, variation in the titer or CQAs of the feed, or equipment errors over long continuous campaigns [94, 104]. Furthermore, downstream mAb processes can often have more than one polishing chromatography step to be carried out in series, such as anion exchange (AEX) followed by cation exchange (CEX) or CEX followed by hydrophobic interaction chromatography (HIC), sometimes requiring in-line conditioning between the steps that also needs to be carefully monitored and controlled over hundreds of cycles [105]. At-line HPLC has been shown to be a powerful tool for making pooling decisions due to the ability of the HPLC to reliably quantify the components in the chromatographic load and elution streams. Rapid HPLC methods along with at-line sampling from streams have been demonstrated as a useful PAT tool in various biopharmaceutical process chromatography applications, including hydroxyapatite chromatography for separation of monomer and aggregated species [104], reverse-phase process chromatography for separation of product analogues [106], and cation exchange chromatography for separation of charge variants of mAbs [94]. In summary, there is a pressing need for real-time analytical tools and controllers for continuous polishing chromatography with the following key elements:

1. Sensor(s) for monitoring the attributes of the process material of either feed or elution stream in real time. These can measure the different variants of mAb directly (such as at-line HPLC, at-line mass spectroscopy, or online spectroscopic sensors) and/or the process parameters (such as pH, conductivity, redox, column packing/symmetry/fouling, etc.).
2. A control strategy for using the measured variables to predict the optimum pooling criteria for the current cycle. This can be a mechanistic model for predicting the elution profile, an empirical model determined by scouting the design space of potential variations, or a statistical/deep learning model trained on historical elutions.
3. An online execution module that can modify the pooling criteria in real time for each chromatography cycle as per the results of the control models and carry out required scheduling based on surge tank level, switch the flow to standby safety columns in the case of column failure, and direct an elution to waste in the case of process errors such as wrong elution gradient.

2.4.3 Control of Continuous Viral Inactivation

Viral inactivation is a critical step in mAb manufacturing as most processes use CHO cells that are susceptible to contamination by adventitious viruses [107]. A low-pH hold step is usually included in downstream processing to inactivate viruses in the process stream and is typically performed after Protein A capture chromatography

[13, 108]. In batch mode, the process material is incubated at pH <4 for 0.5–2 hours. Converting the batch process to continuous mode is challenging as the nature of the viral inactivation step is a hold time in which the process material is kept in a fixed condition over a period of time. Additionally, if the pH adjustment is not performed correctly, there can be adverse effects on the CQAs affected by this step, namely, the viral clearance and the aggregate content [107]. If the pH is not sufficiently low or the hold time is not enough, the viruses would not be properly inactivated. Conversely, if the pH is too low or the hold time is too long, the product molecules aggregate into high-molecular-weight species that not only leads to loss of active product but also can worsen the safety and efficacy of the final drug product [92, 109]. Therefore, it is critical to control the pH and the hold time in each cycle of viral inactivation over long continuous campaigns.

Various enabling technologies for continuous viral inactivation have recently emerged. One option is to use a tubular or coiled flow reactor [12, 14, 110, 111], essentially a long pipe in a twisted configuration with a long flow path that has residence time equivalent to the desired low-pH hold time in the viral inactivation step. The pH of the process material is adjusted to <4 at the inlet and neutralized at the outlet, thus ensuring that each volumetric fluid element has the same low-pH hold time despite flowing continuously through the tubular reactor. Another option is an automated multivessel system [112] that transfers fixed volumes of process material into separate scheduled hold tanks. Another option is to use an additional chromatography column to provide a low-pH hold for the material while it is passing through the column [113, 114]. Essentially, a control strategy for continuous viral inactivation must incorporate the following elements:

1. Sensors for monitoring the pH in real time and ensuring that the pH adjustment (acidification and neutralization) is done correctly and consistently for each cycle.
2. Hardware for facilitating low-pH hold such as a continuous flow reactor or a set of hold tanks with scheduled pumps and valves for pH adjustment and transfer of the process material.
3. Real-time communication with the rest of the unit operations in the continuous train to trigger pH adjustment pumps and fill/empty the reactors/vessels as per the schedule of the adjacent unit operations to prevent any vessels from overfilling or running dry.
4. A safety valve allowing the process material to be directed to a safety tank in case the pH or hold time is outside the prescribed limits for the current process.
5. If possible, models for the aggregation kinetics can be incorporated to predict the impact of pH/hold time deviations and determine whether to discard the material or proceed and handle the variation in the following chromatography step(s).

2.4.4 Control of Continuous Precipitation

Precipitation is one of the simplest, highest yielding, and least expensive methods of separation in manufacturing of biotherapeutics and can be used as an alternative to chromatography [115, 116]. Making changes to the pH, ionic strength, or temperature conditions of the in-process mixture can lead to the conversion of soluble proteins (either the target protein or impurities) to an insoluble state, causing them to precipitate from the solution as solids for easy separation by filtration or centrifugation [117]. The simplest operation is impurity precipitation, with the downside that the precipitants (such as neutral salts, organic solvents, polymers, polyelectrolytes, acids, bases, and affinity ligands) are carried over to subsequent unit operations. The alternative is product precipitation, which leads to high concentration of pure protein but increases the risk of damaging the CQAs of the biotherapeutic protein due to the changes in its structure [28, 118, 119]. Both options have been demonstrated in batch mode for various mAbs and for other biotherapeutics, with the process material kept in a large stirred hold tank, followed by the addition of reactants for the precipitation reaction and subsequent removal of the solids [120–122].

In batch mode, each step from addition of reactants to mixing to solid separation can be carefully monitored offline before proceeding to the next. However, in continuous mode, all these steps must be consistently monitored and controlled over long campaigns [123, 124]. Furthermore, the nature of the precipitation step involves induction of changes in the structure of the molecules in the process mixture, and thus there is a pressing need for monitoring the addition of the reactants and the progress of the reaction to ensure that it is proceeding as planned. There is a high chance of deviations in the CQAs of aggregation and impurity content in the case of a process or feed deviation, as there is a complex interplay between the variables of pH, temperature, time, protein concentration, precipitant concentration, rate of stirring, and rate of reactant addition [121, 123]. Therefore, there is a great need for appropriate instrumentation, sensors, and control strategies. The following elements must be present in a well-controlled continuous precipitation process:

1. Appropriate instrumentation for eliminating the batch-mode hold tank. Similar to the setup described in Section 2.4.4 for viral inactivation, this can be achieved either by switching to a continuous flow tubular reactor or by using a system of scheduled hold tanks. Both options are feasible though the latter is more of a semi-batch approach rather than truly continuous.
2. Appropriate instrumentation for inducing the changes required for precipitation, such as variable control pumps for addition of acid/base/precipitant or heating/cooling jackets or coils, with appropriate scheduling and triggers based on the inflow and outflow from other continuous unit operations.
3. Appropriate instrumentation for continuous solid–liquid separation post precipitation, such as the dead-end filtration skid described in Section 2.4.2 or a continuous centrifuge.
4. In the case of product precipitation, an appropriate setup is required for continuous re-solubilization of the pellets for further downstream processing.
5. Sensors for monitoring the changes induced for causing precipitation in real time, such as pH, conductivity, temperature, redox, and spectroscopic probes calibrated to the precipitant.
6. Model(s) characterizing the relationship between the key reaction variables and the yield and CQAs of the output including concentration, impurity content, and

aggregate percentage to determine the appropriate control actions to be taken in the case of process deviations or feed variability.
7. Control elements such as multi-way valves to recycle/discard incorrectly reacted material and start/stop cleaning and flush cycles.

2.4.5 Control of Continuous Formulation

Tangential flow ultrafiltration is used for the formulation, which is the final step in the downstream process for mAbs. In the formulation step, the drug substance is concentrated to the final concentration and buffer exchanged from the in-process buffers to the final formulation buffer that has been optimized and contains excipients and stabilizers enhancing the drug's stability, osmotic balance, and bioavailability [125, 126]. These two steps of ultrafiltration and diafiltration (UF–DF) are the final steps prior to fill-finish and packaging [127]. Both mAb and excipient concentrations are CQAs, thus necessitating robust control of UF–DF [128, 129]. In batch mode, UF–DF is carried out by recirculating the biotherapeutic material through a fully retentive membrane and discarding the permeate (containing no drug molecule) until the desired volume reduction or diavolumes of buffer exchange are complete [130–132]. This is followed by offline analysis for measuring the concentration of the drug molecule and excipients. However, such a system is not feasible for continuous manufacturing as large tanks are needed to hold the material, and long process times are required for repeated recirculation. There is also the need for offline analysis and manual intervention.

Single-pass tangential flow filtration (SPTFF) technology has emerged as an enabler for continuous UF–DF [17]. Novel membrane modules have been designed consisting of long flow paths to increase the residence time of the process material through the module and optimize the permeate flux, allowing high volumetric concentration factors (VCFs) in a single pass without the need of recirculation [133]. Similar modules with multiple inlets have also been designed to facilitate >7 diavolumes of buffer exchange in a single pass. These modules can be inserted in-line to handle continuously flowing process streams. Various commercial SPTFF modules well suited for ultrafiltration at various scales are available today, including the Cadence series of modules from Pall Corporation [134] and the Pellicon® SPTFF kits from Merck Millipore Sigma [131]. However, though these enabling technologies effectively facilitate in-line UF–DF, they are not enough to guarantee tight control of the CQAs without including real-time sensors and control strategies to overcome issues such as variations in feed concentration, reduced flux due to the membrane fouling, the necessity of scheduling periodic flush and cleaning cycles, and models for characterizing the complex interplay between feed concentration, permeate flux, and pressure [135, 136]. Figure 2.5 illustrates the challenges that arise due to the process variability in continuous operation. A strategy for effective control of continuous UF–DF should include the following elements:

1. Sensors for measuring the concentration of the feed in real time, such as in-line UV or other spectrometric sensors [135, 137].

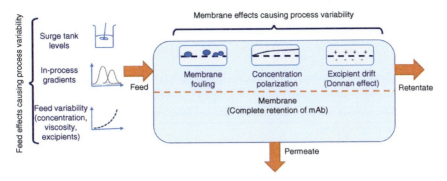

Figure 2.5 UF–DF process variability challenges requiring control strategies for continuous operation.

2. Sensors or calculations for measuring the concentration of the retentate in real time, for example, by calculating VCF based on pressure readings, in-line flow meters, or direct measurements of concentration.
3. Sensors for measuring the concentrations and/or the degree of buffer exchange in real time, for example, in-line conductivity sensors, spectrometric sensors, and at-line HPLC.
4. Models characterizing the relationship between feed concentration, VCF, concentration polarization, and module pressure to determine the appropriate control actions to be taken in the case of feed variability.
5. Models for tracking membrane fouling and module health to trigger cleaning cycles or replacement.
6. Control elements such as variable feed pump, permeate pump/valve, and retentate pump/valve to compensate for flux decline due to the fouling and concentration polarization effects.
7. Control elements such as multi-way valves to recycle/discard incorrectly formulated product and start/stop cleaning and flush cycles.

2.5 Integration Between Adjacent Unit Operations Using Surge Tanks

Surge tanks are hold vessels placed in the process train to absorb the flow rate fluctuation from the upstream unit and to provide a constant flow to the downstream units. It is often said that for well-established continuous process with adjusted flow rates for all unit operations, surge tanks are not necessary as it increases holdup time, increases capital cost, decreases productivity, and increases operational complexities [138]. However, realization of such process is difficult in practice due to the inherently periodic nature of various unit operations such as chromatography and the high chance of deviations caused by fluctuations in the upstream material or equipment errors over long continuous processes of weeks or months. The following considerations should be taken into account when determining whether a surge tank should be placed at a given point in the continuous train [93]:

1. Is flow matching required?
   If the outflow from the previous unit operation is periodic (such as periodic elution pools from multicolumn chromatography), it is beneficial to include a surge tank with mass balance on the outflow rate to ensure a consistent feed stream for the next unit operation.
2. Is there a need to dampen process gradients?
   If the outflow from the previous unit operation contains gradients of concentration, pH, temperature, etc., such as in the case of Protein A elution pools, it is beneficial to include a surge tank with sufficient holdup volume to dampen the gradients.
3. Is the subsequent unit operation very sensitive to variation in feed?
   Various process disturbances or pump pulsation can cause small, temporary variations in feed flow rate and characteristics. In case a unit operation is very sensitive to these (such as single-pass filtration), it is beneficial to include a surge tank.
4. Is there a high chance of equipment breakdown?
   If yes, it is beneficial to have a surge tank prior to the unit operation to hold the incoming process material for the expected duration of replacement or repairs to prevent wastage of valuable drug substance.

Incorporating and automating the level control strategy of surge tanks enable integration of unit operations in which flow is intermittent, such as multicolumn chromatography with periodic elution. In addition, surge tanks help in maintaining low operating pressure throughout the process and process variability [139]. However, the integration of surge tank in the continuous chain for regulating the flow rates to different operations is a complicated operation. The in-line probes immersed in surge tanks help in real-time monitoring and enable the implementation of PAT approaches. Critical control decisions in the case of titer variations, charge variant variations, and membrane fouling can be employed with the help of surge tanks [21, 22, 30, 140]. Figure 2.6 summarizes the approach for designing a surge tank system for a continuous bioprocessing train.

Control strategies for surge tank-based control of continuous processing are required. These strategies are required for dampening of concentration gradient or adjusting flow rates or scheduling unit operations. Here, the residence time of material should be taken into consideration while developing strategies to avoid the disruptions in stability or structure of protein. The agitator speed also plays an important role in protein fragmentation, aggregations, and other structural changes [141]. If the chromatography step is in bind and elute mode, then the product stream varies in concentration during each elution, affecting methodology adopted for the next unit operation [138]. Therefore, there is a need to monitor the material specification in surge tank and accordingly operate consecutive units at optimized conditions. In implementing these, it is also essential to have real-time monitoring of surge tank levels and a central control system with the ability to execute surge tank-based

2.6 Emerging Approaches for High-Level Monitoring and Control of Continuous Bioprocesses | 59

Figure 2.6 Key approach for surge tank-based control system design for bioprocessing.

triggers for starting and stopping the unit operations in the continuous train based on the surge tank level to prevent them from overflowing or running dry [93].

2.6 Emerging Approaches for High-Level Monitoring and Control of Continuous Bioprocesses

The unit operation-specific control guidelines given in the above sections are fundamental building blocks of a well-controlled continuous operation. It is essential to have real-time sensors and controllers in place for each unit operation, each with the objective to maintain and control the relevant CQAs affected by that particular operation. These first-layer control strategies utilize feedforward or feedback control and require extensive process knowledge, expertise, and clear-cut control objectives. Implementation of these strategies demands process model and tools including analyzers, controllers, pumps, and control valves. In addition, a number of controlled and manipulated disturbances need to be taken into account for control system design. Considering process complexity, instead of standard control techniques, advanced process control using emerging computational tools based on artificial intelligence, machine learning, and statistical process control is an attractive orthogonal option for mitigating error propagation in process and delivering consistent quality and productivity. These approaches also provide a global view of the process above the single unit operation level. These techniques can be supervised, unsupervised, or reinforced and take advantage of the large datasets generated over continuous processes to statistically characterize the overall process and make predictions. It is also essential to digitalize the process for automation and data storage in an accessible manner to be used by real-time control models and in future by process characterization or prediction models that leverage extensive historical data [142].

2.6.1 Artificial Intelligence (AI) and Machine Learning (ML) Control

AI tools include expert systems, fuzzy approach, neural networks, case-based reasoning, genetic algorithms, or ambient intelligence. These techniques coupled with computational capability mimic human-like reasoning and decision making in the case of uncertainties. Vast applications in the field of bioprocess control, optimization, monitoring, parameter estimation, prediction, chemometrics, process scale-up/scale-down, and soft sensor development are seen. However, the applications are not equally distributed in all of upstream and downstream operations as the structure and content of data collected are not useful for practical implications. However, combining the datasets from different biopharma companies can cover a wider range of operating space.

In bioprocessing, neural network application for supervision [143], online prediction in fermentation process [144], fault diagnostic in vitrification process [145], rule-based supervisory system for fermentation [146], modeling of antibiotic fermentation [147], two-level control of fed-batch reactor [148], and open-loop control for monitoring yeast production [149] have been published recently. For process intensification of continuous chromatography, reinforcement learning has been used to optimize the process flow rate for cation exchange chromatography for optimal separation of charge variants using a mechanistic model that is solved computationally much faster than using traditional solvers, enabling more effective use for real-time process control in a continuous train [150]. Also, for testing the enzyme tolerance at high temperatures, NN has been implemented [151]. The online adaptation of the neural network is limited due to its time-consuming training. Although a trained ANN predicts well in the design space, the aim is to provide satisfactory control over a broader range. For this, neural network weights need to be updated constantly in real time depending on the perturbations.

Fuzzy control systems provide online decision support to the operator and conduct automatic control. Applications in fault diagnostic and control [152] and pattern recognition [153] in bioprocesses are since long. In combination with the genetic algorithms, fuzzy control system has shown better performance in comparison with the other predictive algorithms for the control purpose [154]. Multivariable fuzzy control [155], hybrid fuzzy control [156], and fuzzy control [157] techniques have been successfully used for bioreactor operations. One of the main benefits of implementing fuzzy control systems includes the process knowledge and expertise written down in a logical form and accessible to everyone and not limited to the concerned person.

As for mammalian bioprocesses, due to the high process time, control strategies having computational time of several minutes are acceptable. Owing to the complexities and variabilities in these processes, nonlinear model predictive control (NMPC) with dynamic models is preferred [158]. This can be designed using the minimum/maximum optimization techniques [159] or artificial neural networks

[160]. Real-time control of quality attributes [161] is feasible with MPC. Single-input single-output (SISO) MPC and multiple-input multiple-output (MIMO) MPC when used for the online estimation and control in the fed-batch reactor give better results than proportional integral controller and feedback/feedforward controller. In addition, MPC implementation for forecasting can be done using the mechanistic model built over correlative process parameter [162].

2.6.2 Statistical Process Control

Statistical quality control is a standardized tool for making systematic decision to ensure, in a cost-efficient manner, that the process is operating at the maximum efficiency and the product shipped to customers meets the desired specifications. SPC compares present performance with the model created using the historical data to detect the root cause of the process deviation. The general framework for SPC implementation is shown in Figure 2.7. Techniques like histogram analysis to summarize the distribution of univariate dataset and scatter plots to reveal relationship or association between variables and control charts (univariate or multivariate) to monitor quality aspects have been used. For visualization of variability be it normal (due to common cause) or abnormal (due to unexpected circumstances), control charts are recommended. These variabilities can be easily predicted using statistical modeling. Techniques like principal component analysis (PCA), independent component analysis (ICA), partial least square (PLS) regression, and neural network (NN) are helpful in creating control charts. Application of PCA and PLS in mammalian cell culture process for characterization, prediction [124], and fault detection [49] is being explored and applied.

The combined approach of neural network for modeling and random forest for variable selection is emphasized for bioprocessing [152]. The presence of uncertainties can be well managed by SPC while keeping the process in control [163]. In addition, Hotelling T^2 statistic is also used for fault detection and real-time monitoring [164]. To determine the datasets corresponding to a particular fault, the distance between the data point and the nearest neighbor in control is calculated. This helps

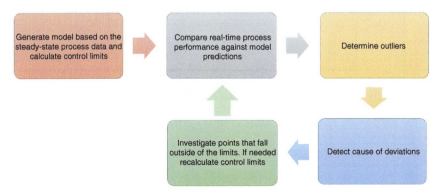

Figure 2.7 Framework for SPC implementation.

in the evaluation of the contribution of individual observation. Squared prediction error charts also ensure process performance within control limits.

2.6.3 Process Digitalization

As a prerequisite to Bioprocessing 4.0, process digitization is required. This concept revolves around connectivity, intelligence, and automation. Connectivity refers to combining equipment and monitoring tool with machine learning approaches, statistical approaches, and digital twins, leading to more efficient and flexible process designs. Digital twins can be built from mechanistic models or statistical models. Mechanistic model-based digital twin can derive deep process insights from the production data, whereas statistical model-based digital twin is beneficial for processes with fundamental and quantitative understanding.

Intelligence involves streamlining data analysis, and developing control logics for optimized process operation and automation helps in communication of data that can be from one equipment to another or across different stage of process development. Figure 2.8 presents an example of process digitalization wherein a supervisory control and data acquisition system is placed for centralized control of the process. The process considered here comprises bioreactor for protein generation, acoustic wave separator for clarification, BioSMB for capture and polishing chromatography, coiled flow inversion reactor for viral inactivation, and inline ultrafiltration–diafiltration unit for final formulation. The process is enabled for real-time data collection using monitoring tools like at-line HPLC and NIR. Here, process-wide control is implemented with a holistic approach to optimize the process parameters as per the set points given to the process. The SCADA system gives feedback to the control system that in turn communicates with the equipment to manipulate the variables to obtain specified output. In addition, SCADA system

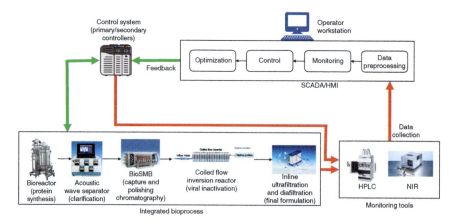

Figure 2.8 Example of centralized control system for an end-to-end integrated process. Red arrow depicts data collection, and green arrow indicates the feedback from SCADA after optimizing and controlling the parameters as per the set points.

with properly designed control logics and optimization techniques can handle system disturbances efficiently.

The process of converting analog to digital depiction and then using advanced technologies for gaining process insights is complex and less understood in the biopharmaceutical industry. Updated systems built in distributed control system are the prime necessity of digitalization. The knowledge of updated technologies and proper training of operators is required. Presently, DCS equipped with plug and play connection and function designed for process scale-up are already present. For data visualization, upcoming technologies like thin client technology for reducing the error probability accompanying mobile equipment and augmented reality as a lookup tool for standard operating procedure help in improving facility efficiency.

2.7 Conclusions

In conclusion, there are unique control challenges that need to be overcome when converting unit operations in mAb manufacturing from batch to continuous mode. The major challenge of continuous processing is handling deviations in real time without the need to stop the process, in contrast to batch mode in which the process is paused after each step for offline analysis and corrections. Therefore, there is a need for a rich and robust set of real-time online or at-line analytical tools coupled with automated control strategies that can monitor process parameters and CQAs either directly or indirectly and, when required, implement control decisions. It is critical for each unit operation to have a set of real-time sensors coupled with real-time control strategies designed to handle variations in the quality attributes of the upstream material and equipment errors and process parameter deviations. Surge tanks are also critical and must be carefully placed for flow matching, gradient dampening, and added safety and flexibility. A set of guidelines are formulated for each unit operation, characterizing the interplay between novel instrumentation, monitoring tools, and control strategies required to ensure smooth continuous operation in the face of process deviations, along with a framework for process digitalization. Finally, emerging approaches are also described that can be implemented orthogonally with individual unit operation-level controls, leveraging computational techniques such as machine learning and artificial intelligence for a global picture of the entire process.

References

1 Konstantin, B.K. and Cooney, C.L. (2015). White paper on continuous bioprocessing May 20–21, 2014 continuous manufacturing symposium. *J. Pharm. Sci.* 104: 813–820.
2 Rathore, A.S., Agarwal, H., Sharma, A.K. et al. (2015). Continuous processing for production of biopharmaceuticals. *Prep. Biochem. Biotechnol.* 45 (8): 836–849. https://doi.org/10.1080/10826068.2014.985834.

3 Gupta, S.K. (2017). Upstream continuous process development. In: *Continuous Biomanufacturing: Innovative Technologies and Methods*. Wiley-VCH.

4 Mehta, S. (2014). Automated single-use centrifugation solution for diverse biomanufacturing process. In: *Continuous Processing in Pharmaceutical Manufacturing* (ed. G. Subramanian), 385–399. Wiley-VCH Verlag GmbH & Co. KGaA.

5 Richardson, A. and Walker, J. (2018). Continuous solids-discharging centrifugation: a solution to the challenges of clarifying high-cell-density mammalian-cell cultures. *Bioproc. Intl.*

6 Hadpe, S.R., Sharma, A.K., Mohite, V.V., and Rathore, A.S. (2017). ATF for cell culture harvest clarification: mechanistic modelling and comparison with TFF. *J. Chem. Technol. Biotechnol.* 92 (4): 732–740. https://doi.org/10.1002/jctb.5165.

7 Karst, D.J., Serra, E., Villiger, T.K. et al. (2016). Characterization and comparison of ATF and TFF in stirred bioreactors for continuous mammalian cell culture processes. *Biochem. Eng. J.* 110: 17–26. https://doi.org/10.1016/j.bej.2016.02.003.

8 Collins, M. and Levison, P. (2016). Development of high performance integrated and disposable clarification solution for continuous bioprocessing. *Bioprocess Int.* 14: 30–33.

9 Hong, J.S., Azer, N., Agarabi, C., and Fratz-Berilla, E.J. (2020). Primary clarification of CHO harvested cell culture fluid using an acoustic separator. *J. Visualized Exp.* 159: e61161. https://doi.org/10.3791/61161.

10 Steinebach, F., Müller-Späth, T., and Morbidelli, M. (2016). Continuous counter-current chromatography for capture and polishing steps in biopharmaceutical production. *Biotechnol. J.* 11 (9): 1126–1141. https://doi.org/10.1002/biot.201500354.

11 David, L., Maiser, B., Lobedann, M. et al. (2019). Virus study for continuous low pH viral inactivation inside a coiled flow inverter. *Biotechnol. Bioeng.* 116 (4): 857–869. https://doi.org/10.1002/bit.26872.

12 Kateja, N., Nitika, N., Fadnis, R.S., and Rathore, A.S. (2021). A novel reactor configuration for continuous virus inactivation. *Biochem. Eng. J.* 167: 107885. https://doi.org/10.1016/j.bej.2020.107885.

13 Klutz, S., Lobedann, M., Bramsiepe, C., and Schembecker, G. (2016). Continuous viral inactivation at low pH value in antibody manufacturing. *Chem. Eng. Process. Process Intensif.* 102: 88–101. https://doi.org/10.1016/j.cep.2016.01.002.

14 Parker, S.A., Amarikwa, L., Vehar, K. et al. (2018). Design of a novel continuous flow reactor for low pH viral inactivation. *Biotechnol. Bioeng.* 115 (3): 606–616. https://doi.org/10.1002/bit.26497.

15 Hebbi, V., Thakur, G., and Rathore, A.S. (2021). Process analytical technology application for protein PEGylation using near infrared spectroscopy: G-CSF as a case study. *J. Biotechnol.* 325: 303–311. https://doi.org/10.1016/j.jbiotec.2020.10.006.

16 Kateja, N., Nitika, N., Dureja, S., and Rathore, A.S. (2020). Development of an integrated continuous PEGylation and purification process for granulocyte colony stimulating factor. *J. Biotechnol.* 322: 79–89. https://doi.org/10.1016/j.jbiotec.2020.07.008.

17 Casey, C., Gallos, T., Alekseev, Y. et al. (2011). Protein concentration with single-pass tangential flow filtration (SPTFF). *J. Membr. Sci.* 384 (1–2): 82–88. https://doi.org/10.1016/j.memsci.2011.09.004.

18 Rucker-Pezzini, J., Arnold, L., Hill-Byrne, K. et al. (2018). Single pass diafiltration integrated into a fully continuous mAb purification process. *Biotechnol. Bioeng.* 115 (8): 1949–1957. https://doi.org/10.1002/bit.26708.

19 Yehl, C.J. and Zydney, A.L. (2020). Single-use, single-pass tangential flow filtration using low-cost hollow fiber modules. *J. Membr. Sci.* 595: 117517. https://doi.org/10.1016/j.memsci.2019.117517.

20 Glassey, J., Gernaey, K.V., Clemens, C. et al. (2011). Process analytical technology (PAT) for biopharmaceuticals. *Biotechnol. J.* 6 (4): 369.

21 Rathore, A.S., Bhambure, R., and Ghare, V. (2010). Process analytical technology (PAT) for biopharmaceutical products. *Anal. Bioanal.Chem.* 398: 137–154.

22 Rathore, A.S. and Winkle, H. (2009). Quality by design for biopharmaceuticals. *Nat. Biotechnol.* 27 (1): 26–34.

23 Walsh, G. (2018). Biopharmaceutical benchmarks 2018. *Nat. Biotechnol.* 36 (12): 1136–1145. https://doi.org/10.1038/nbt.4305.

24 Somasundaram, B., Pleitt, K., Shave, E. et al. (2018). Progression of continuous downstream processing of monoclonal antibodies: current trends and challenges. *Biotechnol. Bioeng.* 115 (12): 2893–2907. https://doi.org/10.1002/bit.26812.

25 Handlogten, M.W., Lee-O'Brien, A., Roy, G. et al. (2018). Intracellular response to process optimization and impact on productivity and product aggregates for a high-titer CHO cell process. *Biotechnol. Bioeng.* 115: 126–138.

26 Huang, Y., Hu, W., Rustandi, E. et al. (2010). Maximizing productivity of CHO cell-based fed-batch culture using chemically defined media conditions and typical manufacturing equipment. *Biotechnol. Progr.* 26: 1400–1410.

27 Kim, J.Y., Kim, Y.-G., and Lee, G.M. (2012). CHO cells in biotechnology for production of recombinant proteins: current state and further potential. *Appl. Microbiol. Biotechnol.* 93: 917–930.

28 Kuczewski, M., Schirmer, E., Lain, B., and Zarbis-Papastoitsis, G. (2011). A single-use purification process for the production of a monoclonal antibody produced in a PER. C6 human cell line. *Biotechnol. J.* 6: 56–65.

29 Fisher, A.C., Kamga, M.-H., Agarabi, C. et al. (2019). The current scientific and regulatory landscape in advancing integrated continuous biopharmaceutical manufacturing. *Trends Biotechnol.* 37: 253–267.

30 Rathore, A.S. (2009). Roadmap for implementation of quality by design (QbD) for biotechnology products. *Trends Biotechnol.* 27: 546–553.

31 US FDA, 2004. *Guidance for industry, PAT-A Framework for Innovative Pharmaceutical Development, Manufacturing and Quality Assurance.* http//www.fda.gov/cder/guidance/published.html (accessed 10 January 2021).

32 Abu-Absi, N.R., Kenty, B.M., Cuellar, M.E. et al. (2011). Real time monitoring of multiple parameters in mammalian cell culture bioreactors using an in-line Raman spectroscopy probe. *Biotechnol. Bioeng.* 108: 1215–1221.

33 Claßen, J., Aupert, F., Reardon, K.F. et al. (2017). Spectroscopic sensors for in-line bioprocess monitoring in research and pharmaceutical industrial application. *Anal. Bioanal.Chem.* 409: 651–666.

34 Teixeira, A.P., Oliveira, R., Alves, P.M., and Carrondo, M.J.T. (2009). Advances in on-line monitoring and control of mammalian cell cultures: supporting the PAT initiative. *Biotechnol. Adv.* 27: 726–732.

35 Xu, J., Rehmann, M.S., Xu, M. et al. (2020). Development of an intensified fed-batch production platform with doubled titers using N-1 perfusion seed for cell culture manufacturing. *Bioresour. Bioprocess.* 7: 1–16.

36 Henry, O., Kwok, E., and Piret, J.M. (2008). Simpler noninstrumented batch and semicontinuous cultures provide mammalian cell kinetic data comparable to continuous and perfusion cultures. *Biotechnol. Progr.* 24: 921–931.

37 Bielser, J.-M., Wolf, M., Souquet, J. et al. (2018). Perfusion mammalian cell culture for recombinant protein manufacturing – a critical review. *Biotechnol. Adv.* 36: 1328–1340.

38 Chen, C., Wong, H.E., and Goudar, C.T. (2018). Upstream process intensification and continuous manufacturing. *Curr. Opin. Chem. Eng.* 22: 191–198.

39 Nikolay, A., Bissinger, T., Gränicher, G. et al. (2020). Perfusion control for high cell density cultivation and viral vaccine production. In: *Animal Cell Biotechnology*, 141–168. Springer.

40 Clincke, M., Mölleryd, C., Zhang, Y. et al. (2013). Very high density of CHO cells in perfusion by ATF or TFF in WAVE bioreactor™. Part I. Effect of the cell density on the process. *Biotechnol. Progr.* 29: 754–767.

41 Konstantinov, K., Goudar, C., Ng, M. et al. (2006). The "push-to-low" approach for optimization of high-density perfusion cultures of animal cells. In: *Cell Culture Engineering*, 75–98. Springer.

42 Wolf, M.K.F., Pechlaner, A., Lorenz, V. et al. (2019b). A two-step procedure for the design of perfusion bioreactors. *Biochem. Eng. J.* 151: 107295.

43 Wolf, M.K.F., Müller, A., Souquet, J. et al. (2019a). Process design and development of a mammalian cell perfusion culture in shake-tube and benchtop bioreactors. *Biotechnol. Bioeng.* 116: 1973–1985.

44 Martínez-Monge, I., Roman, R., Comas, P. et al. (2019). New developments in online OUR monitoring and its application to animal cell cultures. *Appl. Microbiol. Biotechnol.* 103: 6903–6917.

45 Kyung, Y.-S., Peshwa, M.V., Gryte, D.M., and Hu, W.-S. (1994). High density culture of mammalian cells with dynamic perfusion based on on-line oxygen uptake rate measurements. *Cytotechnology* 14: 183–190.

46 Feng, Q., Mi, L., Li, L. et al. (2006). Application of "oxygen uptake rate-amino acids" associated mode in controlled-fed perfusion culture. *J. Biotechnol.* 122: 422–430.

47 Goudar, C.T., Piret, J.M., and Konstantinov, K.B. (2011). Estimating cell specific oxygen uptake and carbon dioxide production rates for mammalian cells in perfusion culture. *Biotechnol. Progr.* 27: 1347–1357.

48 Rigual-González, Y., Gómez, L., Núñez, J. et al. (2016). Application of a new model based on oxygen balance to determine the oxygen uptake rate in mammalian cell chemostat cultures. *Chem. Eng. Sci.* 152: 586–590.

49 Mishra, S., Kumar, V., Sarkar, J., and Rathore, A.S. (2021). CFD based mass transfer modeling of a single use bioreactor for production of monoclonal antibody biotherapeutics. *Chem. Eng. J.* 412: 128592.

50 Andersen, D.C. and Goochee, C.F. (1995). The effect of ammonia on the O-linked glycosylation of granulocyte colony-stimulating factor produced by Chinese hamster ovary cells. *Biotechnol. Bioeng.* 47: 96–105.

51 Ozturk, S.S., Riley, M.R., and Palsson, B.O. (1992). Effects of ammonia and lactate on hybridoma growth, metabolism, and antibody production. *Biotechnol. Bioeng.* 39: 418–431.

52 Zhou, M., Crawford, Y., Ng, D. et al. (2011). Decreasing lactate level and increasing antibody production in Chinese hamster ovary cells (CHO) by reducing the expression of lactate dehydrogenase and pyruvate dehydrogenase kinases. *J. Biotechnol.* 153: 27–34.

53 Kornecki, M. and Strube, J. (2018). Process analytical technology for advanced process control in biologics manufacturing with the aid of macroscopic kinetic modeling. *Bioengineering* 5: 25.

54 Karst, D.J., Steinhoff, R.F., Kopp, M.R.G. et al. (2017a). Intracellular CHO cell metabolite profiling reveals steady-state dependent metabolic fingerprints in perfusion culture. *Biotechnol. Progr.* 33: 879–890.

55 Karst, D.J., Steinhoff, R.F., Kopp, M.R.G. et al. (2017b). Isotope labeling to determine the dynamics of metabolic response in CHO cell perfusion bioreactors using MALDI-TOF-MS. *Biotechnol. Progr.* 33: 1630–1639.

56 Niu, H., Amribt, Z., Fickers, P. et al. (2013). Metabolic pathway analysis and reduction for mammalian cell cultures – towards macroscopic modeling. *Chem. Eng. Sci.* 102: 461–473.

57 Fernandes de Sousa, S., Bastin, G., Jolicoeur, M., and Vande Wouwer, A. (2016). Dynamic metabolic flux analysis using a convex analysis approach: application to hybridoma cell cultures in perfusion. *Biotechnol. Bioeng.* 113: 1102–1112.

58 Glaessgen E, Stargel D. 2012. The digital twin paradigm for future NASA and US Air Force vehicles. *53rd AIAA/ASME/ASCE/AHS/ASC Structures, Structural Dynamics and Materials Conference, 20th AIAA/ASME/AHS Adaptive Structures Conference 14th AIAA*, p. 1818.

59 Chen, Y., Yang, O., Sampat, C. et al. (2020). Digital twins in pharmaceutical and biopharmaceutical manufacturing: a literature review. *Processes* 8: 1088.

60 Zobel-Roos, S., Schmidt, A., Mestmäcker, F. et al. (2019). Accelerating biologics manufacturing by modeling or: is approval under the QbD and PAT approaches demanded by authorities acceptable without a digital-twin? *Processes* 7: 94.

61 Portela RMC, Varsakelis C, Richelle A, Giannelos N, Pence J, Dessoy S, von Stosch M. 2020. *When Is an In Silico Representation a Digital Twin? A Biopharmaceutical Industry Approach to the Digital Twin Concept.*

62 Hutter, S., Wolf, M., Papili Gao, N. et al. (2018). Glycosylation flux analysis of immunoglobulin G in Chinese hamster ovary perfusion cell culture. *Processes* 6: 176.

63 Nolan, R.P. and Lee, K. (2011). Dynamic model of CHO cell metabolism. *Metab. Eng.* 13: 108–124.

64 Sha, S., Huang, Z., Wang, Z., and Yoon, S. (2018). Mechanistic modeling and applications for CHO cell culture development and production. *Curr. Opin. Chem. Eng.* 22: 54–61.

65 Sokolov, M., Ritscher, J., MacKinnon, N. et al. (2017). Enhanced process understanding and multivariate prediction of the relationship between cell culture process and monoclonal antibody quality. *Biotechnol. Progr.* 33: 1368–1380.

66 Zhang, L., Schwarz, H., Wang, M. et al. (2020). Control of IgG glycosylation in CHO cell perfusion cultures by GReBA mathematical model supported by a novel targeted feed, TAFE. *Metab. Eng.* 65: 135–145.

67 Farzan, P. and Ierapetritou, M.G. (2018). A framework for the development of integrated and computationally feasible models of large-scale mammalian cell bioreactors. *Processes* 6: 82.

68 Menshutina, N.V., Guseva, E.V., Safarov, R.R., and Boudrant, J. (2020). Modelling of hollow fiber membrane bioreactor for mammalian cell cultivation using computational hydrodynamics. *Bioprocess. Biosyst. Eng.* 43: 549–567.

69 Camacho, D.M., Collins, K.M., Powers, R.K. et al. (2018). Next-generation machine learning for biological networks. *Cell* 173: 1581–1592.

70 Kotidis, P. and Kontoravdi, C. (2020). Harnessing the potential of artificial neural networks for predicting protein glycosylation. *Metab. Eng. Commun.* 10: e00131.

71 Warikoo, V., Godawat, R., Brower, K. et al. (2012). Integrated continuous production of recombinant therapeutic proteins. *Biotechnol. Bioeng.* 2012 (109): 3018–3029.

72 Xu, S. and Chen, H. (2016). High-density mammalian cell cultures in stirred-tank bioreactor without external pH control. *J. Biotechnol.* 231: 149–159.

73 Walther, J., McLarty, J., and Johnson, T. (2019). The effects of alternating tangential flow (ATF) residence time, hydrodynamic stress, and filtration flux on high-density perfusion cell culture. *Biotechnol. Bioeng.* 116: 320–332.

74 Shekhawat, L.K., Sarkar, J., Gupta, R. et al. (2018). Application of CFD in bioprocessing: separation of mammalian cells using disc stack centrifuge during production of biotherapeutics. *J. Biotechnol.* 267: 1–11.

75 Kelly, W., Rubin, J., Scully, J. et al. (2016). Understanding and modeling retention of mammalian cells in fluidized bed centrifuges. *Biotechnol. Progr.* 32: 1520–1530.

76 Shirgaonkar, I.Z., Lanthier, S., and Kamen, A. (2004). Acoustic cell filter: a proven cell retention technology for perfusion of animal cell cultures. *Biotechnol. Adv.* 22: 433–444.

77 Gorenflo, V.M., Ritter, J.B., Aeschliman, D.S. et al. (2005). Characterization and optimization of acoustic filter performance by experimental design methodology. *Biotechnol. Bioeng.* 90: 746–753.

78 Ryll, T., Dutina, G., Reyes, A. et al. (2000). Performance of small-scale CHO perfusion cultures using an acoustic cell filtration device for cell retention: characterization of separation efficiency and impact of perfusion on product quality. *Biotechnol. Bioeng.* 69: 440–449.

79 Khanal, O., Singh, N., Traylor, S.J. et al. (2018). Contributions of depth filter components to protein adsorption in bioprocessing. *Biotechnol. Bioeng.* 115 (8): 1938–1948.

80 Yigzaw, Y., Piper, R., Tran, M., and Shukla, A.A. (2006). Exploitation of the adsorptive properties of depth filters for host cell protein removal during monoclonal antibody purification. *Biotechnol. Progr.* 22 (1): 288–296.

81 Agarwal, H., Rathore, A.S., Hadpe, S.R., and Alva, S.J. (2016). Artificial neural network (ANN)-based prediction of depth filter loading capacity for filter sizing. *Biotechnol. Progr.* 32 (6): 1436–1443.

82 Lutz, H. (2009). Rationally defined safety factors for filter sizing. *J. Membr. Sci.* 341 (1–2): 268–278.

83 Goldrick, S., Joseph, A., Mollet, M. et al. (2017). Predicting performance of constant flow depth filtration using constant pressure filtration data. *J. Membr. Sci.* 531: 138–147.

84 Yavorsky, D., Blanck, R., Lambalot, C., and Brunkow, R. (2003). The clarification of bioreactor cell cultures for biopharmaceuticals. *Pharm. Technol.* 27 (3): 62–77.

85 Liu, H.F., Ma, J., Winter, C., and Bayer, R. (2010). Recovery and purification process development for monoclonal antibody production. In: *MAbs*, 480–499. Taylor & Francis.

86 Boerlage, S. (2001). *Scaling and Particulate Fouling in Membrane Filtration Systems*. CRC Press.

87 Krupp, A., Please, C., Kumar, A., and Griffiths, I. (2017). Scaling-up of multi-capsule depth filtration systems by modeling flow and pressure distribution. *Sep. Purif. Technol.* 172: 350–356.

88 Pham, C.Y. (2010). Use of depth filtration in series with continuous centrifugation to clarify mammalian cell cultures. Google Patents US 7, 759, 117 B2, filed 27 December 2006 and issued 20 July 2010.

89 Thakur, G., Hebbi, V., Parida, S., and Rathore, A.S. (2020a). Automation of dead end filtration: an enabler for continuous processing of biotherapeutics. *Front. Bioeng. Biotechnol.* 8 https://doi.org/10.3389/fbioe.2020.00758.

90 Bink, L.R. and Furey, J. (2010). Using in-line disposable pressure sensors to evaluate depth filter performance. *Bioprocess Int.* 8 (3): 44–48.

91 Bangtsson, P., Estrada, E., Lacki, K. and Skoglar, H. (2012). Method in a chromatography system. US Patent 20120091063A1, GE Healthcare Bio-Sciences AB, filed 21 June 2010 and issued 19 April 2012.

92 CMC Biotech Working Group (2009). *A-Mab: A Case Study in Bioprocess Development*. Emeryville: CASSS http://casss.org/associations/9165/files/A-Mab_Case_Study_Version_2-1.pdf.

93 Thakur, G., Nikita, S., Tiwari, A., and Rathore, A.S. (2021b). Control of surge tanks for continuous manufacturing of monoclonal antibodies. *Biotechnol. Bioeng.* https://doi.org/10.1002/bit.27706.

94 Tiwari, A., Kateja, N., Chanana, S., and Rathore, A.S. (2018). Use of HPLC as an enabler of process analytical technology in process chromatography. *Anal. Chem.* 90 (13): 7824–7829. https://doi.org/10.1021/acs.analchem.8b00897.

95 Pathak, M., Ma, G., Bracewell, D.G., and Rathore, A.S. (2015). Re-use of Protein A resin: fouling and economics. *BioPharm Int.* 28 (3): 28–33.

96 Pathak, M. and Rathore, A.S. (2017). Implementation of a fluorescence based PAT control for fouling of protein A chromatography resin. *J. Chem. Technol. Biotechnol.* 92 (11): 2799–2807.

97 Hober, S., Nord, K., and Linhult, M. (2007). Protein A chromatography for antibody purification. *J. Chromatogr. B* 848 (1): 40–47.

98 GE Healthcare (2005). *MabSelect SuRe - Studies on Ligand Toxicity, Leakage, Removal of Leached Ligand, and Sanitization*. GE healthcare.

99 Fahrner, R.L. and Blank, G.S. (1999). Real-time control of antibody loading during protein A affinity chromatography using an on-line assay. *J. Chromatogr. A* 849 (1): 191–196.

100 Rüdt, M., Brestrich, N., Rolinger, L., and Hubbuch, J. (2017). Real-time monitoring and control of the load phase of a protein A capture step. *Biotechnol. Bioeng.* 114 (2): 368–373.

101 Chmielowski, R.A., Mathiasson, L., Blom, H. et al. (2017). Definition and dynamic control of a continuous chromatography process independent of cell culture titer and impurities. *J. Chromatogr. A* 1526: 58–69.

102 Flatman, S., Alam, I., Gerard, J., and Mussa, N. (2007). Process analytics for purification of monoclonal antibodies. *J. Chromatogr. B* 848 (1): 79–87.

103 Schiestl, M., Stangler, T., Torella, C. et al. (2011). Acceptable changes in quality attributes of glycosylated biopharmaceuticals. *Nat. Biotechnol.* 29 (4): 310–312.

104 Rathore, A.S., Yu, M., Yeboah, S., and Sharma, A. (2008). Case study and application of process analytical technology (PAT) towards bioprocessing: use of on-line high-performance liquid chromatography (HPLC) for making real-time pooling decisions for process chromatography. *Biotechnol. Bioeng.* 100 (2): 306–316.

105 Carta, G. and Jungbauer, A. (2010). *Protein Chromatography: Process Development and Scale-up*. John Wiley & Sons.

106 Mendhe, R., Thukkaram, M., Patil, N., and Rathore, A.S. (2015). Comparison of PAT based approaches for making real-time pooling decisions for process chromatography – use of feed forward control. *J. Chem. Technol. Biotechnol.* 90: 341–348. https://doi.org/10.1002/jctb.4448.

107 Shukla, A.A. and Aranha, H. (2015). Viral clearance for biopharmaceutical downstream processes. *Pharm. Bioprocess.* 3 (2): 127–138. https://doi.org/10.4155/pbp.14.62.

108 Shukla, A.A., Hubbard, B., Tressel, T. et al. (2007). Downstream processing of monoclonal antibodies – application of platform approaches. *J. Chromatogr. B* 848 (1): 28–39. https://doi.org/10.1016/j.jchromb.2006.09.026.

109 Vázquez-Rey, M. and Lang, D.A. (2011). Aggregates in monoclonal antibody manufacturing processes. *Biotechnol. Bioeng.* 108 (7): 1494–1508. https://doi.org/10.1002/bit.23155.

110 Gillespie, C., Holstein, M., Mullin, L. et al. (2019). Continuous in-line virus inactivation for next generation bioprocessing. *Biotechnol. J.* 14 (2): 1700718. https://doi.org/10.1002/biot.201700718.

111 Klutz, S., Magnus, J., Lobedann, M. et al. (2018). Developing the biofacility of the future based on continuous processing and single-use technology. *J. Biotechnol.* 213: 120–130.

112 Schofield, M. and Johnson, D. (2018). Continuous low-pH virus inactivation: challenges and practical solutions. *Genet. Eng. Biotechnol. News* 38 (15): S13–S15. https://doi.org/10.1089/gen.38.15.12.

113 Bolton, G.R., Selvitelli, K.R., Iliescu, I., and Cecchini, D.J. (2014). Inactivation of viruses using novel protein A wash buffers. *Biotechnol. Progr.* 31 (2): 406–413. https://doi.org/10.1002/btpr.2024.

114 Senčar, J., Hammerschmidt, N., Martins, D.L., and Jungbauer, A. (2020). A narrow residence time incubation reactor for continuous virus inactivation based on packed beds. *New Biotechnol.* 55: 98–107. https://doi.org/10.1016/j.nbt.2019.10.006.

115 Capito, F., Bauer, J., Rapp, A. et al. (2013). Feasibility study of semi-selective protein precipitation with salt-tolerant copolymers for industrial purification of therapeutic antibodies. *Biotechnol. Bioeng.* 110: 2915–2927.

116 Stavrinides, S., Ayazi Shamlou, P., and Hoare, M. (1993). Effects of engineering parameters on the precipitation, recovery and purification of proteins. In: *Processing of Solid–Liquid Suspensions* (ed. P. Ayazi Shamlou), 118–158. Butterworth-Heinemann.

117 Kumar, A., Galaev, I.Y., and Mattiasson, B. (2003). Precipitation of proteins: nonspecific and specific. In: *Isolation and Purification of Proteins* (eds. R.H. Kaul and B. Mattiasson), 225–276. Marcel Dekker.

118 Lydersen, B.K., Brehm-Gibson, T., and Murel, A. (1994). Acid precipitation of mammalian cell fermentation broth. *Ann. N.Y. Acad. Sci.* 745: 222–231.

119 Xiao-Ping, D., Jue, W., Deena, A. et al. (2009). Precipitation of process-derived impurities in non-protein A purification schemes for antibodies. *BioPharm. Int.* 2009 (7): 1–6.

120 Brodsky, Y., Zhang, C., Yigzaw, Y., and Vedantham, G. (2012). Caprylic acid precipitation method for impurity reduction: an alternative to conventional chromatography for monoclonal antibody purification. *Biotechnol. Bioeng.* 109: 2589–2598.

121 Hammerschmidt, N., Hintersteiner, B., Lingg, N., and Jungbauer, A. (2015). Continuous precipitation of IgG from CHO cell culture supernatant in a tubular reactor. *Biotechnol. J.* 10: 1196–1205.

122 Tscheliessnig, A., Satzer, P., Hammerschmidt, N. et al. (2014). Ethanol precipitation for purification of recombinant antibodies. *J. Biotechnol.* 188: 17–28.

123 Kateja, N., Agarwal, H., Saraswat, A. et al. (2016). Continuous precipitation of process related impurities from clarified cell culture supernatant using a novel coiled flow inversion reactor (CFIR). *Biotechnol. J.* 11 (10): 1320–1331. https://doi.org/10.1002/biot.201600271.

124 Li, Z., Gu, Q., Coffman, J.L. et al. (2019). Continuous precipitation for monoclonal antibody capture using countercurrent washing by microfiltration. *Biotechnol. Progr.* 35 (6) https://doi.org/10.1002/btpr.2886.

125 Dimitrov, A.S. (2009). *Therapeutic Antibodies: Methods and Protocols*. Springer.

126 Gottschalk, U. (2009). *Process Scale Purification of Antibodies*. Wiley Online Library.

127 Wang, C., Coppola, G., Gervais, J. et al. (2013). Ultrafiltration and diafiltration formulation methods for protein processing. In: (ed. USPTO), 25. North Chicago, IL: AbbVie.

128 Kamerzell, T.J., Esfandiary, R., Joshi, S.B. et al. (2011). Protein–excipient interactions: mechanisms and biophysical characterization applied to protein formulation development. *Adv. Drug Delivery Rev.* 63: 1118–1159.

129 Steele, A. and Arias, J. (2014). Accounting for the Donnan effect in diafiltration optimization for high-concentration UFDF applications. *Bioprocess Int.* 12: 50–54.

130 Schwartz, L., K. Seeley (2002). *Introduction to tangential flow filtration for laboratory and process development applications*. Pall Scientific & Technical Rep., 33213.

131 Strathmann, H., L. Giorno, E. Drioli, *Introduction to Membrane Science and Technology*, Wiley-VCH Weinheim, 2011.

132 Zeman, L.J. and Zydney, A.L. (2017). *Microfiltration and Ultrafiltration: Principles and Applications*. CRC Press.

133 Lutz, H. and Parrella, J. (2015). Single-pass filtration systems and processes. In: (ed. USPTO), 13. Billerica, MA (US): EMD Millipore Corporation.

134 Subramanian, G., *Antibodies. Volume 1: Production and Purification*, Springer Science & Business Media, 2013.

135 Rolinger, L., Rüdt, M., Diehm, J. et al. (2020). Multi-attribute PAT for UF/DF of proteins – monitoring concentration, particle sizes, and buffer exchange. *Anal. Bioanal.Chem.* 412 (9): 2123–2136. https://doi.org/10.1007/s00216-019-02318-8.

136 Thakur, G., Thori, S., and Rathore, A.S. (2020b). Implementing PAT for single-pass tangential flow ultrafiltration for continuous manufacturing of monoclonal antibodies. *J. Membr. Sci.* 613: 118492.

137 Thakur, G., Hebbi, V., and Rathore, A.S. (2021a). Near infrared spectroscopy as a PAT tool for monitoring and control of protein and excipient concentration in ultrafiltration of highly concentrated antibody formulations. *Int. J. Pharm.*: 120456. https://doi.org/10.1016/j.ijpharm.2021.120456.

138 Steinebach, F., Ulmer, N., Wolf, M. et al. (2017). Design and operation of a continuous integrated monoclonal antibody production process. *Biotechnol. Progr.* 33 (5): 1303–1313.

139 Hernandez, R. (2015). To surge or not to surge? *Genet. Eng. Biotechnol. News* 17 (16): 120–130.

140 Read, E.K., Park, J.T., Shah, R.B. et al. (2010). Process analytical technology (PAT) for biopharmaceutical products. Part I. Concepts and applications. *Biotechnol. Bioeng.* 105 (2): 276–284.

141 Clark, R.H., Latypov, R.F., De Imus, C. et al. (2014). Remediating agitation-induced antibody aggregation by eradicating exposed hydrophobic motifs. *MAbs* 6 (6): 1540–1550. https://doi.org/10.4161/mabs.36252.

142 Rathore, A.S., Nikita, S., Thakur, G., and Deore, N. (2021). Challenges in process control for continuous processing for production of monoclonal antibody products. *Curr. Opin. Chem. Eng.* 31: 100671. https://doi.org/10.1016/j.coche.2021.100671.

143 Glassey, J., Ignova, M., Ward, A.C. et al. (1997). Bioprocess supervision: neural networks and knowledge-based systems. *J. Biotechnol.* 52 (3): 201–205.

144 Breusegem, V.V., Thibault, J., and Chéruy, A. (1991). Adaptive neural models for on-line prediction in fermentation. *Can. J. Chem. Eng.* 69 (2): 481–487.

145 Lennox, B., Rutherford, P., Montague, G.A., and Haughin, C. (1998). Case study investigating the application of neural networks for process modelling and condition monitoring. *Comput. Chem. Eng.* 22: 1573–1579.

146 Paul, G.C., Glassey, J., Ward, A.C. et al. (1996). Towards intelligent process supervision: industrial penicillin fermentation case study. *Comput. Chem. Eng.* 20 (Suppl. 1): S545–S550.

147 Potocnik, P. and Grabec, I. (1999). Empirical modeling of antibiotic fermentation process using neural networks and genetic algorithms. *Math. Comput. Simul* 49: 363–379.

148 Tian, Y., Zhang, J., and Morris, J. (2002). Optimal control of a fed-batch bioreactor based upon an augmented recurrent neural network model. *Neurocomputing* 48: 919–936.

149 Teissier, P., Perret, B., Latrille, E. et al. (1997). A hybrid recurrent neural network model for yeast production monitoring and control in a wine based medium. *J. Biotechnol.* 55 (9): 157–169.

150 Nikita, S., Tiwari, A., Sonawat, D. et al. (2021). Reinforcement learning based optimization of process chromatography for continuous processing of biopharmaceuticals. *Chem. Eng. Sci.* 230: 116171. https://doi.org/10.1016/j.ces.2020.116171.

151 Kumar, S., Dangi, A.K., Shukla, P. et al. (2019). Thermozymes: adaptive strategies and tools for their biotechnological applications. *Bioresour. Technol.* 278: 372–382.

152 Numers, C.V., Nakajima, M., Siimes, T. et al. (1994). A knowledge based system using fuzzy inference for supervisory control of bioprocesses. *J. Biotechnol.* 34 (1994): 109–118.

153 Shi, Z. and Shimizu, K. (1992). Neuro-fuzzy control of bioreactor systems with pattern recognition. *J. Ferment. Bioeng.* 74 (1): 39–45.

154 Causa, J., Karer, G., Núñez, A. et al. (2008). Hybrid fuzzy predictive control based on genetic algorithms for the temperature control of a batch reactor.

Comput. Chem. Eng.: 3254–3263. https://doi.org/10.1016/j.compchemeng.2008.05.014.

155 Cosenza, B. and Galluzzo, M. (2011). Nonlinear fuzzy control of a fed-batch reactor for penicillin production. *Comput. Chem. Eng.* 36 https://doi.org/10.1016/j.compchemeng.2011.07.016.

156 Karer, G., Mušič, G., Škrjanc, I., and Zupančič, B. (2007). Hybrid fuzzy model-based predictive control of temperature in a batch reactor. *Comput. Chem. Eng.*: 1552–1564. doi: 10.1016/j.compchemeng.2007.01.003.

157 Sousa, R. and Almeida, P. (2001). Design of a fuzzy system for the control of a biochemical reactor in fed-batch culture. *Process Biochem.* 37: 461–469. https://doi.org/10.1016/S0032-9592(01)00239-4.

158 Sommeregger, W., Sissolak, B., Kandra, K. et al. (2017). Quality by control: towards model predictive control of mammalian cell culture bioprocesses. *Biotechnol. J.* 12 (7) https://doi.org/10.1002/biot.201600546.

159 Santos, L.O. et al. (2012). Nonlinear model predictive control of fed-batch cultures of micro-organisms exhibiting overflow metabolism: assessment and robustness. *Comput. Chem. Eng.* 39: 143–151. https://doi.org/10.1016/j.compchemeng.2011.12.010.

160 Kovárová-Kovar, K., Gehlen, S., Kunze, A. et al. (2000). Application of model-predictive control based on artificial neural networks to optimize the fed-batch process for riboflavin production. *J. Biotechnol.* 79: 39–52. http://dx.doi.org/10.1016/S0168-1656(00)00211-X.

161 Lyon, D. (2019) Achieving active control of cell culture performance with the aid of machine learning techniques, Abstracts. *74th Northwest Regional Meeting of the American Chemical Society* (16–19 June 2019), Portland, OR, NORM-298.

162 Woodley, J.M. (2019). Accelerating the implementation of biocatalysis in industry. *Appl. Microbiol. Biotechnol.* 103 (12): 4733–4739. https://doi.org/10.1007/s00253-019-09796-x.

163 Birle, S., Hussein, M.A., and Becker, T. (2016). Management of uncertainty by statistical process control and a genetic tuned fuzzy system. *Discrete Dyn. Nat. Soc.* 2016 https://doi.org/10.1155/2016/1548986.

164 Cedeño, M., Rodriguez, L., and Sánchez, M. (2016). Bioprocess statistical control: identification stage based on hierarchical clustering. *Process Biochem.* 51 https://doi.org/10.1016/j.procbio.2016.08.020.

3

Artificial Intelligence and the Control of Continuous Manufacturing

Steven S. Kuwahara

GXP BioTechnology, Tucson, AZ 85741, USA

3.1 Introduction

Continuous processing has many advantages after improvements in efficiency and speed of processing. The automotive, petroleum, chemical, and food processing industries, among others, have been using it and have developed many useful methods and concepts for dealing with the continuous flow of materials and intermediates through their processes. The pharmaceutical industry has been slow to adopt these methods primarily due to the threat of regulatory problems. This is no longer justified as regulatory agencies are strongly encouraging the adoption of more modern and efficient methods by the pharmaceutical industries.[1] The US Food and Drug Administration (FDA) has published a guidance document on process analytical technology [1] that encourages real-time release testing (RTRT) – an important component of continuous processing.

Maintaining control of a continuous manufacturing process can become a major undertaking depending upon the complexity of the process that must be controlled. A simple one- or two-step process can be easily controlled through a computer programmed with feedback loops of the "If-Then" type that will control on–off switches or secondary controllers that operate with step functions. While this type of process can easily be controlled through fairly simple programs based on limited algorithms, a full manufacturing process involves several simple processes that interact with each other and eventually create a web of interactions that will require the use of artificial intelligence (AI) algorithms and programs to accommodate the speed of processing and the interaction of different factors.

When considering the use of AI with continuous processing, it is important to determine the level of intelligence that is needed. As with humans, there are

---

1 Note that the author is located in the United States and works primarily with FDA documents. Thus, the references cited in this chapter are primarily US documents; however it should be noted that similar documents may, and usually do, exist for other jurisdictions, especially in the European Union and Japan. Thus, the reader should seek comparable documents that are applicable to the specific region of the world where the products will be manufactured and used.

Process Control, Intensification, and Digitalisation in Continuous Biomanufacturing, First Edition.
Edited by Ganapathy Subramanian.
© 2022 WILEY-VCH GmbH. Published 2022 by WILEY-VCH GmbH.

different levels and types of "intelligence" that may be encountered. These levels are often the result of human cultural conditioning, physical and mental development, or genetic inheritance. For instance, two equally "intelligent" individuals from different cultures, who speak different native languages, try to converse with each other in a third language that they have both studied. Because of their accents and cultural backgrounds, they may have difficulties in communicating, and both of them leave the conversation thinking that the other is stupid. We also know that children's brains develop at different rates, and a concept that is easily assimilated by an older person may be very difficult to comprehend by a child despite the fact that the child may eventually develop an intelligence equal to that of the older person.

Examples may also be found when thinking about the Turing Test. In 1947, Alan Turing gave a lecture in which he proposed the possibility of an intelligent machine. In 1950 [2], he proposed a test by which a machine might be judged to be intelligent. In this test, a human-A and a computer-B are isolated from each other in separate rooms and communicate using typewritten notes so that differences in vocal intonation would not be a factor. An isolated human-C observer is also present and able to read the messages and also to communicate with the computer-B and human-A. The computer-B is programmed to act like a human even to the point of lying, and human-A is instructed try to prove that B is a computer. If the observer-C cannot distinguish human-A from the computer-B, one may then conclude that the computer is intelligent and possesses AI.

An objection to the Turing Test has been raised in the form of a thought experiment called the Chinese room [3]. The experiment has been dismissed by many as being irrelevant to the discussion of AI, but it contains many elements that bear on the current discussion. The experiment was proposed by John Searle in 1980 and goes as follows: first, consider a person who cannot speak, read, or write Chinese, but is in a room containing a large number of books that contain questions written in Chinese. Each question also comes with an answer also written in Chinese. The person receives a question written in Chinese and then searches through the books until he finds a phrase with the same characters in the same order as the question that was submitted. The person then takes the answer that was associated with the question and submits it as the answer to the question.

Now consider a room that is divided by a wall. On the one side, there is a computer that is capable of recognizing Chinese characters and whose memory banks are programmed to contain all of the information that was in the books that were in the room described previously. On the other side, you have a questioner who is fluent in Chinese and submits printed questions in Chinese through a slot to the other side of the wall. Printed answers are returned to the questioner. The questioner's job is to ask questions that will allow him to decide if he is conversing with a human or a computer. Again, the computer is programmed to act like a human even to the point of lying to the questioner. Thus, a question like "Are you a computer?" will receive a negative answer. If the questioner concludes that he is conversing with a human, the computer will have passed the Turing Test even though it is only programmed with a complex sorting program that does not meet any of the criteria for "intelligence."

There is no simple answer to the question of "What is artificial intelligence?" A good discussion of this question can be found in a paper by John McCarthy [4] that was designed to be read by laymen interested in the subject. This is very important because the popular media have presented AI as some sort of super intelligent (almost magical) entity requiring large computers or robotic devices that act and think like humans. This is incorrect and has led many people to believe that AI can replace humans in all aspects of process management. Additionally, it has led many people in the upper management levels of pharmaceutical companies to believe that the introduction of AI will result in extremely high costs and a disruption to operations that cannot be borne by their companies. Actually, there are limitations to AI at present, and the development of a super intelligence, while possible, should be left to the future and science fiction writers.

Of course, the other problem is that the intelligence and sophistication of the human observer in the Turing Test are important. A young, inexperienced person of moderate intelligence may not be able to tell if they are dealing with a computer, while an older, highly intelligent person with extensive life experiences could, by asking carefully crafted questions. Thus, there could be different levels of AI just as we see different levels of intelligence in people. From the management point of view, an AI system capable of making decisions based on many different considerations may not be desirable. One would think that in situations such as those involving out-of-specification test results (OOS) or out-of-trend test results (OOT), it would be sufficient for an AI system to gather relevant data while leaving a final decision to fail or release a production lot or to significantly alter the process to a management team. Therefore, the AI system needed would not need to be good enough to pass a Turing Test. So, a system good enough to operate in the Chinese room would be sufficient. In fact some of the earliest examples of intelligent machines involved the use of computers that were programmed to recognize a large number of patterns. The computers used to play chess against skilled opponents such as Grand Masters were provided with memories of a large number of previous games that were then used to predict the outcomes of specific moves. This was derived from the investigation of chess masters who used to play against a large number of contestants where they would glance at a board and make a move and then move on to the next player. It was found that these players had played a very large number of practice games or read descriptions of a large number of games and memorized the different patterns of each game. As a result, they could recognize a pattern simply by glancing at a board and would know what the next move should be. This is very much like the operation of the computer envisioned in the Chinese room. In this case it would be sufficient to build a very large memory with a sorting program that could locate the correct pattern and the best response to that pattern.

The subject of AI has become very complex and confused. The popular press and entertainment media have given the public the idea that AI will constitute a super intelligence capable of mimicking a human or, in some cases, exceeding the thinking ability of humans. Different people have different ideas about what AI is. For instance, machine learning systems capable of gathering and analyzing massive amounts of information provided by the Internet of things (IoT) may be considered to be a form of AI, while others may consider lifelike robots capable of answering questions in several different languages to be a form of AI. The fact is that continuous processing really does not require an AI system that is capable of mimicking a human. Given the limitations of humans, this may not be desirable, anyway. In the control of continuous processing for pharmaceutical products, AI may be considered to be a system capable of interpreting information from several process stream analyzers [5] that themselves have functions that were validated [6, 7]. With bioprocessing, where the manufacturing environment may be critical, the AI in control of a continuous processing will need to be aware of the ambient conditions and maintain the process flow in a validated state. This may be done through the use of sensory processes that already exist. These systems may appear to be AI to the average worker, while experienced AI workers may not consider them to be "intelligent." For the present purposes, it will be sufficient to have a system that is "intelligent enough." The AI used here will appear to be super intelligent in that it will be able to monitor several concurrent conditions and activities while integrating the data in a manner that would normally require the interaction of several human supervisors.

Given the factors of complex programming and the cost of constructing and operating a Turing machine, it is unlikely that such a computer would be desirable for the control of a continuous process. The level of AI that would be needed to control a continuous process will not be as high as what would be expected for a Turing machine. A computer programmed to operate at the If-Then and If-GoTo level should be sufficient. An AI system capable of machine learning and making complex decisions or "judgment calls" probably would not be desirable from a management stand point, given the high complexity and regulated condition of the biomanufacturing industry.

For practical purposes, the size of computer needed here and the complexity of the programs needed to operate it in an AI mode could result in a slow response that would be detected by a questioner or observer in a Turing test. In the case of these theoretical discussions, computing speed is often ignored as the speed with which computer processors operate is so high that any delay in response times will be hardly noticeable to a human. However, it could be a problem in high-speed continuous processing. Computers are said to operate in the nanosecond (ns) range at the speed of light. If we consider the fact that the speed of light drops from 300 000 to 125 000 K/s in a diamond. (It is affected by the refractive index of the conductor.) It would still be about 12.5 cm/ns compared with 10 cm/msec for a nerve impulse. Thus, from a human perspective, computer actions appear to be instantaneous, but with a computer-controlled decision-making process, there could be a significant delay if a long-distance connection or a human decision is required.

A decision-making process, especially one that may require accessing a significant amount of data, may need a large amount of wiring. Consider the possibility of a manufacturing facility that may fill a whole city block with the controlling computer located in one corner of the block, but needing to draw on information from sensors located at several places within the block. The situation could become worse if a manufacturing process were being controlled by a computer located at a long distance from another computer (say, at the company headquarters) that was needed to provide input to a control decision. The lag time for a control instruction may be long enough to allow a significant number of defective products to be made.

To avoid these situations, the AI system programmed into the controlling computer should have the required information available in its memory and the authority to make the appropriate decisions, and the computer should be physically located close to the manufacturing area. Any need for a human interaction will also result in a significant delay.

Given the speed at which a continuous processing operation can run, it is quite possible that a significant amount of product that is out of specification could be generated before a decision to stop the manufacturing line was executed. Moreover, given the current and future costs of pharmaceutical products, the cumulative costs of these OOS products will force the company to insist on developing more rapid programs and control mechanisms. In turn, this will increase the cost of developing and maintaining these systems.

A major issue here has already been seen with Laboratory Information Management Systems (LIMS). In this case, the algorithms and programs used for the computer-based management system were initially developed by programmers who had little or no experience in laboratory operations, especially in quality control laboratories. This resulted in a huge amount of conflict between the programmers and the workers who provided the inputs to the LIMS and also those who had to review and evaluate the output data. With continuous processing, it will be critical to include experienced process engineers within the AI development team. It would be especially important to include process engineers who actually have hands-on experience with the batch process and have a basic understanding of the type of AI system. From the software side of the AI system, it will be necessary to have programmers who have worked with continuous processes for other products, even if they were outside of the pharmaceutical industry. A third type of specialist that should be included would be an experienced regulatory affairs or quality control worker who possesses a basic understanding of both AI programming and continuous processing. This person would be especially important for navigating the regulatory requirements that are imposed on pharmaceutical products and processes. In the case of bioprocessing, the product and the process are tightly linked to the point where regulatory agencies often approve products for release to the public on a lot-by-lot basis. Thus, the limits and requirements imposed by regulations would need to be incorporated into the AI controlling the continuous processing system. This would be further complicated by the fact that when operating on the global level, the regulations of several different regions (which sometimes are in conflict with each other) will need to be considered.

To have a system that resembles an AI, the computer would need to be able to make adjustments to the process to meet the needs resulting from the normal variations of inputs, but these changes would need to be made within well-studied limits. These limits would be determined during the process design phase of developing the process. This would be when the early limits of the design space will be defined. Most regulatory agencies have accepted the idea that changes within the design space can be made without further regulatory approval so the early parameters would need to be programmed into the AI, and provisions in the programming would need to be designed to allow future changes of this space after regulatory approval is obtained. At the earliest stages, this would be derived from a batch process. During this phase, early estimates would be made of the critical quality attributes (CQA) of the final and intermediate products and the critical process parameters (CPP) that will allow the different products to meet their CQAs. CQAs and CPPs taken together or in groups will define the multidimensional design spaces [3] for producing a releasable product that is safe and efficacious and meets the requirements of quality by design (QbD). These are also described and defined in a useful guidance document [8].

One of the major control parameters is the quality of the raw material used in the process. A company must define and evaluate the quality of its raw material [9]. This raw material must be held on-site, and the AI system must be aware of the status of each lot of each component so that only approved material may be used in the process. The status of each lot of each raw material (received, quarantined, approved, or rejected) will be defined through an interaction of information from the receiving warehouse and the quality control department. The quantity of material received in the warehouse, the amount removed for testing, including the identity of the containers from which the samples were removed, and the amount taken for sampling will need to be monitored. In addition to these quantities, the amounts dispensed to the processing floor will need to be recorded as it is the standard practice to account for the disposal of all material ordered by the company, especially in the case of highly active or narcotic drugs where regulatory requirements often require accounting for the disposition of these substances down to the gram or milligram level.

The same is true for containers and closures. In addition to monitoring the quantities received, tested, and released for use, in bioprocessing it is common to require cleaned and sterilized containers and closures for use in the filling line. The total number of containers filled will need to be reconciled vs. the number cleaned and sterilized, as well as the number of secondary containers and labels issued.

The labels are also important in that they need to be verified against regulatory requirements for the information contained in them, as well as company design requirements. The direction inserts also need to be checked as they are normally considered to be a part of the labeling and their information content is regulated along with matters as the size of their print and order of presentation. Under normal circumstances, a copy of all labeling would need to be verified by quality assurance (QA) personnel and an exemplar entered into the batch record. With an AI system, one would use a scanner and optical character reader (OCR) program to record the labeling and verify that the wording matches that required by the company specifications.

Because of the need to have material available at all times for the maintenance of continuous processing, the AI system will need to interact with the purchasing function to ensure that orders are placed when inventories decline to critical levels. It will also be necessary to monitor the status of these orders to ensure that material will be available as needed for the process. The AI system will need to be aware of the delivery status of the material and the status of their acceptance test to ensure that only qualified material will be made available to the processing line.

The control of the numbers of raw materials, containers, closures, and labeling is very important as drug counterfeiters like to obtain actual containers, closures, labels, and even active pharmaceutical ingredients (API) and excipients, to use with their products. This adds to the apparent legitimacy of their products and makes it easier to insert them into legitimate supply chains.

Because of these requirements, the AI controller will need to be able to account for the actual distribution and quantities used of all of the material received by the manufacturing function.

Continuous processing will lead to a modified just-in-time (JIT) system. The capital required to secure and maintain such large quantities of raw material, API, containers, closures, and labels will be high enough to require the use of a JIT system. In most cases a slightly modified JIT system is will be employed with small amounts of material being held in stock to accommodate short-term fluctuations in the supply chain. The AI system used here will need to gather and correlate data about purchase orders, receipt, testing, and quantities available to ensure a smooth flow of material for the processing line.

A basic principle of quality management as applied to manufacturing processes holds that "the earlier a problem is found and fixed, the lesser the cost to the company." The basic idea here is that this minimizes the amount of labor and materials that will be lost when the defective material becomes scrap. This is often a difficult principle to maintain when using a continuous process, as the defective material may progress through several additional processing steps before a problem is noticed and the process can be stopped. This, in turn, calls for the development of rapid or continuous measurement and detection systems to continuously verify the quality of the product as it progresses through the manufacturing system. This problem can be solved through the use of relatively simple AI methods.

However, this is also where limitations on AI programs will become apparent. Brian Bergstein [10] has noted that it is very difficult for an AI system to grasp the idea of cause and effect. While it would seem that programming an AI system with appropriate If-Then commands would solve the problem, this requires the programmer to know what the effect will be "if" a particular situation arises and tell the AI how to cope with the occurrence. This means that the programmer will need to have extensive knowledge about the manufacturing system. For instance, if a certain situation arises during the step C to D transition in a process, an experienced worker may "know" that a problem will arise when material at step M produces product N. With a new process, this experienced worker will not exist, and even a machine learning system may not develop an awareness of a problem until a considerable amount of defective product is produced. Fortunately, this will probably occur during a development phase, and the loss of product and time will be assigned to "development costs."

For pharmaceutical purposes, US FDA has defined AI software as medical devices that must follow the device Good Manufacturing Practices (GMP) [11] and the regulations for software validation [12, 13]. This means that an AI system will need to process the information needed for a continuous processing system [14] while remaining in a validated state, itself. As may be seen from the earlier discussion, regulatory acceptance of this type of system will require close cooperation between experienced manufacturing regulators and software regulators and company workers. In turn, this should mean that regulatory approval will require more time than usual and the amount of data provided to regulatory will be large. Difficulties will also arise when changes must be made as these will need to pass through a company's change control system before submission to a regulatory body for acceptance of the change. This would claim for an evolutionary process involving many interactions between manufacturing personnel and regulatory workers during the transition from batch to continuous processing accompanied by the concurrent introduction of AI control.

The data required for manufacturing releasable products will need to be generated during the development process [8]. Initially, there would be a process where the product developers would generate a list of characteristics that would be desirable in the product. This would be during a batch processing stage where the initial CQAs and CPPs would be set, along with the required characteristics of the raw material that will be used in the process. This information could then be programmed into a machine learning system that would be used to modify the numbers as needed during the transition from a batch processing to a continuous processing system. It will be necessary to determine the quality target product profile (QTPP). This is initially done during the early development phase to determine the design criteria for the product. The QTPP should be developed by the upper management of the company to ensure acceptance by all levels within the company and QTPP should include considerations from the patient's point of view as well as the manufacturer's needs. Moreover, this should be programmed into a machine learning system to initially define the limits and targets for the design of the process and will need to be modified as development proceeds, and a better understanding of the product and process is secured. A good machine learning system should be able to make the necessary modifications as it gathers data from the development work.

The QTPP along with quality risk management (QRM) methods will serve as the basis for developing CQA and CPP for the product that satisfy the CQA. These products will be process intermediates and the final product. Initially there will also be sets of CQA to define the quality of the material that may be introduced into the process stream. Furthermore, these will basically be the acceptance criteria for APIs, excipients, solvents, containers, closures, labels, and similar material that will be introduced into the process as needed. When processed under the requirements of the CPP, they will produce intermediates that will meet other CQA and so on down the process until final products meeting the final product CQA are the result.

The first set of CQA and CPP will be designed to meet the requirements of the QTPP and will be the result of paper studies by the product development team. A knowledgeable and experienced group of personnel will be needed, and past experience with other products of similar types will be useful. Data from large-scale laboratory experiments will also be needed. It is understood that the initial set of CQA and CPP will need to be modified as development progresses, and the AI system should be able to monitor results from development studies and propose modifications as the need arises. The reason for the AI system only proposing changes is that changes of the CQA or CPP may require regulatory and management approvals before they can be emplaced, especially when development has progressed to the level of nonclinical studies. It is for this reason that the initial CQA and CPP specification ranges should be set to be as wide as possible. Data from development studies will allow workers and the AI system to narrow acceptance ranges as work progresses.

The AI system will need to be equipped with statistical software of the type required for statistical process control (SPC) charts. These would be of the type such as the moving range charts that are generated by making periodic observations of a CQA during the development process. With a continuous process, it should be possible to continuously monitor the CQA and CPP, if the appropriate testing apparatus is available. It will be necessary to slightly modify the calculations to accommodate differences. The moving range chart (X-mR chart) assumes that the data are normally distributed and requires 25 previous observations before it can be constructed. In very early stages of the work, these observations could be taken from those used for the Pre-Control charts (discussed below), but they should be replaced by 25 observations from later, more established production runs as the implementation of the continuous process proceeds. As with most SPC charts, there are really two charts that are plotted against common points in time.

With a standard chart, these data points are collected discretely by periodic sampling, but in a continuous process these observations could be made at very close intervals so that the resulting chart would appear to consist of a smooth continuous line. The AI system should be fast enough to produce metadata quickly to give results in a continuous line over time and should also be programmed to detect trends and results that surpass alert limits and give alarms and notices to appropriate personnel before failure limits are reached. It is very important to have the moving average compared with the moving range (deviation) data as a large deviation range for the data will invalidate the average. This is because a wide deviation range will indicate the presence of a wide confidence limit that makes any statement about the correct average highly uncertain.

The X chart is constructed by using the individual observations (X_i) in the following manner:

$$\overline{X} = \Sigma X_i / n \text{ (centerline)}$$

where n is the the number of X_i observations.

The upper control limits (UCL) and lower control limits (LCL) are

$$\text{UCL} = \overline{x} + d_2 \left(\overline{mR}\right) \quad \text{LCL} = \overline{x} - d_2 \left(\overline{mR}\right)$$

where d_2 is the control chart constant that is used to convert ranges into standard deviations (s). The reader should consult tables of Control Chart Constants that are found in many textbooks on statistics.

The mR chart is constructed from moving ranges using the differences between successive values of X as follows:

$$mR = |X_2 - X_1|, |X_3 - X_2| \ldots$$

$\overline{mR} = \Sigma mR/_{n-1}$ (centerline) $n-1$ is used as the number of range pairs is one less than the total number of X_i.

The UCL for the moving range is UCL = $D_4\left(\overline{mR}\right)$. However, the LCL for ranges does not exist as its calculation requires a constant known as D_3 that is zero for less than seven replicates. Note that the constants are case sensitive such that D_2, D_3, and d_2, d_3 all exist as different numbers in tables found in standard statistics textbooks. An estimate of the standard deviation (s) of the data can be calculated as

$$s = \overline{mR}/d_2$$

Note that the control limits (UCL and LCL) for control charts are actually calculated from the data. In contrast, the UCL and LCL used for process capability studies are not statistically based, but are assigned as specification limits by the company.

3.2 Continuous Monitoring and Validation

With a continuous process, a continuous validated state will be required. Continuous validation of a continuous process is done after the first two phases of process validation are completed [1]. All of the previously described work will need to be completed, and this will provide the worker with the data needed to set the specifications for the continuous verification of the validated state of the process.

During the early stages of introducing the continuous process, the process stream analyzer and any other monitoring equipment will not have completed the process qualification (PQ) phase of their qualification/validation studies. The process itself will be new, and there will be no sufficient manufacturing runs to provide the replicates needed to initiate monitoring via control charts. In some instances, the control charting may be started by using data from preceding batch processing, but it is best to allow the process to speak for itself. In this situation, the use of a Pre-Control chart (stoplight chart) [15] will be useful. These charts are not dependent on the form of the distribution and work with single point data. The centerline is based on the target specification and allows a determination of whether or not the process is ready for control charting. The target specification (centerline) is chosen by the company and is usually the targeted center point of the specification range that is desired. These values are normally developed in the design and development phase of the process.

3.3 Choosing Other Control Charts

The most commonly used control charts are based on the assumption that the data will be normally distributed. However, this is not true for all types of process data, and a determination of the actual type of distribution should be made during the process development stage. Control charts have been developed for data that follow the binomial and Poisson distributions. Also, multivariate control charts have been generated for use with software packages that may be employed with process stream analyzers simultaneously measuring more than one parameter. The continuous validation of a continuous process may be performed through the use of appropriate control charts that monitor CQAs and CPPs. However, these charts by themselves are not sufficient as a process validation requires attention to more than the CQA and CPP. The status of the manufacturing environment and the supporting validations should be periodically checked, especially with continuous processes that are run constantly and may only shut down once or twice a year for maintenance. AI will greatly assist in this as its system can maintain an awareness of these environmental parameters and alert management if any of them approach unacceptable levels.

3.4 Information Awareness

The continuous validation of the continuous process itself is actually a continual process. At various times, workers will need to obtain a sample from the process stream and test it using the independent primary test method (PTM). This test result (a primary test method result [PTMR]) can then be used in two ways. First, it will be used to confirm the corresponding predicted primary test method result (PPTMR) that will be monitored by the process stream analyzer. Second, the difference between the PTMR and the PPTMR will be used for the control chart that monitors the performance of the process analyzer. PPTMRs from the process analyzer system, themselves, should be monitored using software that can detect trends and violations of control limits. It will be wise to program alert limits into the AI system, as it is not sensible to wait for outright out-of-specification results before taking action on a problem. In this sense, the software should continuously generate a control chart showing the status of the process stream. This will produce a continuous verification of the validated state of the process as required for phase 3 of a process validation [15–17].

The AI must retain an awareness of the status of factors such as the environment of the manufacturing area where the process is operating. Matters such as particulate matter in the air, bioburden levels and presence of objectionable organisms in raw material and containers and closures, and cleanliness levels in the processing area must be known. This will usually be the result of communications between the laboratory information system (LIMS) and the AI. If aseptic processing is required, monitoring of the aseptic status of the processing area is needed along with knowledge about the sterilization procedures. In reality, an AI system must be programmed to monitor all of the factors needed to be checked and provide alerts to workers. The speed of AI systems in providing information for making simple decisions is advantageous and should be used to provide a rapid response to problems. For instance, a computer may receive continuous input from a laser particle monitor that checks the particle count of the air in an aseptic processing area. This computer can be programmed to alert workers when a certain particle level is noted and to immediately stop the process if an extreme level is surpassed.

The problem here is that in most pharmaceutical companies that are manufacturing products via batch processing, an LIMS already exists for the quality control laboratories and related functions. This will need to be modified to work with the AI system as the conversion from batch to continuous processing occurs [18]. In the past there have been many problems related to programming languages and data processing systems that have impeded the connection of a LIMS to a company's data handling system. It will be important for management to ensure the compatibility of the existing LIMS and the AI that is being introduced. Similar problems may be expected if the AI is required to interact with an existing data management system.

3.5 Management and Personnel

The use of AI with continuous processing will require the ability of employees to rapidly respond to failures or alarm/alert conditions where the AI system will be set to warn workers of an impending mechanical failure or out-of-specification event. Qualified quality control and manufacturing management personnel will need to be available at all times to make final decisions on what actions need to be taken. This is because a continuous process can produce a large amount of product in a short period of time. If a failure is noted or an alarm is raised, personnel will need to quickly correct the problem to minimize the amount of product that could be lost. It will be very important for the AI system to be able to note the point at which a failure is noted and to identify all material that was manufactured after the failure occurred. Some workers have assumed that the AI system would immediately stop a continuous processing line if these conditions occur, but it must also be realized that the stoppage of a continuous processing line also generates costs from the delay in manufacturing. Also, some material in a continuous process cannot be held for very long before it can no longer be used in the next processing step, and this material would be lost.

It should be clear that using AI to control continuous processes will require the formation of a team of individuals with an in-depth knowledge of several areas of manufacturing. For a company that does only continuous processing, this team should be a part of its normal management structure. However, if batch processing and continuous processing are both done, separate teams at appropriate management levels will be needed. The continuous processing team will need the authority to quickly request help from various specialists and the ability to request help from consultants. As a result, the head of the team should be at the upper levels of company management.

There is a human element also involved here. Company management may be very reluctant to allow an AI system to fail a lot or to stop a processing line. Since an AI program will be unforgiving and operate on logic alone, factors such as costs and production needs will not have an impact on go/no-go decisions. These factors are known to affect human decisions about processing, and management, particularly upper management, may be reluctant to surrender them to a computer. This human factor may prove to be a major stumbling block for a company that wishes to introduce AI to a continuous process. Company personnel, especially junior and middle management, may fear that the adoption of AI will endanger their jobs, and consequently they will impede the adoption process. The people involved in introducing AI into a continuous process must find ways to assure existing employees of the security of their positions.

A related problem here arises from the fact that AI requires the use of specialized methods and programming languages, and a company's existing information technology (IT) personnel may not possess the information and training to introduce or even maintain an AI-controlled system to an existing continuous processing system. If a company wishes to convert an existing continuous process into an AI controlled system, it may be necessary to introduce new employees and hire consultants to perform the conversion and maintain the AI system. If, on the other hand, the company is starting from a batch process and intends to convert the process into an AI-controlled continuous process, it might be easier to form a new processing team by retraining existing IT personnel and using existing workers to develop the continuous process from the batch process having characteristics they are already familiar with. In all cases it will be necessary to provide the workers with a basic understanding of the AI system and also the basic characteristics of the process. From a quality management point of view, it is vital to have all of the workers understand not only their jobs but also the activities that occur at stages before and after they perform their work. This will lead to a more cohesive team that will know how their jobs have an impact on manufacturing the product.

Past experience has shown that when workers only know the requirements for performing their own specific jobs, they may make decisions that have serious consequences for work that is done at subsequent stages. There have been many instances where workers at stage C have made apparently minor changes that resulted in changes in the product at stage D. This was serious enough in batch processing, but with continuous processing a considerable amount of product and time may be lost while dealing with the problem. This is why it is very important to carefully consider the limits on the ability of an AI to make changes.

The other personnel problem that may easily arise is that lower ranking employees, especially those who perform repetitive or relatively simple tasks, will be very reluctant to comply with the adoption of an AI system due to the fear of losing their jobs. There are already many examples of relatively simple robotic systems replacing receptionists and bank tellers and self-driving automobiles and trucks replacing taxi and delivery truck drivers. Similar lurid stories put forth by irresponsible reporters can easily panic personnel with lower levels of training or education. It will be important for companies to provide retraining and other forms of reassurance to workers to prevent interference with the introduction of AI systems.

It is in the performance of batch reviews that AI will be very useful. As a practical matter a company should want to have individual lots of products. This is because it can be very useful to be able to isolate specific units of products when dealing with customer complaints, out-of-specification product reviews, product liability lawsuits, and similar incidents where it saves a considerable amount of money if a relatively small, isolated group of products needs to be reviewed or investigated. Even with continuous processing, it is advantageous to be able to isolate lots of products rather than whole production batches. Each of these lots or batches will need to undergo a review of the data derived from the manufacture of the lot to verify that the lot was produced under conditions meeting the requirements of the various design spaces before it could be distributed to the public. This review can be performed rapidly by an AI system operating under human supervision.

In defining these lots, one may follow the definitions given in the regulations for GMP in the production of drugs or biologics. A *batch* means a specific quantity of a drug or other material that is intended to have uniform character and quality, within specified limits, and is produced according to a single manufacturing order during the same cycle of manufacture [19]. Similarly, a *lot* means a batch, or a specific identified portion of a batch, having uniform character and quality within specified limits. Otherwise, in the case of a drug product produced by continuous process, it is a specific identified amount produced in a unit of time or quantity in a manner that assures it has uniform character and quality within specified limits [20]. This is also related to the definition of lot numbers [21] also called *lot number, control number, or batch number* that means any distinctive combination of letters, numbers, or symbols or any combination of them, from which the complete history of the manufacture, processing, packing, holding, and distribution of a batch or lot of drug product or other material can be determined. However, in the case of biologics [22], "lot means that quantity of uniform material identified by the manufacturer as having been thoroughly mixed in a single vessel." Clearly the definition is highly dependent on the definitions proposed by the manufacturer and accepted by the regulators.

The requirements for Master Batch/Lot Records are listed in the US Code of Federal Regulations [23]. In this case the master record must be programmed into the AI, and a provision must be made to define the lot as described above. For instance, with continuous processing, this cannot be continued forever. There will need to be a point at which processing will need to stop, and time is allotted for maintenance and repairs of the processing system. This will be especially important for the production of sterile products by aseptic processes. Thus, it will be necessary to have stopping points for processing. This might be a week or a month, in which case the product produced in one week or month could be identified as a batch with each day's production labeled as a separate lot. Given the speed of operations and the quantity that can be produced in a given time by continuous processing, control with an AI will be very useful.

The AI will also need to have connections to not only the manufacturing line but also the QC laboratory's LIMS system and other data points where critical data are produced. This will allow the AI to create individual lot records based on the master record. Based on the US regulations, the master record will have sections for the following:

(a) The name and strength of the product and a description of the dosage form.
(b) The name and weight or measure of each active ingredient per dosage unit or per unit of weight or measure of the drug product and a statement of the total weight or measure of any dosage unit.
(c) A complete list of components designated by names or codes sufficiently specific to indicate any special quality characteristic.

(d) An accurate statement of the weight or measure of each component, using the same weight system (metric, avoirdupois, or apothecary) for each component. Reasonable variations may be permitted, however, in the amount of components necessary for the preparation in the dosage form, provided they are justified in the master production and control records.
(e) A statement concerning any calculated excess of component.
(f) A statement of theoretical weight or measure at appropriate phases of processing.
(g) A statement of theoretical yield, including the maximum and minimum percentages of theoretical yield beyond which investigation according to §211.192 is required.
(h) A description of the drug product containers, closures, and packaging materials, including a specimen or copy of each label and all other labeling signed and dated by the person or persons responsible for approval of such labeling.
(i) Complete manufacturing and control instructions, sampling and testing procedures, specifications, special notations, and precautions to be followed.

The usual procedure will require the AI to provide all of the information required above. The ability of the AI to monitor all of the important data generating points will be important as this speed will be needed when producing lots by continuous processing. In normal processing each lot will need to have a corresponding lot record with all data entered, and this information will require checking by the QA workers. Normally, this will need a little over the equivalent of 16 hours working time, but an AI should be able to produce the lot record in a matter of minutes. The information that will need to be verified is given in the US regulations [24] as follows:

(a) An accurate reproduction of the appropriate master production or control record, checked for accuracy, dated, and signed.
(b) Documentation that each significant step in the manufacture, processing, packing, or holding of the batch was accomplished, including (i) dates; (ii) identity of individual major equipment and lines used; (iii) specific identification of each batch of component or in-process material used; (iv) weights and measures of components used in the course of processing; (v) in-process and laboratory control results; (vi) inspection of the packaging and labeling area before and after use; (vii) a statement of the actual yield and a statement of the percentage of theoretical yield at appropriate phases of processing; (viii) complete labeling control records, including specimens or copies of all labeling used; (ix) description of drug product containers and closures; (x) any sampling performed; (xi) identification of the persons performing and directly supervising or checking each significant step in the operation or, if a significant step in the operation is performed by automated equipment under 21CFR211.68, identification of the person checking the significant step performed by the automated equipment; (xii) any investigation made according to 21CFR211.192; and (xiii) results of examinations made in accordance with 21CFR211.134.

The examinations noted in (xii) are basically for discrepancies in the testing results as compared with the specifications, especially apparent failures to meet specifications, and in (xiii) are for problems related to the final drug products, such as discrepancies between the counts of actual units produced vs. the numbers of containers or closures that were delivered to the filling line.

The completion of these lot records and their independent review often require a large amount of personnel time due to the discrepancies that are found during the review and the inability to locate the documentation to resolve the problems. Yet, these records are vital to the company. Once the lot is released for distribution to the public, these records will remain as the only documentation that can describe the events that led to the production of the particular lot. The speed and reliability of a validated AI system will allow the reduction of personnel time and provide for the fast preparation of complete records.

The use of continuous processing coupled with AI offers many advantages that will lead to the reductions in unit production costs. These will include reduced final product testing as a result of the continuous monitoring and testing permitting the use of PAT principles and the resulting ability to release products with a minimum amount of final product testing allowing reductions in inventory and a more rapid recovery of production expenses. The advantage is greatest for companies that intend to produce a product over a long term. Careful planning combined with knowledgeable management, willing to work closely with regulators, will be beneficial for stockholders and patients.

References

1 FDA (2004) *Guidance for industry PAT – a framework for innovative pharmaceutical development, manufacturing, and quality assurance*. CDER, CVM, ORA; FDA; U.S. DHHS, September.
2 Turing, A.M. (1950). Computing machinery and intelligence. In: *Mind*, vol. LIX, 236, 433–460. https://doi.org/10.1093/mind/LIX.236.433.
3 Searle, J. (1980). Minds, brains and programs. *Behav. Brain Sci.* 3 (3): 417–457. https://doi.org/10.1017/S0140525X00005756.
4 McCarthy, J, (2007) *What is Artificial Intelligence?* http://www-formal.stanford.edu/jmc/
5 ASTM. *Standard practice for validation of the performance of process stream analyzer systems*. ASTM Designation D3764-15.
6 ASTM. *Standard practice for sampling a stream of product by variables indexed by AQL*. ASTM Designation: E2762-10 (Reapproved in 2014).
7 ASTM. *Standard guide for establishing a linear correlation between analyzer and primary test method results using relevant ASTM standard practices*. ASTM Designation: D7235-14.
8 FDA (2009), *Guidance for industry*. ICH Q8(R2) Pharmaceutical Development. Annex. CDER, CBER; FDA; USDHHS. November.
9 U.S. Code of Federal Regulations (2020) *Testing and approval or rejection of components, drug product containers, and closures*. Title 21, Chapter I, Part 211, Subpart E, paragraph 211.84.

10 Bergstein, B (2020) What AI still can't do. *MIT Technology Review*, (19 February).

11 U.S. Code of Federal Regulations (2020) *Quality systems regulations*. Title 21, Chapter I, Subchapter H – Medical Devices, FDA, DHHS, Part 820.

12 Guidance for Industry and FDA Staff. (2002) *General principles of software validation*. Center for Devices and Radiological Health (CDRH), CBER, FDA, USDHHS, January.

13 U.S. Code of Federal Regulations (2020) *Electronic records: electronic signatures*. Title 21, Chapter I, Subchapter A – Medical Devices, FDA, DHHS, Part 11.

14 Kuwahara, S.S. (2018). Continuous validation for continuous processing. In: *Continuous Biomanufacturing: Innovative Technologies and Methods* (ed. G. Subramanian), 533–547. Wiley-VCH, Weinheim, Germany.

15 Center for Drug Evaluation and Research (CDER), Center for Biologics Evaluation and Research (CBER), and the Center for Veterinary Medicine (CVM), FDA, USDHHS. (2011). *Guidance for Industry: Process Validation: General Principles and Practices*.

16 ASTM. (2007). *Standard guide for specification, design, and verification of pharmaceutical and biopharmaceutical manufacturing systems and equipment*. ASTM Designation E 2500-07.

17 ASTM. (2008). *Standard guide for application of continuous quality verification to pharmaceutical and biopharmaceutical manufacturing*. ASTM Designation E 2537-08.

18 Durivage, M.A. (2015). *Practical Engineering, Process, and Reliability Statistics*, 134. Milwaukee, WI: American Society for Quality, Quality Press.

19 U.S. Code of Federal Regulations (2020) *Quality systems regulations*. Title 21, Subchapter C – Drugs: General, Part 210.3 Definitions. (b)(2).

20 U.S. Code of Federal Regulations (2020) *Quality systems regulations*. Title 21, Subchapter C – Drugs: General, Part 210.3 Definitions. (b)(10).

21 U.S. Code of Federal Regulations (2020) *Quality systems regulations*. Title 21, Subchapter C – Drugs: General, Part 210.3 Definitions. (b)(11).

22 U.S. Code of Federal Regulations (2020) *Quality systems regulations*. Title 21, Subchapter F – Biologics, Part 600, Biological Products: General, Subpart A – General Provisions. 600.3 Definitions, (x).

23 U.S. Code of Federal Regulations (2020) *Quality systems regulations*. Title 21, Subpart J – Records and Reports. 211.186 Master production and control records.

24 U.S. Code of Federal Regulations (2020) *Quality systems regulations*. Title 21, Subpart J – Records and Reports. 211.188 Batch production and control records.

Part II

Intensified Biomanufacturing

4

Bioprocess Intensification: Technologies and Goals

William G. Whitford

DPS Group, Strategic Consulting Group, 959 Concord St #100, Framingham, MA 01701, USA

4.1 Introduction

Biomanufacturing has evolved into a remarkably dynamic discipline. Considering new product types and applications, new manufacturing processes and platforms, and new facility designs and construction methods, it is difficult to even catalog the developments arising. It was not too long ago that the field consisted of a very few protein biologicals, adoptive immunotherapies, and vaccines produced in batch mode. Today, such completely new therapeutic modalities as personalized medicine and advanced therapy medicinal products (ATMP) challenge every step of product development, filing, and manufacturing.

Initially, cost control and efficiency promotion were not top priorities in biopharmaceutical manufacturing but now have been the focus. The science and technology required to accomplish these can be difficult to catalog because of the diverse industry sectors. For example, the need to produce large quantities of products can be due to either dosage mass or global demand. For some entities, there is a pressure to quickly develop a functional first-generation process. Only later does attention turn to process capacity, robustness, consistency, efficiency, manufacturing flexibility, and sustainability [1]. On the other hand, there are viral-vectored and cell therapies that first require intensification of the legacy platform to become economically effective or schedule efficient. Of course, these development goals are carefully balanced with the goals of change minimization and process finalization.

Bioprocessing presents unique engineering considerations because complex living systems present challenges such as being composed of many interacting metabolic pathways. The number and complexity of potential process parameters to consider have previously thwarted true comprehensive monitoring and open-loop model predictive control. Bioprocessing also presents challenges such as high labor and material costs and a demand for extensive experimental efforts. It has been observed that for reasons including regulatory and quality constraints, bioprocess engineering has been slow to incorporate new power from industry digitalization ("4.0"), automation, standardization, miniaturization, and operations integration.

Process Control, Intensification, and Digitalisation in Continuous Biomanufacturing, First Edition.
Edited by Ganapathy Subramanian.
© 2022 WILEY-VCH GmbH. Published 2022 by WILEY-VCH GmbH.

Describing advances relevant to the field of bioprocess intensification (BI) in particular is a challenge for several reasons. First is due to the sheer number of specific and unique developing sciences and technologies demonstrating value in this arena. These range from the many omics advances in cell biology to the revolution in digital technologies. Second is the growing number of bioproduct types ranging from 3D bioprinted diagnostic constructs to nucleic acid vaccines to the many ATMPs. Third are the innovative expression systems to consider, from transient plant-based expression to ciliate (*Tetrahymena thermophila*). Fourth is the at-scale application of each product type in any number of diverse manufacturing systems and operations modalities. These range from plant design to continuous manufacturing approaches to developments in process design and control. Next, there is sometimes confusing overlap in the common definitions and applications of these developing technologies. Many technologies that can powerfully support each other are not necessarily dedicated to supporting some common application. For example, digital biomanufacturing can support both advanced enterprise resource planning (ERP) and continuous processing, as well as activities completed unrelated to BI (Figure 4.1). Then, the highly regulated nature of biopharmaceutical production makes the update and improvement of processes more difficult than some other manufacturing environments. Finally, considering that some published technologies are still in late research phase while others are commercially available and ready to be implemented, there is a need to reveal in just what ways these many new applications are actually powerful or practical [2]. This all makes comprehensive or categorical statements difficult including all biological products or manufacturing technologies.

Today, we see a blast in the development of such new protein biologicals as monoclonal antibody (mAb) fragments, designed ankyrin repeat proteins (DARPins), single-domain antibodies (dAbs), and bispecific monoclonal antibodies

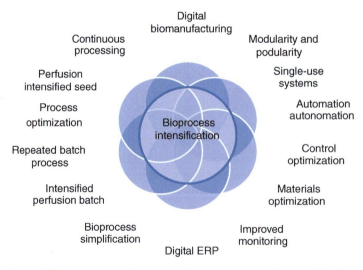

Figure 4.1 Many new initiatives work together in various combinations, sometimes supporting bioprocess intensification. Source: Whitford, original work '20.

(bsMAbs), to name a few. In fact, the number of extremely divergent entities, from exosomes to nanoparticle-vectored mRNA vaccines, obviate any quick review or easy generalizations regarding advanced biomanufacturing techniques. BI enablers include a wide array of technologies and vendors, improving distinct product types and supporting diverse analytic, diagnostic, and therapeutic fields [3]. It is truly beyond the scope of any publication to examine process intensification for each permutation of the above.

The traditional protein biological manufacturing is being challenged by such platforms as bacterial, yeast, and plant-based expression systems (to name a few). Animal cell expression processes are advancing through such innovations as fixed-bed reactors, perfusion technologies, and modular facilities. These provide such improvements as unique bioreactor capability, continuous manufacturing, and shortened process trains. Also, each of these processes is now supported by such developments as the multi-attribute method (MAM), new probes and bioreactor sampling devices, and multiplexed, integrated analytics. New facility design technologies include such revolutionary improvements as prefabricated modular and "podular" designs, modality agnostic suites, and shared services plants.

Beyond this, our capabilities in biomanufacturing process/product development (PD) are being advanced by the increased process understandings provided by developments in all of the omics fields, a number of automated and high-throughput analytical tools, and categorical advances in genetic engineering (synthetic biology). Here, we see gains from next-generation sequencing to new gene isolation and cloning techniques to more specific and efficient gene editing such as CRISPR/CasX.

In a category by itself are elements of the so-called Industry 4.0 or digital manufacturing. Beginning with process analytical technologies (PAT) and quality by design (QbD), at the end of the last century an effort to introduce modern science- and risk-based approaches to biomanufacturing was established. At that time, some newer monitoring and analytical and control technologies were just beginning. However, just as in smartphone technology, no one envisioned the types of advances we are now seeing. From cloud and edge computing, the industrial Internet of things (IIoT), and artificial intelligence (AI) to digital twin (DT)-type *in silico* process modeling and predictions, "Biomanufacturing 4.0" or "digital biomanufacturing" (DB) is changing the face of biomanufacturing (Table 4.1). All these initiatives support such visions as heightened product consistency and quality; real-time product release; continued/ongoing process verification; ERP/enterprise asset management (EAM) tools such as manufacturing execution systems (MES), electronic batch records (EBR), laboratory information management systems (LIMS), and asset performance management (APM); and bioprocess intensification.

There are usually distinct (even unique) parameters to be identified in any specific biological product manufacturing process. There are, however, many commonalities between an HEK293 culture producing recombinant erythropoietin, an AAV-vectored gene therapy, and a multi-peptide subunit-based epitope vaccine against COVID-1. Therefore, unless otherwise identified, it will be assumed here that the mainly upstream process addressed is an established animal cell line, producing of an enzyme or antibody, in advanced regulation, for pharmaceutical

Table 4.1 Digital biomanufacturing will support many systems related to bioprocess intensification.

Process development	Machine vision
Process optimization	Speech recognition
Control optimization	Integrated analytics
Process prediction	Improved monitoring
Soft sensors	Knowledge management
Risk analysis	ERP: APM, MES, EBR, LIMS
Process troubleshooting	Continued process verification
Supervisory process control	Automation, autonomation
Mixed and dynamic process models	Real time process analytics
Quality control from materials to product	Real-time product release
Process and supply chain management	Cloud/edge computing
Virtual and augmented reality	Advanced data governance

Source: Whitford, original work '19.

employment, using currently popular facility design, equipment, and control. Also, as there are so many new materials, processes, and technologies proposed – in development or being generally employed. Only those proven practical at this time will be discussed here.

While some average or generalized comparative measurements are presented, accurate values through modeling of any particular scenario as compared with any other are best accomplished by professionals trained in such endeavors, such as *BioProcess Technology Consultants (BPTC), BioProcess Technology Group* (BPTG), *Latham BioPharm Group*, or others [4].

4.2 Bioprocess Intensification

4.2.1 Definition

Each new technology and advancement in the field presents its own unique set of features to the operations they affect. These features include development ease and speed, process power and economy, and product yield and quality. It is also apparent that significant changes can have unintended consequences, for good or ill, and such tradeoffs maybe in areas not in the original design. Most of the many improvements in biomanufacturing can be corralled under the commonly employed headings of "Factory of the Future," "Next-gen," or "Smart Bioprocessing." Some authors include advances such as increases in product quality or regulatory ease under the heading "Bioprocess Intensification". When taking a comprehensive look at manufacturing, one sees that there is a bit of muddling and even contradiction in what particular processes, aspects of a technology, or material development should properly be designated "process intensification." While the biomanufacturing industry is addressing many distinct goals such as improved manufacturing flexibility, product quality and cost-effectiveness, the terms used in identifying each of the initiatives commonly employed to accomplish them currently overlap at best. This ambiguity and lack of precision can cause inefficiency and errors across an industry or within an individual company [5].

The Cambridge University Press dictionary defines "intensification" as "The fact of becoming greater, more serious, or more extreme, or of making something do this." For modern industry in general, process intensification has been defined as "Any chemical engineering development that leads to a substantially smaller, cleaner, safer, and more energy efficient technology" [6]. Consequently, under some current (sometimes very broad) usage, precisely what aspects of biomanufacturing have been improved when employing the phrase "process intensification" can include facility design, equipment and process control, capital costs, a reduction in suite classification, and environmental stress. For bioprocesses, these activities could include increases in cell density, manufacturing flexibility, environmental sustainability, product quality, and more. In a broad usage, there is virtually no improvement in which the facility, material, process, or equipment that could not be regarded by some to be part of an "intensification" effort, and some authors have been employing it in this way.

However, many in the biomanufacturing community use the term "process intensification" only when referring to an improvement in the productivity or economy of the process, such as to optimize an existing process to increase output via one of many metrics, as in more product in a shorter time or with less cost. One group sees it as follows: "Process intensification is used to get more out of the process, whether it's by producing more product upstream or retaining more product downstream" and "take an existing process and optimize it to increase output: more product in a shorter time, with fewer steps, and from a smaller working footprint" [7]. Yet, another defines bioprocess intensification from the other direction but, to the same conclusion, as "smaller footprint/volume" systems that "yield the same desired results and performance" [8]. A similar conclusion was reached, albeit in a different field, when agricultural intensification was defined as an "increase in agricultural production per unit of inputs" [9].

Acceptance of this usage would indicate that, for example, a plant modernization or a method of reducing an undesirable product glycoform may be desirable – but would not then be referred to as an "intensification" of the process. In any event, it is suggested that our industry might consider clarifying this by establishing an industry standard for such biomanufacturing terms, such as has been accomplished for many other terms in the *ISPE Glossary of Pharmaceutical and Biotechnology Terminology* [10]. Some limit the scope of the definition of the initiatives examined to only those that increase productivity but, due to the overlap in either the definition or application of each new empowering technology, may adopt a different approach to their categorizing or grouping. For example, one author categorically distinguishes (and contrasts) perfusion-based continuous processing from "intensified bioprocessing," whereas others might consider continuous processing one means of it [11].

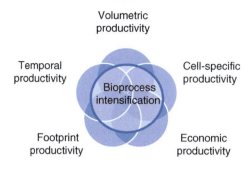

Figure 4.2 Bioprocess intensification can increase productivity in many distinct metrics, each of which can influence other metrics. Source: Whitford, original work '20.

In this chapter, the meaning of process intensification is limited to getting more product out of a process or producing more product by any metric, e.g. per cell, time, volume, footprint, or cost (Figure 4.2). Whether it is by producing more product upstream or retaining more product downstream, intensifying the process requires changes in a manufacturing plant, equipment, materials, control, or process. It therefore has the potential for significant impact upon the molecule, and for this reason these activities are most efficiently done during early phase development. They can still be pursued later on – or even postlaunch – but this would require demonstration of a lack of adverse effect on the identity, quality, purity, and potency of the biological product and formal regulatory activities. While many developments are improving yield per volume or time downstream, such as acoustic wave cell separation and tangential flow filtration (TFF), space here permits only an introduction to those that are primarily powerful upstream.

4.2.2 New Directions

The "Factory of the Future," "Next-gen," and "Smart Bioprocessing" designators, as they are often implemented, include many distinct and recently initiated technologies (Table 4.2). In fact, remarkable innovations in nearly all biopharmaceutical entities, processes, platforms, modes, equipment, materials, and facilities have been seen. It is important to note that (i) individual process development groups or teams can focus upon achieving different applications or end goals from a particular technology, (ii) as above, application of one intensification effort can have consequences in more than one distinct metric of production efficiency gain, and (iii) a developing technology can easily apply to more than one type of process improvement goal. An example of the latter is that the reduction or stabilization of a transition metal concentration in a culture medium formulation might increase the volumetric productivity of bioreactor operation, although it was designed for an entirely different primary application – improving product quality. Therefore, it may be more efficient, and clearer, to use the phrase "bioprocess intensification" *only* when referring to developments immediately *intended* to provide an increased productivity of some type. In this way we could reserve the designators "Factory of the Future," "Next-gen," and "Smart Bioprocessing" as phrases that include technologies supporting all kinds of process and product improvements (Table 4.2). However, which particular terms are restricted in their scope – and which are allowed to remain more

Table 4.2 Identification and organization some major programs and technologies currently implemented in upstream biomanufacturing.

Factory in a box	Next-gen, factory of the future, and smart bioprocessing			Intensified biomanufacturing
	Continuous	Biopharm 4.0		
Future proofing	Connected and closed	PAT/MAM		Train simplification
Modular, podular	Contiguous, not closed	APM/OPM		Footprint productivity
Single-use systems	In-line fluids conditioning	Adaptive plant		Temporal productivity
Hybrid SU systems	Straight-through processing	Cloud/AI/IIoT		Economic productivity
Standardized vs. free	Perfusion-based continuous	Improved monitoring		Volumetric productivity
Integrated, enterprise	Pseudo- and quasi-continuous	ERP: MES, EBR, LIMS		Cell-specific productivity
Shared service plant	Intensified perfusion continuous	Model predictive control		Improved clonal expression
Platform agnostic suites	Integrated continuous processes	Real-time product release		Improved process/medium
Entity and mode flexible	Repeated fed (or intensified) batch	Automation, autonomation		Many bioprocess simplifications
Shortened process train	Continuous but unjoined operations	Integrated, real-time analytics		Perfusion-intensified seed (N-1)
Prefabricated cGMP facilities	Enterprise continuous biomanufacturing	Continued process verification		

Note: Many programs support or otherwise closely relate to each other in some essential way or have been initiated in another context and now are being applied to biomanufacturing. Some have been described and defined in the literature by authors of disparate backgrounds; thus the same initiatives have been named and/or described using different terminology. Many terms have been used to refer to different groupings or organizations of elements from the same overall set of component technologies.
Source: Whitford, original work '19.

comprehensive – is not as important as codifying in some way the decisions that are made.

Regardless of the particular types of process or product improvements expected from an intensification effort, the scope of the effort can range from unit operations to entire processes – even extending to entire facility design and build engineering. This increasing scope of application directly correlates to the degree of organizational impact and subsequent effort required to successfully implement and underscores the importance of terminology during communication for goal setting and alignment. It must be acknowledged that while we will here address the process intensification aspect of the new technologies introduced in this chapter, biomanufacturers are increasingly finding other benefits in them as well, such as flexible manufacturing, futureproofing, modality agnostic suites, heightened product quality, or increased sustainability.

4.2.3 Sustainability Synergy

A concern for the Earth and calls for manufacturers to work toward more sustainable processes and facilities are increasing. Both suppliers and biomanufacturers are discovering that, in contrast to green initiatives incurring exorbitant costs, they can sometimes provide actual financial savings. For example, many programs under the "process intensification" umbrella not only increase productivity by some unit of measure but will also reduce environmental stress. Employing perfusion to increase the density of cell culture to skip an N-x cycle saves not only time and expense but also an entire cycle of single-use (SU) materials (Table 4.3). Engineering a clone to secrete more product in a given volume of medium can increase not only productivity per dollar spent on media but also productivity per consumable piece or energy consumed. The employment of prefabricated and modular components in facility design will be shown in Section 4.3 to support greater productivity per some aspect of CapEx or OpEx. However, they can also support a more sustainable process in a few

Table 4.3 Operational states (modes) of stirred tank bioreactors demonstrating means of intensification.

Batch: Cell seed and all nutrients employed are added at beginning of the run, with no medium or feed added or removed until culture is harvested as a unit

Fed batch: Mid-operation supply (various ways) of nutrients (i.e. substrates or primary metabolites) during operation, with entire culture harvested as a unit

Perfusion concentrated batch: Perfusion-based media exchange while retaining all cells and product for up to 10× the product concentration of a batch harvest

Perfused continuous: Equivalent volumes of media simultaneously added and removed (with or without product) with some of the cells retained in bioreactor

Intensified perfused continuous: Retaining all the cells during early perfusion to amplify cell density to predefined level, followed by near-steady-state operation

Dynamic perfusion: Production medium perfusion under initial cell-specific rate control, switched to a constant feed rate with very lean medium during harvest

High productivity harvest: Perfusion with nutrients (medium) to maintain peak cell performance in the production reactor providing a more controlled harvest

Repeated batch: Incomplete harvesting of material from culture and replacing with production medium allowing another culture expansion without cleaning

Repeated concentrated batch: Using perfusion to amplify both cell and product concentrations for intensification of production using the repeated batch mode

Intensified seed train: Perfusion to raise n–x reactor concentration (and reduce seed-train steps) to accelerate progression (reduce time) to production reactor

Intensified cryostock: Greatly increasing the amount of cells in frozen working stock units to accelerate progression and/or reduce n–x cultures in a seed train

Source: Whitford, original work '19 and Data from Mirasol [12].

ways. In reducing the mass of construction materials, there is less to manufacture and ultimately dispose of. Increased facility flexibility and equipment or product agnostic suits provides a better likelihood of "futureproofing" the construction, and installing suites with integral modular air-handling/HVAC ensures proper capacity and can obviate redundant systems. Similar serendipity is apparent from other advances, such as automation, as well.

4.3 Intensification Techniques

4.3.1 Enterprise Resource Management

It is appropriate to begin with programs that improve performance and reliability of existing processes, equipment, and facilities [13]. APM and enterprise resource management (ERM) initiatives are now supplied in a wide range of components and implementations, all of which are designed to provide better and more enterprise control over production, materials management, and maintenance [14]. By using advanced data analysis to optimize the activities within of existing facilities, APM improves the performance of a plant while minimizing risk and cost. It includes condition monitoring, predictive maintenance, and equipment integrity management and supports many approaches to operational excellence for improving efficiency by increasing operational performance and minimizing downtime.

APM and ERM are now being greatly improved through Industry 4.0-type computing, cloud/edge techniques, IIoT, and AI [13]. Integrated process historian databases store and analyze data used to mitigate operational risks. MES collect real-time data on many of the conditions in the production process, supporting automated monitoring and tracking of parameters associated with improved manufacturing efficiency [14]. Distributed control systems (DCSs) provide independent control for each aspect of a user-defined control area, providing better monitoring, performance tracking, and failure risk reduction. APM programs lean heavily upon EAM systems to help manage the overall maintenance and organization of a facility's assets [15].

APM and operations performance management (OPM) can contribute to increasing the productivity of a plant and are being driven by suites of software tools designed to streamline industrial processes, optimize throughput, and increase productivity across multiple sites. Modern software leverages real-time and historical analytics results to provide a portfolio-wide view of each piece of equipment, unit operation, overall business performance, and, more importantly here, actionable insights to improve these processes and equipment performance. Modern APM/OPM software begin by aggregating data from both operational technology (OT) and information technology (IT) sources. In applying newly developed modeling and analysis, they are able to deliver holistic operational awareness to drive intelligent decision making and support the productive implementation of such newer developments as digital twins, augmented reality, and operation autonomation [16].

4.3.2 Synthetic Biology and Genetic Engineering

To date, mammalian cells have dominated other protein expression systems in approved recombinant protein biologicals. They have the capacity and posttranslational (PT) systems to produce large and complex recombinant proteins with acceptable characteristics. However, many developments in genetic engineering and so-called "rational design" are supporting the optimization of many classical cell platforms, including prokaryotes, to provide more efficient expression of quality product. Advances in metabolic engineering (e.g. synthetic biology) can improve such factors as peak cell density, high-density culture duration, cell-specific production, product quality, and homogeneity and reduce the emergence of nonproducer mutants [17, 18]. This manipulation of outputs (e.g. transcripts) or gene contexts (e.g. epigenetic marks) is occurring in products from protein biologicals to vaccines to cell and gene therapies. Systems biology combines data generated from multiple 'omics approaches (e.g. transcriptomics, proteomics, and metabolomics) to create a more holistic understanding of cells. Such work is revealing information regarding multiple interactions between genes, transcripts, and proteins [19].

Meetings and symposia, such as the American Institute of Chemical Engineers (AiChE) *International Mammalian Synthetic Biology Workshop*, present some of the newest technologies in this field. A few of the approaches to dynamic control of cell behavior are applicable to the development of null cell lines providing improved characteristics in final producer constructs. For example, new techniques can assist in the control of transcription factors in tuning of cell-to-cell variability in gene expression [20]. Codon optimization is an expression technology used to increase recombinant protein secretion. The degeneracy of the genetic code provides opportunity for codon optimization to increase protein expression in some systems by as much as 1000-fold. However, in mammalian cells, assumptions underlying this are tricky [21].

Accurate, efficient, and robust methods of gene editing have been facilitated by RNA interference (RNAi), ribozyme engineering, zinc finger nucleases (ZFNs), transcription activator-like effector nucleases (TALENs), and recombinase-mediated cassette exchange (RMCE). CRISPR/CasX is now being used to modulate the level or activity of genes encoding products involved in complex metabolic pathways reflected in functional behaviors of individual cells or assemblies of cells in tissues and organs. These approaches can be used to manipulate the regulation of transcription and affect the structural and functional features of chromatin to reveal how genetic material is organized and utilized within a cell. This has been powerful in elucidating the relationship between native 3D structures in the genome and the expression of heterologous genes in engineered cells [18].

Cloning, cell line development, and gene editing are being greatly advanced through such radically new technologies as microfluidics [22, 23]. One new approach uses a "laser tweezers" type of technology (but actually more gentle) that directs millions of light-actuated pixels to move individual cells so they can be isolated, cultured, assayed, and exported. Suppliers can describe software algorithms that automatically identify single cells and then gently direct them into one of over 14 000 wells in a microfluidic chip in less than 30 minutes. As fed clonal cultures proliferate, the software images each individual well to continuously calculate growth rates. The phenotype of cells is reported in hours to days instead of weeks in more classical systems. Once colonies of interest are identified, the light patterns move them into position for collection. The increased footprint, materials, and temporal productivity is enormous [22].

4.3.3 New Expression Systems

The majority of new biopharmaceutical active ingredients in the market are recombinant proteins, and of those, the vast majority are from nonhuman mammalian cells [24]. Nevertheless, many disparate systems in development hold out promise for identified products, and a few of the most popular and successful are mentioned here [18]. Human cell lines enhance the expression of proteins with human-like PT modifications [25]. However, the cultivation of these cells on a commercial scale remains problematic and the more robust lines can contain animal virus sequences. *Escherichia coli* is the preferred prokaryotic host for recombinant proteins due to its low cost, well-known biochemistry and genetics, rapid growth, and good productivity [26]. Disadvantages include a lack of human PT modifications, inclusion body (IB) formation, codon bias, and endotoxin issues. While newer genetic engineering techniques have improved producer strains, further technological development is still required.

Yeasts have become popular expression hosts due to their rapid growth, easy genetic manipulation, cost-effective growth medium requirements, available complete genome sequences, and ability to provide engineering controllable PT modifications [27]. *Pichea pastoris* and *Saccharomyces cerevisiae* are the most common for recombinant biopharmaceutical production. Several gene targets that might enhance acceptable protein production have been identified; however, the platform currently suffers from a low yield of properly glycosylated protein. *P. pastoris* (also known as *Komagataella phaffii* or *K. pastoris*) is popular for properly folded, functional secreted proteins and high cell densities [28]. Despite being genetically engineered to perform human-like N-glycosylation, final moiety sequences remain an issue [29].

Insect cells have been used for the expression of various recombinant proteins for over 30 years. The baculovirus expression vector system (BEVS) is one means of the production of recombinant proteins in insect cells. Engineering of acceptable glycosylation patterns from BEVS- and CRISPR-based systems for virusless engineering of insect cells promises the use of these cells as alternatives for the manufacturing of therapeutic proteins and vaccines [30].

Transgenic plants present advantages such as low cost, safety, ease of scale-up, stability, and ability to produce *N*-glycosylated proteins. While studied and considered for many years, a biotechnology company in the development of a plant-derived quadrivalent virus-like particle (VLP) recombinant influenza vaccine recently announced that its vaccine candidate for COVID-19 induced a positive antibody response only 10 days after a single dose in mice. The commercialization of an innovative influenza vaccine manufactured using the metabolic capability of plants would be a milestone in the advancement of these systems. It is anticipated that flu vaccine products using VLPs from plants could become available soon [31].

4.3.4 Bioprocess Optimization

Nominal operating parameters can be optimized to enhanced productivity for bioreactor cultivation in any chosen operation mode or expression platform. The improvement of standard operating parameters can result in enhanced volumetric productivities through higher cell densities or cell-specific production. Such strategies have resulted in an increase in plant yield.

Many factors contribute to enhance the bioprocess in this regard, including improvements to the process technology, cell line, nutritional environment, reactor environment control parameters, feeding and supplementation schedules, the reactor control scheme or algorithm, and the production mode chosen (Table 4.3) [12].

Process improvements have made it possible to reach higher titers in upstream bioprocesses through different types of intensification scenarios [32]. Standard control parameters such as temperature, pH, agitation, aeration, dissolved oxygen, and CO_2 are the first choices for adjustment to achieve enhanced productivity. Others, such as feed timing and components, temperature shifts, gas exchange rates and shifts, mixing rates, and hydrodynamic shear, can also be examined. Proximate goals include increased cell densities, enhanced specific and volumetric productivities, increased fraction of quality product, and more robust or reproducible performance. Such process optimization has been common in fed-batch production bioreactors using non-perfusion seed cultures [33]. Improvements in the fed-batch approach have followed four significant developments: (i) improved monitoring of primary and secondary nutrient levels; (ii) design of experiment (DOE), i.e. Plackett–Burman approaches to optimizing the feed; (iii) algorithm-determined feed rates; and (iv) better design of feedforward-PID-feedback control. One example of qualitative advances includes reports of overcoming the density effect that had for decades limited the cell density of infection (CDI) in the baculovirus infection of insect cell cultures during VLP in vaccine manufacturing [34].

Improvements in the seed train are an example category of a recently successful bioprocess optimization. Approaches include larger seed freeze volumes and cell mass through "high cell density cryopreservation," eliminating early culture expansion steps and shorting its duration. Intensified perfusion in the seed train can (i) also support skipping an n-X culture cycle, saving materials and time, (ii) allow seeding the production reactor at higher densities providing faster and/or higher final production reactor peak cell densities, and (iii) provide greater yields from the production reactor as less nutrient complement is consumed in cell growth to reach peak densities.

Scale factors in the process describe those culture behaviors that vary as the size of the reactor changes. Some values are carefully maintained as the process is scaled up. Others are intentionally altered for reasons of economy or trade-off in other parameters, while even other value changes are obligate as reactor size increases. Many scale values are relatively linear or proportional. However, values can vary for parameters such as reactor power input, heat dispersion, and mass transfer [35]. The large number of relevant scaling parameters makes the effort define the optimal design space multidimensional.

4.3.5 Bioprocess Simplification

The goal of bioprocess simplification is to minimize process activities and extent while maintaining productivity and value [36]. Early on, this simplification can contribute to what is referred to as the "manufacturability" of a product. This is a key consideration in whether or not a product will be commercialized. Bioprocess simplification can improve manufacturability through improved robustness, consistency, efficiency, sustainability, and, most relevant here, productivity.

While late-stage and postlaunch process changes increase in expense and implementation difficulty, they should nevertheless be considered – especially when synergies or serendipities appear. Many examples of these multifactor improvements exist, such as in the amplifying of cell densities in the seed train to eliminate an N-x culture step. Among other advantages (see Section 4.3.4), this can shorten the physical extent of the seed train, materials consumption, and culture duration. Benefits of this simplification include cost savings in reducing the time and equipment required, risk reduction in eliminating an entire process step that can go wrong, and reduced environmental stress in eliminating the clean-in-place (CIP)/steam-in-place (SIP) requirement of a classical reactor or the plastic liner from a single-use reactor.

Another example of such process streamlining exists in release testing. There are many advances in process analytics, from high-level initiatives such as PAT to the procedures being specified in the MAM [37]. It is possible to significantly shorten product release wait time for many entities by replacing cell-based assays – which can take weeks – with significantly faster *in vitro* assays. Microbial testing has traditionally been required throughout biopharmaceutical manufacturing processes, but standard methods can take weeks – too long to be used in some cell therapy approaches. New methods are dramatically reducing the testing time. For example, the commercially available polymerase chain reaction (PCR)-based mycoplasma testing kits are continually improving. Supporting the use in Current Good Manufacturing Practices (cGMP) manufacturing, quantitative external DNA controls have become available as certified reference materials from the American Type Culture Collection (ATCC).

Other examples include streamlining of media and buffer preparation in, for example, automated buffer in-line conditioning systems [38] and materials handling and staging. As in other classifications of PI, there is a significant degree of overlap or exclusion in naming this type of activity. For example, we often consider particular implementations of bioprocess outsourcing, standardization, and automation as process simplification. Personalized medicine, treatments for rare diseases, and cell therapy's pace may provide their own particular imperatives for simplification. These products are typically more challenging and costly to manufacture, and in some cases, simplification not only can contribute to BI but also could be critical in bringing the product to market.

4.3.6 Continuous Bioprocessing

In continuous bioprocessing (CB), raw materials continuously flow into equipment and are processed on an ongoing basis into an intermediate or final product. In biopharmaceutical manufacturing, this process generally begins in a bioreactor. CB occurs at a single location without interruption and proceeds for variable lengths of time – from days to months. Drivers to the incorporation of CB in pharma include its acceptance by the regulatory agencies, anticipated lower-mass "next-gen" products, growth of price-sensitive or price-controlled markets, and trends toward local manufacturing. So, CB is being promoted in biomanufacturing for many reasons, such as operating at similar scales for product and process development and clinical and commercial production. It also provides highly flexible facilities that can respond to changing product mass demands and works well with enablers of a flexible infrastructure such as SU systems and smaller bioreactor sizes [39]. Together, this provides a reduced footprint and capital investment (and potentially operating expenses) while simplifying technology transfer activities. Concerns for CB include performance reliability (incidence, and consequence of failure), validation complexity, the requirement of automation, advanced process monitoring and control, multiple reconnection, and economic justification. However, for many bioprocesses, such previous or perceived limitations are being alleviated [39, 40]. Integrated continuous bioprocessing allows even smaller facilities and equipment footprints while maintaining faster PD and control response [41].

Currently, a majority of CB upstream processes in cell-based manufacturing are variations of large-scale perfused cultures, usually operating in some type of chemostat. In perfusion mode, cells are retained through immobilization, isolation, or concentration in some way to allow older culture medium (perfusate) to be withdrawn and replaced by fresh medium. This is accomplished in a variety of fluidized and packed bed, centrifugal concentrators, gravity-based (conical and inclined ramp) settlers, spin filters, ultrasonic resonators/filters, and crossflow membrane, and diverse (internal and external) hollow fiber-based systems [39, 40]. Depending upon the type of cell, nature of the cell retention, and type of culture density control employed, a variety of cell densities, reaction volumes, and production states can be achieved. In intensified perfusion, cultures are driven to unusually high densities and maintained as a continuous operation. This approach has become prominent of late and provides tremendous increases in footprint and temporal productivity. Several creative solutions to the earlier challenges encountered (especially in scale-up) have been engineered for both adherent and suspension cultures. While perfusion bioreactors are highlighted here, many other equipment and processes supporting integrated, continuous, and straight-through processing exist, both up- and downstream. This includes revolutionary technologies such as continuous buffer and culture media production/conditioning and multicolumn, or simulated moving bed, chromatography [42, 43].

Many have published on comparisons of cost of goods (CoGs) per gram in batch vs. CB. Generally, continuous bioprocesses can provide more cost-effective options, except at smaller scales (<2 kg/year). Sources of cost savings include reductions in capital requirements (from reduced classified space requirements, bioreactors, centrifuges, and filter holders) and reductions in consumables [44, 45]. The least economically efficient aspect of many continuous culture processes is the supply of primary metabolites through non-optimized media exchange.

A perfusion-based production culture provides other values for therapeutic products from mammalian cells. By both supporting cells secreting product from cultures at a consistent state and by decreasing the product's bioreactor residence time, perfusion supports higher harvested product quality profiles [39, 46]. Factors in the optimization of the perfusion process involve medium composition, cell density, feeding strategy, cultivation time, perfusion rate, and culture duration [47, 48].

4.4 Materials

4.4.1 Media Optimization

Hundreds of individual compounds are employed in the various cell culture media formulations, and even more when researchers are examining their specific effects. These components comprise a wide variety of molecular types and physical states. Because clonal derivatives inherently present their own metabolic phenotype, optimizing a medium to a particular clone and production format can easily increase production levels two- to fivefold. Even in well-established generic null cell lines, the desire for increased production efficiency or product quality often establishes another demand for customization of the basal media and feeds. Factors to research in the application of any particular material include validation of the source material with incoming materials tests (including for identified transition metals); methods of intermediate processing; the nature, amount, timing, duration and means of supplementation; and secondary effects of the supplementation on other aspects of the process.

Primary factors determining optimum feeding of cultures include (i) avoiding toxicity to initial low-density seed cultures, (ii) maintaining desired PT processing in high-density cultures, (iii) prolonging cell viability and avoiding apoptosis, (iv) promoting longevity of the culture over short-lived peak densities, (v) promoting high-quality product accumulation over mere quantity, and (vi) encouraging product accumulation over simple cell culture mass.

The controlled supplementation of larger-scale systems has evolved into a discipline of its own. The goal of increasing the harvest mass of quality product can involve such means as increasing the number of cells (or biomass) or increasing the secretion of product per cell. This often requires in-process optimization of the in situ culture medium gases, nutritional substrates, vitamins, growth factors, or cell metabolism-altering chemicals. Concentrates of nutrients (primary metabolites) and vitamins are now commonly employed to either provide additional potency to existing liquid media, to operate in the fed-batch mode, or to optimize for one of many perfusion-based platforms.

The nutrients and factors required for the various goals in media and feed optimization vary greatly, depending upon the product and process. For example, enhancement of cell growth can compete with rate of protein expression or virus production. It is therefore important to distinguish and balance goals such as maximizing peak cell densities and their duration and recombinant expression rates and accumulated product quality. Optimal composition of a medium for the same cell line employed in batch, fed-batch, perfusion-enhanced fed-batch, and perfusion-based continuous operations can vary greatly. The fact that, due to the complexity of cellular metabolic pathways, many optimal media component levels are interdependent upon others often adds another layer of complexity.

Along with other process parameters, media and feed metabolite, cofactor, and growth factor composition have been shown to directly affect cells' PT efficiency and accuracy. Medium composition can alter a cell culture's growth rate and robustness but can also affect the fidelity and variability of product glycoforms [49].

Exciting new technologies are readily available to help in the optimization of production media. In one case, commercially available spent media assays enabled the quantitative measure of different classes of metabolites and amino acids in one 20-minute liquid chromatography–mass spectrometry (LC–MS) run. This spent media analysis was used as part of a MS clone selection package and verification of sequence variant misincorporations observed in peptide map analysis. They also monitored amino acid and metabolite profiles of different clones and applied statistical methods to look for trends to correlate with titer values [50].

4.4.2 Variability

Materials variability is currently being comprehensively addressed by the industry [51]. For example, the BioPhorum Development Group (BPOG) is an industry-wide consortium of suppliers and therapeutic manufacturers enabling networking and the sharing of common practices in the development and manufacturing of biopharmaceuticals. The "raw material risk management" workstream has been studying trace metal variation and analytical characterization of cell culture media and supplements. It has launched products such as a new raw material risk assessment tool aimed at helping industry identify and prioritize the challenging question of material fit. Such industry goals like identifying and reducing materials variability in trace metals promise to improve bioproduction quality and quantity [52].

4.5 Digital Biomanufacturing

As the concept of BI is multifaceted and growing, so are the science, techniques, and applications comprising "digital" approaches in biomanufacturing (Table 4.4). While applications in this arena for biomanufacturing are behind, e.g. the automobile industry, they are growing rapidly and are becoming major contributors to

Table 4.4 Similar digital biomanufacturing concepts presented in many forums.

Digital transformation of industries (DTI)
Digital plant maturity model (DPMM)
Cyber/physical production systems
Next-gen biomanufacturing
Enterprise adaptive control
Smart factory technology
Digital biomanufacturing
Fully automated facility
Factory of the future
Smart bioprocessing
Biopharm 4.0
Industry 4.0
Society 5.0

Source: Whitford, original work '20.

intensification efforts [53]. Such basic goals as EBR and electronic standard operating procedures (eSOPs) have been eclipsed by even more powerful mainstream tools [54]. Demonstrating the power of these "4.0" approaches in biopharma is how a developer of messenger RNA (mRNA)-based therapeutics won the *2019 ISPE Facility of the Future* award. This highly automated enterprise established the first human tests to suppress epidemic COVID-19 by delivering a batch of clinical vaccine to the National Institute of Allergy and Infectious Diseases (NIAID) 42 days from sequence identification. It credits this success to its cloud-based IT systems, robotics, automation, and artificial intelligence-based cGMP environment [55].

4.5.1 Data

The sources, transmission, storage, and general governance or curation of data in biomanufacturing are maturing at a rapid pace. New probes, sampling technologies, monitoring instrumentation, analytical procedures, and high-throughput, multiplexed, and multiply singleplexed analytics are supplying an ever-increasing amount of process data. However, manufacturers are beginning to realize that all the data in the world is useless unless it is efficiently – and accurately – consolidated, formatted, stored, accessed, and employed [56]. Data can now be collected from PD, past production runs, remote manufacturing sites, and even related digital publications. It can originate from scale-down or actual production equipment; in- on-, at-, or offline analytics; electronically interfaced instruments or manually entered results in real time, or archived data lakes. It is now becoming obvious that these data can be especially useful in activities including PD, trial design, and results of examinations, process control, event prediction, excursion management, and phase IV studies.

One of the current challenges for the biomanufacturing industry is the number of poorly managed and isolated data repositories that commonly exist [57]. For years, many of the results of development and operational analytics were unusable, as they commonly remained in various degrees of accumulation at the site of generation, uniquely formatted in outdated media, and/or even remained unexamined. However, this is all changing now [56]. Manufacturers are realizing not only the value of the data being generated, but also of the equipment and algorithms becoming commercially available to receive and manage immediate data and retrieve, recover, massage, clean, and blend historical data [58]. Finally, curators of data are becoming adept at evaluating our "data lakes" for their statistical value as applied to the problem at hand and processing them to data ponds of the most valuable and significant portions of the whole.

4.5.2 Bioprocess Control

Not too long ago, while such bioreactor process control elements as tank heaters were automatically PID controlled through output variables like temperature, some control variables including glucose or glutamine levels were manually maintained. Progress has accelerated for years in the development of control strategies, optimization algorithms, and software frameworks for control systems. Current systems often employ supervisory control and data acquisition (SCADA) often requiring some human–machine interface (HMI). Newer, closed-loop bioreactor control for many variables is now being accomplished with adaptive and model-based controllers (see Section 4.6) [59, 60]. They provide a significant and flexible benefit in two basic ways. First, they provide not only for optimized constraint of bioreactor operation but for a constraint of the control signal itself within optimal ranges. Second, they modify their action in real time to not only the result of their control activity but to other changes in the systems. Their power here is derived from the fact that they fully recalculate the optimal next step in each monitoring cycle of their operation. While the math for some such systems has existed for some time, a number of developments in rather independent fields have actually enabled them. Hardware speed and algorithm diversity now support both linear and nonlinear systems in such iterative activities. More accurate and dynamic base models are being developed due to the more statistically valuable data being supplied by additional monitoring capability and to the increased cell culture systems knowledge. However, parameter estimation can still be based upon a model-free or model-based algorithm.

Unlike many manufacturing processes, the nature of biological manufacturing dictates that many of the process parameters required for optimal control must be inferred, calculated, or correlated to dependent but measurable phenomena. Much work is now occurring in the arena of generating measurements of physical, chemical, and physiological process parameters, process control actuators, and means of data gathering and processing [61]. Our newer adaptive controllers are much better in determining changes in dynamic output variables required to satisfy a desired input. Also, better input measurements are now provided by many devices in Section 4.8. Finally, both bioreactor type and process control are important factors to consider in successful PD and optimization [60].

4.5.3 Digital Twins

The concept of the digital twin is growing in popularity in many arenas, from manufacturing to automobile driving (as of this writing driverless trucks have begun delivering goods in the United States). DTs are virtual models of physical systems or processes [62]. They convert a physical instrument and related activity to an *in silico* model providing a digital transformation of the process. In one example, DTs support the analysis, optimization, prediction, and even control of a manufacturing processes in real time. They employ large amounts of historical data and measured parameters from a number of sources and times. Eventually they will merge mechanistic modeling and machine learning based control to support a true implementation of the "factory of the future."

A bioreactor DT in upstream bioproduction mirrors the process taking place within a bioreactor in a computer program. It can be used for predicting cell metabolism, cell growth, product quality and – particularly relevant here – product titer under dynamic process conditions (see Section 4.6). It can be used to produce virtual experiments in PD or support a higher level of process control during a production run. By definition it is not an external, contained reference tool. It is an integral component of advanced process development, optimization, and control [63, 64]. A DT facilitates more efficient biomanufacturing and the generation of results supporting both strategic and tactical process and product development. It is a resident, learning, and comprehensive reference model and can be designed to supply process knowledge supporting development and response to operational queries regarding past, current, and future process states and product outcomes.

Bioreactor DTs can operate in the multidimensional design space of the dynamical cell culture processes using learned, both historical and monitored, data. They can be considered an ultra-multiplexed *in silico* model of a particular cell culture and bioreactor's behavior. They can predict optimal culture media composition alterations, bioproduction feeding strategy, and other process parameter or strategy alterations supporting (among many others) the goals of high cell density and improved culture productivity. Bioreactor DTs map the results of bioreactor operation within a design space rapidly and systematically. They can potentially simulate an entire complex production run in seconds [65].

As with many innovations, the bioreactor DT concept is currently in a state of comprehensive yet asymmetric growth. The algorithm's supporting statistics and math; the sources, amount, and quality of data employed; the performance parameters modeled; and the applications they support are increasing monthly. On the other hand, there is a wide divergence between what has been developed academically and what is available commercially [65].

4.5.4 Artificial Intelligence

Artificial intelligence refers to the simulation of human intelligence in machines. Its practical power is the ability to "rationalize" and come up with non-anticipated solutions to complex problems with unpredictable data. The applications for AI are endless and include activities in process development, optimization, and control. AI makes appropriate development/prediction/control decisions from systems of advanced monitoring, big data processing capabilities, and industrial connectivity [66].

Intensification applications for biomanufacturing include incorporating the IIoT by utilizing newer data technologies. The use of so-called "SMART" technologies and autonomation in both process development and manufacturing process control will be achieved through developments such as advanced monitoring techniques, machine learning, and AI. Systems supported by AI will mimic cognitive functions, be able to learn incrementally from data of many sources (e.g. historical and remote) and types (e.g. unstructured and multidimensional), and accommodate new information (including empirical data) as it becomes available. This promises real-time predictions, classifications, problem solving, decision making, recommendations, and control. Observability analyzers will review systems and determine what data is valuable and what is redundant, irrelevant, or corrupt. *In silico* modeling (see Section 4.5.3) will soon employ AI in the control of the entire process from supply chain management to the bioprocessing train to documentation of final product attributes. Such activities will ultimately provide full autonomation of model-based experimental design in product and process development - and learning, adaptive, model-based, closed-loop control systems in manufacturing [67].

4.5.5 Cloud/Edge Computing

Cloud functions solve the increasing pressure on IT to open up bandwidth for data from customers, partners, and PD groups. Other enabling activities such as AI and machine learning are dependent upon an increasing amount of data and activity speed. In fact, many manufacturing concerns now run on multi-cloud environments. Edge computing is used to process more time-sensitive data, providing the runtime required for applications employing it. Application programming interfaces (API) are tools used in the creation of software and are becoming increasingly dependent upon cloud and edge computing. They are a documented interface whose adoption is essential to support digital transformation by allowing communication through a web browser, mobile application, or device beyond the firewall [68].

4.6 Bioprocess Modeling

The term "model" can actually apply to such mundane activities as a process flow chart. Today, advanced mathematical models increase our understanding of systems biology, metabolic engineering, flux balance analysis, messenger systems, and bioprocesses in general. Distinct aspects of biomanufacturing benefit from many model applications, from bioprocess development to bioreactor mixing dynamics to cell growth and virus replication. The complexity of cell processes, including metabolic networks and reaction pathways, makes cell process modeling a rather formidable goal. Mechanistic models are simplifications of the systems for which the true and optimized operating parameters are sometimes not known [69]. Recent advances are, in fact, promoting a more thorough handling of existing dynamical models to extend their power for biotechnology research, PD, and process control [70]. The ultimate goal for biomanufacturing might be to provide a comprehensive and universal algorithm (or AI-based machine learning) for all mammalian cell-based processes and applications to build their model upon. Interestingly, while this would greatly simplify the development of model-based applications, we know it is not possible. Even with a better understanding of mammalian cell culture and lots of significant and orthogonal monitoring data, the "no free lunch" theorem would dictate the need for many underlying algorithms for the many purposes required and the now many products and production modes. The newest approach to the selection of the underlying machine learning algorithm best suited to your need is called model-based machine learning. Here, the objective characteristics, or set of assumptions, of your problem determine a "model" that will help you to choose the best underlying algorithm (machine learning, e.g. deep neural network) to use. Finally, another advantage of cellular process model's incorporating mechanistic principals is reported to be their ability to predict the systems behavior beyond the region of experimentation, e.g. in the scale-up to manufacturing levels.

Examples of progress include a hybrid (semi-parametric) model of fed-batch processes for protein production reported to be superior in predicting many different process variables as compared to simple statistical models [60]. A basic economic model of for mAb production is advertised as valuable for any production platform [71]. A scale-down model of a 4000-l culture was produced for foot and mouth disease vaccine production. Here, computational fluid dynamics (CFD) modeled the hydrodynamic environments inside the bioreactors [72].

An important discipline in the development of models in biomanufacturing is knowledge management. Cell biologists, operations personnel, model developers, and biomimicking professionals need to work together in the development of powerful, accurate, and robust models. Thankfully, there are now many consortia working to support sharing of such bioprocess knowledge between researchers, suppliers, and manufacturers [73, 74].

4.7 Automation and Autonomation

Automation is becoming increasingly important in bioprocessing. Driven by PAT, and with the goal of a knowledge-based and optimally controlled processes, many solutions for automation in PD and even operations are appearing. One such tool is automated sampling. Developers are looking for complete automated analytics for bioreactor culture systems. Valuable criteria for this initiative include small sample size, high-throughput support, automated analysis, support of multi-bioreactor computer systems, availability of an expanded assay suite, options to rerun a test (if required), dual sample streams (cell containing and cell free), the ability to integrate multiple analyzers, configuration options for sample dilution, and the ability to sample for multiple analytes through a single sampler.

Automation can be considered for PD or in manufacturing operations. In bioprocessing the most dramatic accomplishments so far have occurred in development [75]. We have now seen online optimal experimental redesign in robotic parallel fed-batch cultivation facilities. Here, integrated frameworks for the online experimental optimization of parallel nonlinear dynamic processes promise to reduce the effort required to provide precise estimates of a kinetic growth model's parameter sets. Such a system used two fully automated liquid-handling robots. One contained eight mini-bioreactors, while the other was used for automated at-line analysis. This second robot supported the immediate use of the available data in the modeling environment, allowing online, real-time redesign of experiments. The value of parallel, concurrent recomputation of the experiment is significant, as compared with sequential methods [75]. Currently available are modular compact enclosures containing intensified, automated biological manufacturing process trains and small multi-bioreactor systems using robotic arms for the sampling process in PD [76].

Some "flexible factory" systems provide automation enhancing the process line with centralized data management and streamlined production in a 21 CFR Part 11 compliance-ready environment. Such systems improve operational efficiency at a process, plant, or global facility level [77]. A fully automated "GMP-in-a-box" concept for patient-scale cell therapy manufacturing is in final stages of development. This closed, automated, highly flexible cell therapy manufacturing platform employs SU, highly customizable cassettes. It supports fully enclosed end-to-end manufacturing of cell therapy products with key unit operations, including isolation, activation, transduction, expansion, and harvest. Technical features include monitoring and control of temperature and gases, pH and DO, and information logging and control with EBRs [78].

All this increasingly supports the integration of analyzers to enable online analytics from instruments such as cell counters, enzyme-based bioanalyzers, and HPLC systems for use in operations. Systems can be constructed al la carte or as complete integrated solutions - from reactors to samplers, analyzers, and SCADA-linked open platform communications (OPC). The 3rd Global Bioprocessing, Bioanalytics and ATMP Manufacturing Congress convened in May 2020 and addressed just such innovative solutions and technologies powering this transformation [79, 80].

We are mostly familiar with autonomous robotics by their adoption in, for example, the automotive manufacturing industry. An autonomous robot has more human-like flexibility than the fixed, dedicated robotics that simply move the same type of object from one place to another. While their adoption within biopharma has been slow, collaborations between industry and academia herald the beginnings of autonomous robotics employment in this highly regulated industry. An example here is autonomous robots for environmental monitoring within pharmaceutical facilities. These were chosen as a first step, with the expectation that their capabilities will be eventually broadened [81].

Eventually, as with other industries, these robots will carry out multiple tasks that fulfill particular defined requirements. Flexible and adaptable autonomous robots are envisioned being able to perform many of the repetitious tasks in both development lab and bioprocess operations on the manufacturing floor [82].

4.8 Bioprocess Monitoring

Improvements in process monitoring can be categorized in many ways. One is the development of means to monitor new operational process parameters vs. improvements to the technology in monitoring existing process parameters. Bioprocess monitoring can develop data in PD to feed process control systems in manufacturing or to monitor manufacturing operations to realize visions such as real-time product release and concurrent process validation. Another way to organize improvements affecting monitoring is by their proximity to the culture as in (i) offline, where a sample is assayed remotely often at a later date; (ii) at-line, manual sampling and transfer for timely local analysis; (iii) online, employing an automated sampling interface delivering a sample to an analyzer(s) for near real-time analysis; and (iv) in-line, the measurement device directly connected to the bioreactor, as with integral probes, providing real-time values. It can also be based upon the type of parameter measured, e.g. cell state, primary/secondary metabolite presence or level, expressed product concentration or characteristics, or bioreactor physical parameters. We can also sort the improvements based upon the means of their sensing, e.g. electrochemical, capacitance (impedance), surface plasmon resonance, mass based, optical, thermistor, or piezoelectric based. Even the subcategory "biosensors" can be further subdivided into the catalytic, where the interactions such as with immobilized substrate-specific enzymes result in the formation of a new reaction products, or affinity, where the analyte binds directly onto the transducer surface. Finally, we must consider the emerging maturity of automated samplers and the multiple singleplexed distribution of samples used in online sampling [83].

It is nearly impossible to even catalog all of the relevant emerging probes and analytical technologies in the development today. To begin, there have been steady improvements in all the tools based on spectrometry: near- and mid-infrared fluorescence and Raman. We also see a continual evolution of the quantity and quality of results from biosensor technologies. Microarray technology enabling high-throughput matrix-assisted laser desorption/ionization (MALDI) (MS) can reveal timely details of N-glycopeptides of IgG1 produced in a perfusion cell culture [84]. LC-MS metabolomics can be used to demonstrate the basal metabolic states of Chinese hamster ovary (CHO) cells in fed batch, revealing their state during entity production [85]. Quantitative MS is being explored in vaccine manufacture to assess parameters such as antigen content as an early indicator vaccine potency.

Capillary electrophoresis mass spectroscopy (CE–MS) instrumentation is just now supporting development and production workflows and methods.

Innovative in-line monitoring directly supports the goals of process state prediction, real-time release, reduced lead times, and improved product quality. Benefits here include improved economy, such as by drastically reducing the costs associated with holding large inventories and reducing cost of quality [86].

4.0 capabilities are enabling the concept of software-based sensing to become of practical value valuable in bioprocessing. So-called soft sensors or virtual sensors are a combination of data from physical sensors, analytical devices, and other hardware with computer-driven mathematical models that create real-time and categorically novel information about the process [87].

4.9 Improved Process and Product Development

There are many emerging components to the improved development capabilities in the world of bioprocessing, many of which are mentioned in this chapter. The more immediate goals of this include developing techniques to improve design engineering speed and capability. The ultimate intent is to reduce costs, risk, and time to market while increasing product quality, consistency, safety, and efficacy. Product and process development tools include those supporting entity development, clone selection, and improved bioprocess operations. PD today increasingly includes goals such as making operations more sustainable and flexible. Toward BI, they include processes becoming less expensive to operate, of reduced footprint and with increased efficiency and productivity.

Some designate these initiatives as "computer-aided biology" (CAB), defined as an orchestration of tools supporting or replacing human activity in biological research (Figure 4.3). CAB organizes the development activities into the *digital* and *physical* domains. The digital outlined in Section 4.5 includes advances from more powerful hardware to AI. It highlights advances including software for modeling biological systems and new methods of importing, curating, structuring, and analyzing experimental data from the literature, other models, and wet lab experiments [88]. The physical includes systems supporting the automated incorporation of DoE-type or *in silico* model-determined experimental designs into actual laboratory experiments. It also specifies that this wet-lab experimentation is now being powerfully enabled by research lab automation (see Section 4.7). So, in CAB, we see machines improving upon humans in both the *in silico* design and organization, as well as in the setup and execution, of the real-world PD experiments.

4.9.1 Design of Experiments

The design of experiments approach is now *de rigueur* in optimizing bioprocess parameters in recombinant protein production [89]. Because of the cell metabolism-based non-regular interaction of operational parameters and culture media components, approaches such as full and fractional factorial design, Taguchi

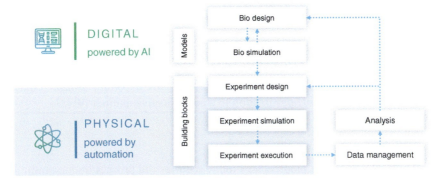

"COMPUTER-AIDED BIOLOGY: Delivering biotechnology in the 21st century", Synthace Ltd, 2018, P.Crane, T.Fell, M.Gershater, S. Ward, M. Watson & R. Wiederhold

Figure 4.3 The computer-aided biology (CAB) landscape, connecting the digital word with the physical, powered by AI and automation. Source: Fell et al. [2].

orthogonal arrays, and the response surface methodology (RSM) are particularly powerful.

4.9.2 QbD and PAT

The basic principles behind QbD are the need for (i) a deep understanding of both the product and the production mode and (ii) an understanding of the production risks and how to control the process to mitigate those risks. It involves defining and monitoring the critical process parameters (CPPs) and critical quality attributes (CQAs) during processing. PAT systems direct the continuous and timely monitoring of material and product characteristics throughout processing. As the influence of process parameters on product levels and CQAs can be determined in an actionable timeframe, an understanding of what control parameters will provide acceptable product quality and quantity is developed. The multidimensional relationship between process inputs and CQAs defines a process "design space" [90], providing value in PD, process control, and product release. Finally, high-throughput screening methods are providing the tools to truly implement the QbD promises made over a decade ago.

4.9.3 High-Throughput Systems

After optimizing a null cell line, selecting the best recombinant clone for stockpiling has generally employed small-scale cultures in microtiter plates, test tubes, tissue culture flasks, and shake flasks. As selection screening techniques have advanced, the criteria for selection have too. Beyond high expression levels, researchers now establish such criteria in a culture behavior and robustness in the production media, behavior in the actual production mode envisioned, performance in scale-down conditions, and a product quality profile throughout the culture's duration. The culture parameters being addressed in popular mini- and microculture systems now include media composition, pH, temperature, agitation and sheer, gasification, cell density, inducers, induction times and feeding approaches, and even perfusion rates. In fact, surveys report that over 80% of biopharmaceutical industry is already using micro- and mini-scale bioreactor systems that offer a small-scale, high-throughput solution for accelerating clone/media selection and process development. High-throughput bioprocess development (HTPD) techniques for these are becoming commercially available for the best process optimization in a cost-effective manner [91].

4.9.4 Methods

Many report that improves bioprocesses today requires the process understanding provided by mathematical models (see Section 4.6). Multi-paradigm numerical computing software allows matrix manipulations, plotting of functions and data, implementation of algorithms, creation of user interfaces, and comprehensive interfacing. They support adaptive modeling from online and at-line data to provide model-based designs with recurrent parameter fitting making the developmental process efficient. Intelligent experimental facilities enable efficient mode-based DoEs supporting multivariate optimization of fed-batch bioprocesses [92, 93].

A thorough knowledge of bioreactor parameters at various scales helps in the successful scale-up of robust production processes (see Section 4.6). The important parameters for scale-up that are critical to efficient cell growth, viability, and protein production are well reported. It is less known that the means are emerging to elucidate cellular genetic heterogeneity (point-to-point values) that will support new metabolic engineering and synthetic biology-based solutions. They promise to provide such practical value as reducing the population of lower-producer mutants in scaled-up bioprocesses increasing product quality and yield [17, 94].

4.9.5 Commercialized Systems

An increasing number of commercially available equipment and services supporting many of the advanced process and product development discussed here are emerging [95, 96]. Many suppliers have been addressing PAT, QbD, or the digital transformation to one degree or another for years. We now see the appearance of a more complete repertoire of data sourcing and handling, process monitoring, advance sensing and miniaturization, and automation. These systems offer new possibilities to monitor and control metabolic activity, process robustness, and related parameters through electrochemical, impedance, and spectroscopy-based probes and related to fluid flow and gradient formation. They address advanced high-throughput analysis, device integration and automation, integrated robotic cultivation, and multiple single-variate and multivariate at-line analytics [97].

Such intelligent facilities promise improved DoEs and adaptive model-based designs with recurrent parameter fitting, making both the product-related and operations development much more efficient. Their parallel screening of product formation and multivariate system optimization implies support of the most modern platforms of production [75]. Finally, SMART soft- and single-use sensors and enhanced data transfer point to the ability to supplement previously established process knowledge and support their self-learning algorithms. This all enables reduced PD effort, integration of various sources of process knowledge, and real-time estimation of product quality.

4.10 Advanced Process Control

Advanced process control (APC) actually has a rather special definition and refers to a number of technologies used to improve basic existing process control systems. Basic process controls facilitate essential operation, control, and automation requirements. APCs are, rather by definition, added subsequently, over time, and provide identified enhancement of the efficiency or economy of the process. Surprisingly, this definition is not generally understood by many biotechnologists.

It is apparent that the terms "basic" and "advanced" are in the long-term relative. Also, we see some debate and evolution in the field of bioproduction as to just what technologies fall under each term. Currently, APC refers to such technologies as feedforward, inferential, and multivariable model predictive control (MPC). This can be implemented at the process local or supervisory control computer level. MPC operates on key-independent and key-dependent process variables and the dynamic relationships between them. As it can control multiple variables simultaneously, nonlinear MPC incorporates dynamic models that have varying, non-linear process gains and dynamics. Soft sensors in APC provide virtual or inferential measurements from models that use true measured variables to estimate some that are difficult or impossible to measure. Intelligent controls use AI computing such as neural networks, Bayesian probability, fuzzy logic, machine learning, evolutionary computation, and genetic algorithms. The newest theoretical developments in bioprocess control (including APC 2.0) are addressing such components [98]. So, we see overlap between APC and the developing fields of process sensors, bioanalytics, and digital biomanufacturing. The AiChE conducts PD2M Advanced Process Control and Future of Pharmaceutical Manufacturing Workshops that address this field and its application in biopharmaceutical processing [99].

4.11 Bioreactor Design

The bioreactors used for biopharmaceuticals production include stirred tank, airlift, bubble column, fixed and moving bed, hollow fiber, and fluidized-bed bioreactors [45]. SU bioreactors have been steadily improving in many respects since their introduction some 20 years ago [100]. Many of the improvements affect factors such as speed and convenience of setup. However, some have influenced their performance and reliability, and these can be regarded as supporting BI. Each of the many suppliers have, through the years, improved the impeller design, providing improved mixing and mass transfer. Similar performance improvements have been accomplished by redesign of the tank (improving mixing and turndown) and sparge apparatus configurations. The composition and structure of the films comprising the SU liners have been improved to avoid leachables and extractables that were a problem in the past. The diversity of applications for SU containers requires film that provides a wide variety of performance attributes such as mechanical strengths, flexibility, transparency, biocompatibility, and suitable gas barrier properties, to name but a few. The right balance of chemical composition and film architecture is critical for achieving desired performance across many applications. It is popular to employ a platform film used for their entire portfolio of SU products, thus enabling a single qualification for multiple SU bags in a bioprocess [101].

4.12 Single-Use Systems

The biopharmaceutical industry now incorporates significant levels of SU technology and systems in the majority of cell culture-based production processes. This is due to the remarkable benefits the technology affords to the majority of popular production platforms and implementations, including advances such as SU, and SU applicable, probes. SU upstream processing results in faster product to market, less capital and operating costs, smaller footprint, and greater flexibility [102]. The most important reasons for their extensive adoption include faster install and turnaround times, lower capital and utility costs, and reduced concern for cross-contamination. Multi-laminate sterile bags, with either impellers or external rocker-type platforms, are now the most popular means of preparation of media and process fluids and the culture of cells in disposable, SU, and presterilized assemblies [103].

SU systems can increase economic productivity by many metrics. Well-published factors influence the exact value of savings, and space prevents their comprehensive portrayal. However, estimates of reduced initial investment costs are in the area of 40% lower than for a traditional stainless-steel facility. Another factor in the CapEx savings is provided by a delay in the facility build start date, vs. a stainless-steel plant, of as much as three years. Quick-to-deploy SU facilities support such improvements of up to 60% faster to market. An increase of up to 30% in OpEx savings is supported by reducing labor, quality, regularity, and materials costs associated with such process steps such as sterilization, cleaning, and maintenance of stainless-steel systems. SU facilities can present up to a 30% smaller footprint (increasing footprint productivity) and be 5 times faster in changeover. For example, a stainless-steel facility producing 15 batches per year may produce 19 batches per year if configured in SU, increasing footprint and temporal productivity. It reduces the amount of time the staff spends on preparation, setup, validation, and documentation along with related costs. By greatly reducing the need for clean water, steam, and cleaning chemicals, they also reduce the time, costs, and complexity of automation and changeover/cleaning validation. Project lead time for implementation of single-use production plants is reported to be reduced by at least eight months compared with stainless-steel reactors, and project testing, validation, and implementation costs are lower. As the product flow path is replaced after each batch, the risk of product cross-contamination between batches is virtually eliminated, heightening overall batch productivity [39, 104, 105]. While not a focus here, one exciting new development downstream is the appearance of SU high-performance purification media [106].

Following the success in the electronics industry, a current initiative involving SU is the "plug-and-play" initiative. Here, the struggle is considering all the advantages of individualized entrepreneurial systems, pieces, and suppliers vs. those of standardization, interchangeability, and security of supply. The field of SU in biomanufacturing has grown so rapidly because of the competition between suppliers and the dynamic interaction between systems designers and component distributors. However, all of this creative development comes at a cost: we are now in a state of employing many suppler-specific films, container dimensions, connectors, and even flow paths. There is hope that AI will be powerful in aiding the harmonization of all the divergent parts and applications into a "greatest common factor" and "least required adaptations" needed to provide compatible and interchangeable components between processes and suppliers. In any event, the promise of platform and supplier interchangeability is driving action in this direction.

A concept related to plug and play is that of the flexible factory. Some SU suppliers are designing their systems such that they can be employed in uniform or hybrid implementations allowing optimization of up- and downstream processes while reducing operational costs. Flexible factories offer faster deployment, multiproduct processing, process agnostic manufacturing, and accelerated production with flexible, scalable processes. Automation enhances a flexible factory process line with centralized data management and streamlined production in a 21 CFR Part 11 compliant-ready environment. Finally, this automated platform helps drive operational efficiency at a process, plant, or global facility level [77].

4.13 Facilities

The factory-in-a-box concept is yet another development revolutionizing the biomanufacturing industry [107]. Prefabricated modular and podular designs are providing much beyond such well-described benefits as futureproofing and supporting the employment of SU systems. Values relevant to intensification include their enabling more efficient and standardized process flow and shared service designs. Many aspects of these facilities contribute to their shortening time to build and time to run, such as the ability to overlap unit and suite start times, work in parallel, and avoid the sequential nature of on-site construction [108].

Three exciting and disruptive approaches in this regard can be roughly categorized as prefabricated (i) modular facilities, (ii) podular suites, and (iii) modular operations units. They provide, in respective ways, reduced footprint, capital expense, time to market, service needs, and operating expenses. Each can support multiproduct and simplified process trains that are more inexpensive to build and operate while offering increased flexibility and heightened environmental sustainability. Entire manufacturing facilities composed of prefabricated and modular units provide for the rapid build and deployment of cGMP process-need compliant

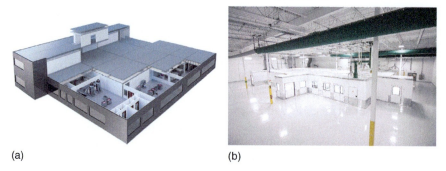

Figure 4.4 Modular biomanufacturing solutions incorporating single-use production equipment enable intensification goals. (a) Picture here is a GEHC KUBio™ modular mAb facility providing a FlexFactory™ single-use production environment. Source: General Electric Company (GE), https://www.gehealthcare.com/. (b) Modular facilities, suits, and isolation chambers now support research, manufacturing, and fill/finish. Depicted here is a cleanroom POD® cluster by G-CON Manufacturing. Source: G-CON MANUFACTURING, INC, www.gconbio.com.

design, materials, space allocation, workflow, equipment, utility, and HVAC for research and manufacturing (Figure 4.4). They efficiently house every component of modern biomanufacturing in suites of application-specific classification and biosafety level. Recent capabilities expansion includes specialized cell and gene therapy support [109]. Prefabricated podular cleanrooms can be rapidly installed in unclassified (gray space) environments. These independent, controlled suites efficiently provide streamlined capacity from research to commercial scale [110]. Some prefabricated, modular, and controlled environment chambers combine principles of process intensification, automation, and in-line chaining in self-contained ISO 14644-1-classified operating modules. They provide upstream, downstream, and aliquoting capacity that drastically reduces facility footprint, capital, and operational expenditures and are especially suited for applications such as viral vaccines, gene therapies, and oncolytic viruses [111].

This basic build economy is furthered by the purchaser being in complete proximate control of the build process and not having to rely on such uncontrolled factors such as weather conditions, personnel density, local labor influences or disputes, or similar interruptions. The laborers, tradespersons, and materials procurement involved in the prefabricated elements are dedicated, trained, and experienced in biotech and pharma facilities and suites. Benefits of the larger systems are that there is no need to establish (i) prepared work areas with temporary parking, access, or staging constructions; (ii) personnel gowning and preparation facilities; (iii) technology-dependent new, unique, or specialized training; or (iv) temporary security programs. Prequalification reduces glitches and the need to revisit or change things at the site, as many prefabricated systems can be factory acceptance-tested before shipment, insuring fully functional systems at the time of installation.

Operationally, pre-fabs can have compact and autonomous air-handling/HVAC systems that allow scale out without interrupting existing processes. In addition, if there is an excursion within one unit or suite, it can be shut down and run on a vaporized hydrogen peroxide (VHP) cycle without disrupting any other suites. So, modular/podular facilities not only heighten facility adaptability but also can reduce costs for a manufacturer by reducing the time and expense in changeover between processes [112]. As they are often designed to support platform and entity-agnostic processes, they enable flexible production modes, multiproduct facilities, and shortened process trains.

By swapping out and adding bioprocessing modules, biomanufacturers can modify functionality and adjust capacity quickly and economically [108]. Each of these factors can contribute to a greater manufacturer's running productivity per, at least, time or CapEx.

So, they start off as more cost-effective than on-site approaches as they provide improved net present value (NPV) and internal rate of return (IRR). They continue with even more specific values, such as relocatability and clonability, that their vendors are more than happy to discuss. Finally, the more standardized designs can be built in a matter of months now, and mass-produced and preloaded components (further reducing the timeline and costs) are envisioned for the future.

Qualified vendors of facility design and build support include many of the intensification initiatives mentioned above, providing for the rapid deployment of highly flexible and adaptable facilities. This includes compliant design, materials, space allocation, workflow, equipment, utility, ventilation and climate control for research, PD, and manufacturing activities. They enable process innovation and high-throughput manufacturing with both product and process flexibility. This includes features establishing flexible manufacturing spaces supporting the rapid changeover of either existing product campaigns or new development candidates. Such futureproofing will also address repurposing suits to address development in the product mix.

Integrating laboratory information management systems with enterprise resource planning and control systems enables process flexibility and technology transfer in a highly automated landscape. The orchestration of digital technologies such as cloud-based IT systems robotics, automation, and artificial intelligence supports all operational aspects of the facility. This includes procurement, process, and quality systems – thus delivering a fully integrated manufacturing and supply chain. Advanced data historian and electronic production records support not only improved product testing and release but also advanced process development, characterization, and intensification techniques. Popular design features supporting expanding pipelines in variable platforms include the following:

- Single-use equipment.
- Digital equipment tracking.
- Shared services plants.
- Ballroom suite design.
- Modality agnostic suites.
- Flexible suite equipment and process flow support.
- Modular and podular operations units.

Multiple, flexible cGMP suites for as follows:

- Production of such materials as culture media and buffers
- Production of both drug substance and sterile drug product
- Planned visibility between manufacturing spaces and auditors
- Digitally enabled spaces
- Storage of single-use materials
- Temperature-controlled warehousing for incoming and outgoing products
- Packaging, labeling, and storage of vialed products
- Quality control laboratories
- Pilot-scale manufacturing space for research and tech transfer
- Such controlled utilities as graded water

4.14 Conclusion

Biomanufacturing has evolved into an era of new product types and applications, new product development approaches and tools, new manufacturing processes and support technologies, and new facility designs and construction methods. There is growing interest in improving process capacity, robustness, consistency, efficiency, flexibility, and sustainability. While exact definitions and scope are difficult, some consider bioprocess intensification as a means of improving production efficiency. In choosing how, what, and when to intensify, those benefits must always be carefully weighed against the potential trade-off risks in not only economy and time but also most importantly product safety and efficacy. Bioprocessing presents such unique engineering considerations because complex living systems are dependent upon many interacting metabolic pathways. It has been slow to incorporate many of the newer manufacturing paradigms because of complexity, the highly regulated nature of the business, the sheer number of developing sciences and technologies, the number of their applications in manufacturing operations, and the need to demonstrate just how these new scientific and academic achievements will be practical, powerful, and robust in improving manufacturing efficiency. Nevertheless, there are now many verified technologies and commercial products being successfully applied to the goal of bioprocess intensification.

Abbreviations and Acronyms

3D	three dimensional
ATMP	advanced therapy medicinal products
AI	artificial intelligence
AiChE	American Institute of Chemical Engineers
APC	advanced process control
API	application programming interfaces
APM	asset performance management
ATCC	American Type Culture Collection

ATMP	advanced medical therapy product
BEVS	baculovirus expression vector system
BI	bioprocess intensification
BPOG	BioPhorum Development Group
BPTC	BioProcess Technology Consultants
BPTG	BioProcess Technology Group
bsMAb	bispecific monoclonal antibody
CAB	computer-aided biology
CapEx	capital expenditure
CB	continuous bioprocessing
CFD	computational fluid dynamics
CFR	Code of Federal Regulations
cGMP	Current Good Manufacturing Practices
CHO	Chinese hamster ovary
CIP	clean in place
CO_2	carbon dioxide
CoGs	cost of goods
COVID-19	coronavirus disease 19
CPPs	critical process parameters
CQAs	critical quality attributes
CRISPR	clustered regularly interspaced short palindromic repeats
dAbs	single-domain antibodies
DARPins	designed ankyrin repeat proteins
DB	digital biomanufacturing
DNA	deoxyribonucleic acid
DO	dissolved oxygen
DoE	design of experiments
DPMM	digital plant maturity model
DSCs	distributed control systems
DT	digital twin
DTI	digital transformation of industries
E. coli	*Escherichia coli*
EAM	enterprise asset management
ERM	enterprise resource management
EBR	electronic batch records
ERP	enterprise resource planning
eSOPs	electronic standard operating procedures
GE	general electric
HMI	human–machine interface
HPLC	high-performance liquid chromatography
HTPD	high-throughput bioprocess development
HTS	high-throughput screening
HVAC	heating, ventilation, and air conditioning
IB	inclusion body
IgG1	immunoglobulin G 1

IIoT	internet of things
IRR	internal rate of return
ISPE	International Society for Pharmaceutical Engineering
IT	information technology
K. pastoris	*Komagataella pastoris*
kg	kilograms
l	liter
LC-MS	liquid chromatography–mass spectrometry
LIMS	laboratory information management systems
mAb	monoclonal antibody
MALDI	mass spectrometry (MS)
MAM	multi-attribute method
MBML	model-based machine learning
MES	manufacturing execution systems
MPC	multivariable model predictive control
mRNA	messenger RNA
NDS	new drug submission
NIAID	National Institute of Allergy and Infectious Diseases
NPV	net present value
OPC	Open Platform Communications
OpEx	operating expense (or expenditure)
OPM	operations performance management
OT	operational technology
P. pastoris	*Pichea pastoris*
PAT	process analytical technologies
PCR	polymerase chain reaction
PD	process development
PD2M	Pharmaceutical Discovery, Development and Manufacturing Forum
PID	proportional–integral–derivative
PT	posttranslational
QbD	quality by design
RMCE	recombinase-mediated cassette exchange
RNA	ribonucleic acid
RNAi	interference RNA
RSM	response surface methodology
S. cerevisiae	*Saccharomyces cerevisiae*
SCADA	supervisory control and data acquisition
SIP	steam in place
SMART	scalable, modeled, adaptive, rational, tested
TALENs	transcription activator-like effector nucleases
TTF	tangential flow filtration
VHP	vaporized hydrogen peroxide
VLP	virus-like particle
ZFNs	zinc finger nucleases.

Acknowledgment

Alain Fairbank is enthusiastically thanked for her assistance with the critical review, editing, and references.

References

1 Larroche, C., Sanroman, M.A., Du, G. et al. (2018). *Current Developments in Biotechnology and Bioengineering: Bioprocess, Bioreactors and Controls*. Elsevier.
2 Tim Fell, Sean Ward, Markus Gershater, Mark Watson, Peter Crane & Robert Wiederhold (2018). Synthace: computer-aided biology. https://synthace.com/computer-aided-biology-whitepaper (Last accessed 1 August 2020).
3 DPS Group Consulting, Engineering, and Construction Management. https://www.dpsgroupglobal.com/novel-therapies (Last accessed 3 July 2021).
4 BDO's BioProcess Technology Group (BPTG). https://www.bdo.com/industries/life-sciences/bioprocess-technology (Last accessed 1 August 2020).
5 Ratz D. (2017). *Terminological Consistency: Quality Through Precision*. Amplexor: Luxembourg, https://blog.amplexor.com/terminological-consistency-quality-through-precision.
6 Joyce, P.C. (2019). *What Is Process Intensification and How Can You Implement It?* St. Louis, MO: Epic Systems https://www.epicmodularprocess.com/blog/what-is-process-intensification.
7 Smith S.F. (2019). Process intensification: getting more from less. *Medicine Maker* (17 January 2019).
8 Montgomery, S.A., Scott, C., and Rios, M. (2017). Chapter 8: The manufacturing perspective – current approaches to bioprocess intensification. *Bioprocess Int.* 15 (6) https://www.scribd.com/document/356374327/Ch-8-the-Manufacturing-Perspective-Current-Approaches-to-Bioprocess-Intensification.
9 The Food and Agriculture Organization (FAO) agency of the United Nations Agricultural Intensification, The Organization acknowledges the contribution of Peter Kenmore, Clive Stannard and Professor Paul B. Thompson to the preparation of this publication. http://www.fao.org/3/j0902e/j0902e03.htm (Last accessed 1 August 2020).
10 ISPE: North Bethesda, MD, 2019 https://ispe.org/glossary.
11 Continuous or Intensified Bioprocessing? The Gold Standard for Improved Productivity. (2018), http://www.global-engage.com/life-science/continuous-or-intensified-bioprocessing-the-gold-standard-for-improved-productivity/ (Last accessed 3 July 2021).
12 Mirasol, F. (2019). The search for bioprocess productivity improvement. *Pharm. Technol.* 43 (5): 16–22.
13 Smart Innovators: Asset Performance Management Software 2019 Rachel Umunna, https://research.verdantix.com/report/smart-innovators-asset-performance-management-software.

14 Predix Asset Performance Management, www.ge.com/digital/applications/asset-performance-management (Last accessed 1 August 2020).
15 The Institute of Asset Management, https://theiam.org/ (Last accessed 1 August 2020).
16 Everything You Need to Know About Operations Performance Management, https://www.ge.com/digital/blog/everything-you-need-know-about-operations-management-opm-power-generation (Last accessed 3 July 2021).
17 Rugbjerg, P. and Sommer, M.O.A. (2019). Overcoming genetic heterogeneity in industrial fermentations. *Nat. Biotechnol.* 37 (8).
18 Tripathi, N.K. and Shrivastava, A. (2019). Recent developments in bioprocessing of recombinant proteins: expression hosts and process development. *Front. Bioeng. Biotechnol.*
19 Hong, J., Lakshmanan, M., Goudar, C., and Lee, D.Y. (2018). Towards next generation CHO cell line development and engineering by systems approaches. *Curr. Opin. Chem. Eng.* 22: 1–10. https://doi.org/10.1016/j.coche.2018.08.002.
20 Gutiérrez-González, M., Latorre, Y., Zúñiga, R. et al. (2019). Transcription factor engineering in CHO cells for recombinant protein production. *Crit. Rev. Biotechnol.* 39 (5): 665–679. https://doi.org/10.1080/07388551.2019.1605496.
21 Mauro, V.P. (2018). Codon optimization in the production of recombinant biotherapeutics: potential risks and considerations. *BioDrugs* 32 (1): 69–81.
22 Berkeley Lights, https://www.berkeleylights.com (Last accessed 1 August 2020).
23 Bioelectronica, www.bioelectronica.com (Last accessed 1 August 2020).
24 Walsh, G. (2018). Biopharmaceutical benchmarks 2018. *Nat. Biotechnol.* 36: 1136–1145.
25 Swiech, K., Picanço-Castro, V., and Covas, D.T. (2017). Production of recombinant coagulation factors: are humans the best host cells? *Bioengineered* 8 (5): 462–447.
26 Gupta, S.K. and Shukla, P. (2016). Advanced technologies for improved expression of recombinant proteins in bacteria: perspectives and applications. *Crit. Rev. Biotechnol.* 36 (6): 1089–1098.
27 Huertas, M.J. and Michán, C. (2019). Paving the way for the production of secretory proteins by yeast cell factories. *Microb. Biotechnol.* 12 (6): 1095–1096.
28 Werten, M., Eggink, G., Cohen Stuart, M.A., and de Wolf, F.A. (2019). Production of protein-based polymers in *Pichia pastoris*. *Biotechnol. Adv.* 37 (5): 642–666.
29 Li, P., Anumanthan, A., Xiugong, G. et al. (2007). Expression of recombinant proteins in *Pichia pastoris*. *Appl. Biochem. Biotechnol.* 142: 105–124. https://doi.org/10.1007/s12010-007-0003-x.
30 Yee, C.M., Zak, A.J., Hill, B.D., and Wen, F. (2018). The coming age of insect cells for manufacturing and development of protein therapeutics. *Ind. Eng. Chem. Res.* 57: 10061–10070.
31 Medicago's New Drug Submission accepted for scientific review by Health Canada: An important step for Medicago towards commercialization of its innovative influenza vaccine (prnewswire.com). (2019) https://www.prnewswire.com/news-releases/medicagos-new-drug-submission-accepted-

for-scientific-review-by-health-canada-an-important-step-for-medicago-towards-commercialization-of-its-innovative-influenza-vaccine-300929093.html (Last accessed 1 August 2020).

32 Lim, A.C., Washbrook, J., Titchener-Hooker, N.J., and Farid, S.S. (2006). A computer-aided approach to compare the production economics of fed-batch and perfusion culture under uncertainty. *Biotechnol. Bioeng.* 93 (4): 687–697.

33 Yongky, A., Xu, J., Tian, J. et al. (2019). Process intensification in fed-batch production bioreactors using non-perfusion seed cultures. *mAbs* 11 (8).

34 Cao, W., Cao, H., Yi, X. et al. (2019). Development of a simple and high-yielding fed-batch process for the production of porcine circovirus type 2 virus-like particle subunit vaccine. *AMB Express* 9: 164. https://doi.org/10.1186/s13568-019-0880-8.

35 Villadsen, J., Nielsen, J., and Lidén, G. (2011). Scale-up of bioprocesses. In: *Bioreaction Engineering Principles*. Boston, MA: Springer.

36 Veyrat P. (2016). 5 awesome ideas for business process simplification, https://www.heflo.com/blog/bpm/business-process-simplification-ideas.

37 Rogstad, S., Yan, H., Wang, X. et al. (2019). Multi-attribute method for quality control of therapeutic proteins. *Anal. Chem.* 91 (22): 14170–14177.

38 Andreas Castan, Thomas Falkman, Eric Fäldt, and Annika Fors. Integration of continuous upstream and downstream operations in mAb production - Bioprocess Development Forum. http://www.processdevelopmentforum.com/ppts/posters/Bioproductionposter_FINAL.pdf (Last accessed July 3, 2021).

39 Whitford, W.G. (2015). Single-use systems support continuous bioprocessing by perfusion culture. In: *Continuous Processing in Pharmaceutical Manufacturing* (ed. G. Subramanian). Wiley-VCH Verlag GmbH & Co.

40 https://www.gelifesciences.com/en/us/shop/cell-culture-and-fermentation/stirred-tank-bioreactors (Last accessed 1 August 2020).

41 Godawat, R., Konstantinov, K., Rohani, M., and Warikoo, V. (2015). End-to-end integrated fully continuous production of recombinant monoclonal antibodies. *J. Biotechnol.* 213 (10): 13–19.

42 https://cdn.gelifesciences.com/dmm3bwsv3/AssetStream.aspx?mediaformatid=10061&destinationid=10016&assetid=19357 (Last accessed 1 August 2020).

43 https://www.gelifesciences.com/en/us/shop/chromatography/chromatography-systems/akta-pcc-and-bioprocess-pcc-continuous-chromatography-systems-p-06265.

44 Schofield M., Hummel, J., Krishnan M., and Bisschops, M. (2018). The Continuous Way, The Medicine Maker (19 February 2018).

45 Hummel, J., Pagkaliwangan, M., Gioka, X. et al. (2018). Modeling the downstream processing of monoclonal antibodies reveals cost advantages for continuous methods for a broad range of manufacturing scales. *Biotechnol. J.*

46 Rivas-Interián, R.M., Guillén-Francisco, J.A., Sacramento-Rivero, J.C. et al. (2019). Concentration effects of main components of synthetic culture media on oxygen transfer in bubble column bioreactors. *Biochem. Eng. J.* 143 (15): 131–140.

47 Gagnon, M., Nagre, S., Wang, W. et al. (2019). Novel, linked bioreactor system for continuous production of biologics. *Biotechnol. Bioeng.* 116 (8): 1946–1958.

48 de Bournonville, S., Lambrechts, T., Vanhulst, J. et al. (2019). Towards self-regulated bioprocessing: a compact benchtop bioreactor system for monitored and controlled 3D cell and tissue culture. *Biotechnol. J.* 14 (7).

49 Whitford, W.G., Lundgren, M., and Fairbank, A. (2018). Chapter 8 – Cell culture media in bioprocessing. In: *Biopharmaceutical Processing* (eds. G. Jagschies, E. Lindskog, K. Łącki and P. Galliher), 147–162. Amsterdam, The Netherlands: Elsevier.

50 Sargent, B. (2020). Rapid Spent Media Analysis of Metabolites During Cell Line Development for Better Product Quality Outcomes, https://cellculturedish.com/rapid-spent-media-analysis-of-metabolites-during-cell-line-development-for-better-product-quality-outcomes/ (Last accessed 3 July 2021).

51 David Raw. The evolution of supply chain security – increasing our focus on raw material variability, https://www.cytivalifesciences.com/en/us/solutions/bioprocessing/knowledge-center/supply-chain-raw-material-variability (Last accessed 3 July 2021).

52 Kara S. Quinn. Proactive prioritization: a new tool for assessing raw material risk, https://www.biophorum.com/a-new-tool-for-assessing-raw-material-risk/ (Last accessed 3 July 2021).

53 Cahill, A. The top 7 digital transformation trends shaping 2020. https://blogs.mulesoft.com/biz/trends/digital-transformation-trends-shaping-2020/ (October 21, 2019).

54 Whitford, W. (2017). *The Era of Digital Biomanufacturing*. BioProcessing International.

55 DPS Group and TRIA Congratulate Moderna on ISPE 2019. Facility of the future category award, https://www.dpsgroupglobal.com/news/2019/4/dps-group-and-tria-congratulate-moderna-on--ispe-2019-facility-of-the-future-category-award/.

56 Manzano, T. and Langer, G. (2018). Getting ready for Pharma 4.0: data integrity in cloud and big data applications. *Pharm. Eng.*: 72–79.

57 James Rutley, Adam Paton, Peter Crane, James Arpino, Nuno Leitao, Steve Brown, Vishal Sanchania, Adam Tozer, Chris Grant, Michael Sadowski, Markus Gershater, Bioprocessing 4.0: Integrating Data & Lab Automation in Upstream Bioprocessing, https://synthace.com/s/DATA_small-93b2.pdf (Last accessed 1 August 2020).

58 Ren, S., Zhang, Y., Liu, Y. et al. (2018). A comprehensive review of big data analytics throughout product lifecycle to support sustainable smart manufacturing: a framework, challenges, and future research directions. *J. Cleaner Prod.* 210 https://doi.org/10.1016/j.jclepro.2018.11.025.

59 Pörtner, R., Platas Barradas, O., Frahm, B., and Hass, V.C. (2017). Advanced process and control strategies for bioreactors. In: *Current Developments in Biotechnology and Bioengineering Bioprocesses, Bioreactors and Controls*, 463–493.

60 Narayanan, H., Luna, M.F., von Stosch, M. et al. (2019). Bioprocessing in the digital age: the role of process models. *Biotechnol. J.*

61 Wang, B., Wang, Z., Tao, C., and Xueming, Z. (2020). Development of novel bioreactor control systems based on smart sensors and actuators. *Front. Bioeng. Biotechnol.* 8.

62 Digital Twins digitize assets and processes to enable better industrial outcomes, https://www.ge.com/digital/applications/digital-twin (Last accessed 1 August 2020).

63 Vincent Price, J., Shivappa, R., Barnthouse, K. et al. Application of a genome-based predictive CHO model for increased mAb production and Glycosylation control. In: *Cell Culture Engineering XVI*.

64 Robinson, A.S., Venkat, R., and Schaefer, G. (2018). Conference Program. In: *Cell Culture Engineering XVI*, ECI Symposium Series (eds. A. Robinson, R. Venkat and E. Schaefer). J&J Janssen https://dc.engconfintl.org/ccexvi/253.

65 Nargund, S., Guenther, K., and Mauch, K. The move toward Biopharma 4.0. In: *Genetic Engineering and Biotechnology News*, vol. 39, 6.

66 Mohammadi, V. and Minaei, S. (2019). *Artificial Intelligence in the Production Process from Engineering Tools in the Beverage Industry*, Volume 3: The Science of Beverages, 27–63.

67 (2017). *Technology Roadmapping*. London, UK: BioPhorum https://www.biophorum.com/phorum/technologyroadmapping.

68 Keerthi Iyengar, Somesh Khanna, Srinivas Ramadath, and Daniel Stephens. (2017). What it takes to really capture the value of APIs. https://www.mckinsey.com/business-functions/mckinsey-digital/our-insights/what-it-really-takes-to-capture-the-value-of-apis.

69 Teresa Baumann.(2020). GE Healthcare Life Sciences and GoSilico mechanistic modeling software, https://gosilico.com/gosilico/blog/ge-and-gosilico-enter-comarketing-agreement-on-chromx/ (Last accessed 3 July 2021).

70 Anane, E., Lopez Ca, D.C., Barzb, T. et al. (2019). Output uncertainty of dynamic growth models: effect of uncertain parameter estimates on model reliability. *Biochem. Eng. J.* 150: 107247.

71 Mir-Artigues, P., Twyman, R.M., Alvarez, D. et al. (2019). A simplified techno-economic model for the molecular pharming of antibodies. *Biotechnol. Bioeng.* 116 (10): 2526–2539. https://doi.org/10.1002/bit.27093.

72 Li, X.R., Yang, Y.K., Wang, R.B. et al. (2019). A scale-down model of 4000-L cell culture process for inactivated foot-and-mouth disease vaccine production. *Vaccine* 37 (43): 6380–6389. https://doi.org/10.1016/j.vaccine.2019.09.013.

73 Knowledge Management. Technology Roadmapping. BioPhorum: London, UK, 2017, https://www.biophorum.com/knowledge-management.

74 BSCT: Biomanufacturing Science and Technology Consortium to Advance U.S. Manufacturing of Biopharmaceuticals, https://www.uml.edu/research/bstc/ (Last accessed 1 August 2020).

75 Cruz Bournazou, M.N., Barz, T., Nickel, D.B. et al. (2017). Online optimal experimental re-design in robotic parallel fed-batch cultivation facilities. *Biotechnol. Bioeng.* 114 (3): 610–619. https://doi.org/10.1002/bit.26192.

76 Jonathan, B. and Frank, B. (2006). Miniature bioreactors: current practices and future opportunities. *Microb. Cell Fact.* 5: 21. https://doi.org/10.1186/1475-2859-5-21.

77 FlexFactory: gain efficiency and flexibility with a configurable single-use biomanufacturing platform, GE Healthcare, FlexFactory | Cytiva (cytivalifesciences.com) (Last accessed 3 July 2021).

78 The Cocoon™ Platform, https://www.lonza.com/products-services/pharma-biotech/bioprocess-systems/patient-scale-cell-manufacturing.aspx (Last accessed 1 August 2020).

79 Liv Sewell, 2020 9 Specifications to Guide Successful Automated Sampling in Bioprocessing, http://www.global-engage.com/life-science/9-specifications-to-guide-successful-automated-sampling-in-bioprocessing/.

80 Automated Facility. Technology Roadmapping. BioPhorum: London, UK, 2017, https://www.biophorum.com/automated-facility.

81 Dave Wolton, 2020, Cleanroom Technology, https://www.cleanroomtechnology.com/news/article_page/Aseptic_manufacturing_The_robots_are_coming/161887 (Last accessioned 1 August 2020).

82 Beri, R.G., Wolton, D., and Helmut, C. (2019). Opportunities for modern robotics in biologics manufacturing. *Bioprocess Int.* 17 (4).

83 MAST platform: system for transferring samples from bioreactors to analytical devices, http://mastsampling.com/mast-platform/ (Last accessed 1 August 2020).

84 Hajduk, J., Wolf, M., Steinhoff, R. et al. (2019). Monitoring of antibody glycosylation pattern based on microarray MALDI-TOF mass spectrometry. *J. Biotechnol.* 302 (20): 77–84.

85 Vodopivec, M., Lah, L., and Narat, M. (2019). Metabolomic profiling of CHO fed-batch growth phases at 10, 100, and 1,000 L. *Biotechnol. Bioeng.* 116 (10): 2720–2729.

86 (2017). *In-Line Monitoring and Real-Time Release. Technology Roadmapping*. London, UK: BioPhorum https://www.biophorum.com/in-line-monitoring-and-realtime-release.

87 Randek, J. and Mandenius, C.F. (2018). On-line soft sensing in upstream bioprocessing. *Crit. Rev. Biotechnol.* 38 (1): 106–121. https://doi.org/10.1080/07388551.2017.1312271.

88 Carbonell, P. (2019). Getting on the path to engineering biology. In: *Metabolic Pathway Design. Learning Materials in Biosciences*. Cham: Springer.

89 Shekhawat, L.K., Godara, A., Kumar, V., and Rathore, A.S. (2019). Design of experiments applications in bioprocessing: chromatography process development using split design of experiments. *Biotechnol. Progr.* 35 (1): e2730.

90 (2017). *Process Technologies. Technology Roadmapping*. London, UK: BioPhorum https://www.biophorum.com/process-technologies.

91 Baumann, P. and Hubbuch, J. (2016). Downstream process development strategies for effective bioprocesses: trends, progress, and combinatorial approaches. *Eng. Life Sci.* 17 https://doi.org/10.1002/elsc.201600033.

92 Process intensification and continuous bioprocessing with single-use devices, https://www.biotech2019.ch/ (Last accessed 1 August 2020).

93 Gerben, Z., Kai, T., Michael, K., and Miriam, M. (2019). Design considerations towards an intensified single-use facility. In: . https://doi.org/10.1002/9781119477891.ch14.

94 Larroche, C., Sanromán, M.Á., Du, G., and Pandey, A. (eds.) (2017). *Current Developments in Biotechnology and Bioengineering Bioprocesses, Bioreactors and Controls*. Elsevier.

95 BioIntelligence Technologies, https://biointelligence.com/products/bioanalyst/ (Last accessed 3 July 2021).

96 Insilico Biotechnology, https://www.insilico-biotechnology.com/ (Last accessed 1 August 2020).

97 DataHow AG, https://www.datahow.ch/ (Last accessed 1 August 2020).

98 Zhang, B., Sun, X., Liu, S., and Deng, X. (2019, 7272387). Recurrent neural network-based model predictive control for multiple unmanned quadrotor formation flight. *Int. J. Aerosp. Eng.*: 18.

99 PD2M Future of Pharmaceutical Manufacturing Workshop, https://www.aiche.org/conferences/future-pharmaceutical-manufacturing-conference/2021 (Last accessed 3 July 2021).

100 Eibl, R. and Eibl, D. (eds.). *Single-Use Technology in Biopharmaceutical Manufacture*, 2e. Kindle Edition.

101 Sargent, B. (2017). Blog: Fortem: A platform film built for bioprocess, https://cellculturedish.com/video-fortem-a-platform-built-for-bioprocess/.

102 Boedeker, B., Goldstein, A., and Mahajan, E. (2018). Fully disposable manufacturing concepts for clinical and commercial manufacturing and ballroom concepts. *Adv. Biochem. Eng./Biotechnol.* 165: 179–210.

103 Purposeful design and development of a next-generation single-use bioprocess film, GE Healthcare, https://www.gelifesciences.com/en/us/solutions/bioprocessing/knowledge-center/purposeful-design-and-development-of-a-next-generation-single-use-bioprocess-film (Last accessed 1 August 2020).

104 Biomanufacturing Investment Calculator, GE Healthcare, https://www.geappdash.com/Biomanufacturing_Investment_Calculator/#/calculator/solution (Last accessed 1 August 2020).

105 Langer, E.S. and Rader, R.A. (2014). Continuous bioprocessing and perfusion: wider adoption coming as bioprocessing matures. *Bioprocess. J.* 13 (1).

106 Fibro chromatography. https://www.gelifesciences.com/en/us/solutions/bioprocessing/products-and-solutions/downstream-bioprocessing/fibro-chromatography?extcmp=--sem&gclid=EAIaIQobChMIlr7n2P_v6AIVA56fCh2f7wLKEAAYASAAEgK9TvD_BwE (Last accessed 1 August 2020).

107 (2017). *Modular and Mobile. Technology Roadmapping*. London, UK: BioPhorum https://www.biophorum.com/modular-and-mobile.

108 Mike May, *Modular Bioprocessing Makes Adaptability a Snap GEN October 1*, 2019, https://www.genengnews.com/insights/modular-bioprocessing-makes-adaptability-a-snap/.

109 KUBio prefabricated Manufacturing facilities for monoclonals: The off-the-shelf modular solution for maximum flexibility and efficiency, GE Healthcare. https://www.gelifesciences.com/en/us/solutions/bioprocessing/products-and-solutions/enterprise-solutions/kubio/kubio-mab-manufacturing-facility (Last accessed 1 August 2020).

110 G-CON manufacturing prefabricated cleanrooms called PODs. https://www.gconbio.com/ (Last accessed 1 August 2020).

111 Univercells NevoLine™ biomanufacturing platform. https://www.manufacturingchemist.com/news/article_page/Univercells_introduces_the_NevoLine_Upstream_Platform/169777 (Last accessed 2 July 2021).

112 Single-use modular facilities, GE Healthcare. https://www.geappdash.com/Biomanufacturing_Investment_Calculator/#/challenges/ge-solutions (Last accessed 1 August 2020).

5

Process Intensification Based on Disposable Solutions as First Step Toward Continuous Processing

Stefan R. Schmidt

COO, BioAtrium AG, Operations, Lonzastrasse K02, 3930 Visp, Switzerland

5.1 Introduction

With the approval of Humulin, the first recombinant protein for therapeutic use in humans, in 1980 an unprecedented success story began. Since that time the pipeline and the number of approved protein therapies increased significantly. Currently the class of monoclonal antibodies and their derivatives from Fc-fusion proteins to bi- and multi-specific versions dominate the field. The whole market segment of recombinant therapeutic proteins has a high single-digit compound annual growth rate. This justifies the huge interest in optimizing the corresponding manufacturing processes. Most of these molecules are produced by mammalian cells requiring a long growth and expansion interval in large bioreactors. The subsequent downstream process (DSP) is characterized by a number of specific unit operations like chromatography and filtration to purify the therapeutic protein. In comparison with traditional chemical synthesis processes for small molecules, the overall manufacturing is more complicated. Figure 5.1 gives an overview on the market demand, the manufacturing cost and selling price, regulatory requirements, and the currently applied level of process intensification for products generated by biotechnology. It is fairly obvious that basic products such as bulk chemicals are treated as commodities being prepared in huge quantities at low cost and have a high level of process intensification, whereas pharmaceuticals are complicated high-cost/price products with huge regulatory requirements and a low level of process intensification. However, what does the term "process intensification" really mean? Using it as search term in the PubMed database, the number of hits exponentially increases from a handful of hits annually at the shift of the century to more than 100 hits in the last years (Figure 5.2). Let us have a closer look at definitions in paragraph 5.1.1.

5.1.1 Theory and Practice of Process Intensification

Process intensification is being used primarily in the chemical industry since almost 50 years and was only recently adopted by biotechnology. Therefore, it is no surprise

Process Control, Intensification, and Digitalisation in Continuous Biomanufacturing, First Edition.
Edited by Ganapathy Subramanian.
© 2022 WILEY-VCH GmbH. Published 2022 by WILEY-VCH GmbH.

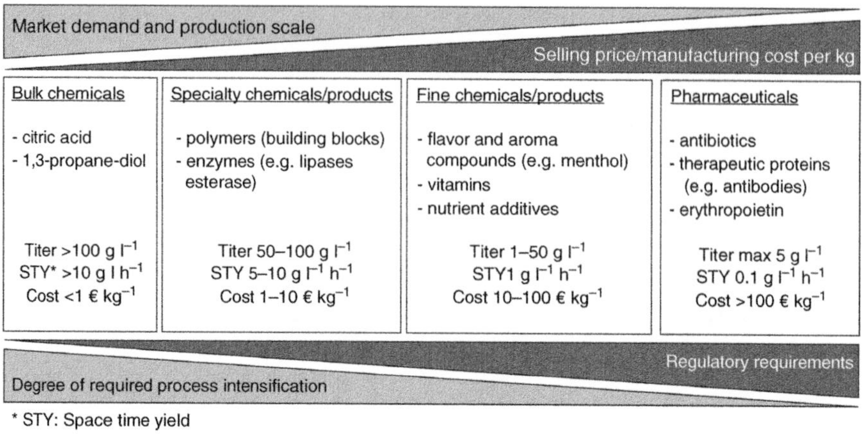

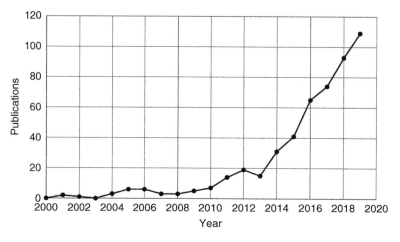

Figure 5.1 Biotechnology products and their degree of process intensification.

Figure 5.2 Literature quantification of process intensification.

that most of the definitions are related to chemical processes. A famous quote is the following: "Process intensification consists of the development of novel apparatuses and techniques that, compared to those commonly used today, are expected to bring dramatic improvements in manufacturing and processing, substantially decreasing equipment-size/production-capacity ratio, energy consumption, or waste production, and ultimately resulting in cheaper, sustainable technologies" [1]. Interestingly the term time saving does not appear here. In general, the authors differentiate between equipment and methods, which is sensible and also true for biologics manufacturing. Looking for a similar definition focusing on biologics only, the following quotation can be useful: "Many employ the term 'bioprocess intensification' to refer to systems for producing more product per cell, time, volume, footprint or cost. This need is being driven by two emerging priorities: cost control and process efficiency" [2].

The essence of several definitions is to improve processes by combining, controlling, and enhancing chemical or transport phenomena. Interestingly, more recent reviews on process intensification claim the need for modeling, which actually fits nicely with current digitization strategies. In the chemical industries, there is a trend toward modular facilities, something that is resembled in bioprocessing as well [3].

Many approaches are focusing on the end results of smaller, more efficient plants and the technical solutions to get there. The current state of the art has been reviewed recently in great detail [4]. Taking examples from the numerous definitions, four larger topics can be isolated that are also relevant for biotechnological processes. The most obvious is the combination and integration of process steps to reduce the overall number of unit operations, thus leading to the increased overall yield. Particularly the targeted enhancement of phenomena has a huge potential in biotechnology where living cells generate the final product. Still intracellular processes can mostly be assessed only indirectly through quantifiable phenomena. Rearrangements of process steps by uncoupling dependencies or frontloading of activities supports the improved process output. Figure 5.3 summarizes the different approaches. Overall, the different or combined process intensification strategies should lead to simpler, more robust, and more efficient processes. However, the most effective approach, eliminating unnecessary and unproductive process steps, is often the hardest to achieve.

By implementing these approaches, intensified processes should on the one hand deliver higher productivity and capacity and better flexibility. On the other hand, this should also result in a decreased complexity, a smaller plant footprint, fewer byproducts, less energy consumption and waste generation, and in summary lower investment and manufacturing costs (Table 5.1).

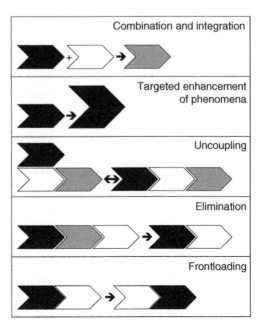

Figure 5.3 Process intensification strategies.

Table 5.1 Expected outcome of process intensification.

Increased	Decreased
Productivity	Complexity
Capacity	Footprint
Titer	Energy consumption
Flexibility	Waste

5.1.2 Current Bioprocessing

Modern bioprocesses relying on living cells can be divided into two large parts: the upstream process (USP) where cells are present and the DSP where cells are absent. The intermediate step, harvesting, usually is counted as the last USP step as it serves to generate a cell-free fluid (CFF) that is transferred to the DSP. Conventional batch USP typically starts with the inoculum, the thawing of a cell bank and its expansion in several steps during the so-called seed train to finally arrive in the main reactor that is used to further increase the cell count while simultaneously expressing the recombinant protein. Most of the current processes require a supplementation of the cells with nutrients, resulting in a so-called fed-batch cultivation. The USP is stopped when the viability of the cells decreases and the cell count becomes stagnant. Cells are separated from the secreted product by sedimentation or filtration or a combination of both. The DSP part usually starts with a capture step that aims to quantitatively adsorb the target protein on chromatography resins. As there exists no absolute selectivity, additional purification steps are required. These steps are either further chromatography or filtration unit operations. When working with mammalian cells, minimally two orthogonal virus reduction steps must be part of the DSP. This is often the combination of virus inactivation through the exposure to either low pH, organic solvents or detergents, and a virus removal close to the end of the process by nanofiltration. Finally, the bulk drug substance is filled into primary containers that are either bottles or bags and stored under controlled conditions. A schematic overview of a traditional antibody process can be seen in Figure 5.4.

5.1.3 General Aspects of Disposables

For all unit operations in USP and DSP, a wide range of technical equipment is utilized. Traditional processes have relied heavily on stainless steel facilities with large bioreactor vessels and hard-piped connections. Due to the huge pipeline of recombinant therapeutic proteins, the demand for clinical and commercial manufacturing is enormous. Both categories differ in the quantities that need to be supplied and the frequency of changes between campaigns. In the last decade disposable equipment became more important that facilitates quick changeovers. However, the history of disposable tools in protein production reaches much farther into the past. Already

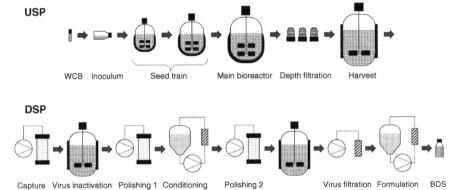

Figure 5.4 Traditional antibody manufacturing process.

in 1991 the first prepacked columns were available, and at about the same time membrane chromatography was introduced. This is soon followed by the first USP applications, for example, the famous wave bioreactor designed in 1996. After some time, the equipment was scaled up and successfully entered the mid-scale manufacturing market, where disposable instruments are widely adopted and are currently state of the art. The various single-use process equipment ranging from bags to bioreactors and filtration have been summarized many times [5].

The benefits and limits of disposables in the manufacturing landscape cover the usual triangle of cost, quality, and time but also the additional parameter of safety. A summary of that analysis can be seen in Figure 5.5. On the one hand, benefits of disposable systems themselves already represent some of the expected outcomes of process intensification. For instance, they require lower investments, reduce the plant complexity, have a smaller footprint, and allow a high degree of flexibility. On the other hand, single-use equipment suffers from quality challenges like outsourcing of quality control, extra efforts for leachable and extractable assessment, risk of leaks, inventory costs, and dependency on external supply [6].

5.2 Technical Solutions

Based on the depicted standard process in Figure 5.4, several entry points for the introduction of disposable solutions for process intensification can be identified.

5.2.1 Process Development

When working with a new recombinant protein, the first step certainly is process development. Already here process intensification can be implemented. The utilization of miniaturized, automated, single-use solutions fulfills the goal of increased productivity by eliminating manual steps, decoupling of preparation work, and reducing the cost. Furthermore, the ability to run up to 24 experiments in parallel enables factorial approaches for design of experiment (DoE) strategies and

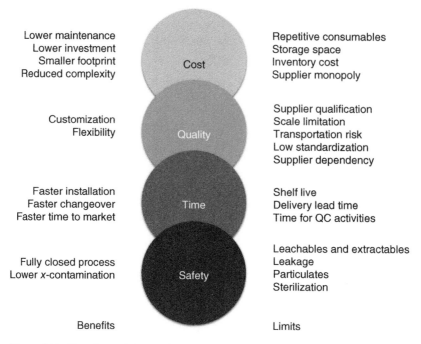

Figure 5.5 Benefits and limits of disposables.

results in further time savings. This has been evaluated for 24 runs with the ambr® mini-bioreactor system in comparison with classical 10 and 2.7 l glass bioreactors. Obviously the glass bioreactors require significantly more time for preparation and cleaning, while the 250 ml single-use bioreactors come presterilized and ready to use. Another important contributor to labor effort and duration of the process for glass reactors is the manual sampling for in process controls that is done automatically for the ambr. Due to the larger volume, clearly the material costs are highest for the 10 l bioreactor. Interestingly the 10 l material costs are still higher than those for the ambr 250 despite the extra cost for the single-use bioreactor. Overall the ambr systems delivers the best values for all three parameters, duration, labor, and material cost as displayed in Figure 5.6.

Similar to USP activities, also DSP development can benefit from process intensification and single-use applications. The central tool for protein purification is chromatography. As mentioned previously, prepacked, miniaturized columns have a long history. However, only recently the full automation including gradient elution, delivering an unmatched resolution, has been established on a robotic platform. Comparing the output of a single-column system with a multichannel liquid-handling instrument, it was demonstrated that similar results could be obtained in 1/3 of the time [7].

5.2.2 Upstream Processing Unit Operations

The main goal of all upstream improvements is to increase volumetric productivity that can be achieved by higher viable cell densities (VCD), enhancing cellular

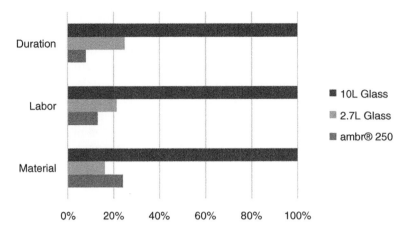

Figure 5.6 Economic performance of different bioreactors in process development.

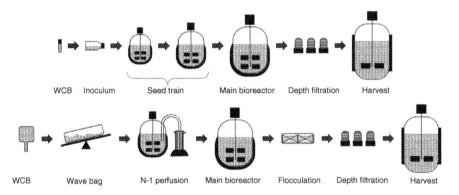

Figure 5.7 Process intensification in USP.

productivity and optimizing bioreactor utilization. Furthermore, it would be advantageous to decrease the complexity by eliminating or combining process steps. For USP there are multiple entry points to introduce process intensification through single-use applications. Interestingly many approaches are amenable to all scales and can even be introduced into existing plants and processes as depicted in Figure 5.7. This is a clear opportunity for a gradual move toward continuous processing.

5.2.2.1 High-Density, Large-Volume Cell Banking in Bags

The first step, the inoculum, starts with thawing an aliquot of a cell bank with stably transfected cells that have the recombinant gene integrated in their genome. Traditionally the cell bank is stored in small vials in the presence of a cryopreservant in liquid nitrogen. Expanding the small number of cells present in the Cryovial is a time-consuming process. The limitations of this procedure are clearly related to the small volume of starting material. An obvious time-saving improvement would be to begin the expansion with a larger cell number. This becomes feasible by utilizing bags as proven for blood transfusion instead of 1.5 ml vials, thus increasing the starter volume by a factor of up to 100.

The first example of that approach dates back to the early 1990s when hybridoma cells were stored at 1.5×10^8 cells/ml in 25 ml bags. This allowed starting the seed train with a 10 times higher cell density at 500 ml [8].

About 10 years later, 50–100 ml cryobags containing $20–40 \times 10^6$ CHO cells/ml were directly introduced into a bioreactor at 2 l starting volume. Further expansion to 12 l in the same bioreactor saved approximately 30 days or 60–70% of the whole process time from thawing to the inoculum of a production reactor when compared to a conventional method [9].

A further decade later, frozen accelerated seed train for execution of a campaign (FASTEC) was developed. Cells from traditional 1.5 ml working cell bank (WCB) ampoules are cultivated in a perfusion bioreactor to accumulate a high-density cell mass of up to 1.1×10^7 cells/ml. The large cell volume is mixed with freezing medium and then aliquoted into 150 ml freezing bags. These bags can be stored in liquid nitrogen at conditions similar to conventional cell banks. Thawing these bags allows to skip several seed train steps, thus saving time and efforts [10].

5.2.2.2 Seed Train Intensification
5.2.2.2.1 Wave Bag Expansion
The next step after optimizing the starting volumes and densities of cell banks is the expansion of the culture volume during the seed train. Traditionally this includes a number of manual steps to go from a vial to shake flasks or spinners and finally the production reactor. The intermediate manual operations could be eliminated if a vessel was found that allows dynamic increasing the culture volume without stepwise transfers of serially growing quantities.

The vessel enabling that approach was designed in the mid-1990s and is called wave bioreactor. It allows smooth and gentle mixing while maintaining a high level of aeration [11]. In an example starting with a 5 ml cryotube containing a cell bank with up to 1×10^8 cells/ml and direct seeding into a 20 l wave bag with continuous volume expansion, nine days of operations could be saved [12]. Current processes frequently start in a 25 l wave bag with a cultivation volume of 2–3 l to be extended to 15 l over the course of several days. A further benefit of using the wave system is that it decreases the contamination risk of open handling as all transfers can be made through sterile coupling or welding of tubes. Wave bioreactors as all other disposable instruments come presterilized avoiding any cleaning efforts.

5.2.2.2.2 N-1 Perfusion
The last step before the main production reactor is the so-called N-1 step, critical for defining the cultivation duration in the main reactor. The higher the cell count at the end of N-1 to inoculate the finale stage, the shorter the runtime in the last bioreactor. Shorter durations will ultimately lead to more batches per time interval and thus a higher productivity of the facility, fulfilling one of the goals of process intensification.

As before in the case of increasing VCD of cell banks by perfusion, the same principle can be applied for the N-1 step. Ideally the initial seeding density for the main reactor should be above 2×10^6 cells/ml. A very recent example started the last reactor with 20×10^6 cells/ml with cells from N-1 perfusion in a single-use bioreactor. Here the initial growth in the main reactor is replaced by an N-1 perfusion step, resulting in the comparable final product concentrations in a shorter time corresponding to a higher volumetric productivity [13]. The same group demonstrated previously that higher VCD at the start of the main reactor and identical cultivation duration leads to higher protein titers. Interestingly this could also be obtained without perfusion but by optimizing the feed strategy and an enriched culture medium [14]. Comparing the high feed and the N-1 perfusion, the product concentrations increased by a factor of 4 and 8, respectively, in contrast to the original fed-batch process [15]. Looking at process time savings, a reduction from 14 to 8 days could be achieved in another example through N-1 perfusion and 25 times higher seeding cell densities. That resulted in similar protein titers in a shorter timeframe [16]. Another case study demonstrated a time saving of 5 days for a N-1 perfusion based high seed process, achieving the same output at 12 days instead of 17 days [17].

The improved cellular performance by seeding with higher VCD can be explained on the molecular level by an earlier shift of the gene expression pattern from growth toward production-associated genes [18].

Although many of the examples from the literature did not apply exclusively single-use equipment, the general advantages remain the same. Replacing the conventional fed batch at N-1 by perfusion to start with elevated VCDs, either a higher final product titer in the same cultivation timeframe or a shorter cultivation duration for the same product concentration can be obtained, as shown in Figure 5.8. Nevertheless, when directly comparing a fully disposable process with intensified seed train, N-1 perfusion, and perfusion in a 500 l disposable main reactor, a similar output as in a traditional stainless steel USP setup can be generated [19]. Even at commercial manufacturing scale of a Chinese hamster ovary (CHO) antibody process, N-1 perfusion was successfully introduced at 3000 l, feeding a 13 500 l main reactor, hereby reducing the cultivation time in the main bioreactor by 20% [20].

Typical parameters to optimize the N-1 perfusion are the inoculum density [21], the perfusion rate [22], and the media composition [23]. A comparison of some so far applied settings can be seen in Table 5.2.

A wide range of different single-use bioreactors is nowadays available for perfusion and continuous processes. The preference for disposable solutions is also related to the modest reactor volumes that are required to deliver a similar productivity as much larger conventional fed-batch steel bioreactors [24]. The footprint reduction is repeated in the downsized utilities that no longer need to offer clean steam. However, the space savings are partly reclaimed by enlarged media preparation capabilities.

5.2.2.3 Cell Retention and Harvest

Having just discussed the different approaches to intensify USP with higher cell densities through perfusion, it is worthwhile to focus now on the available

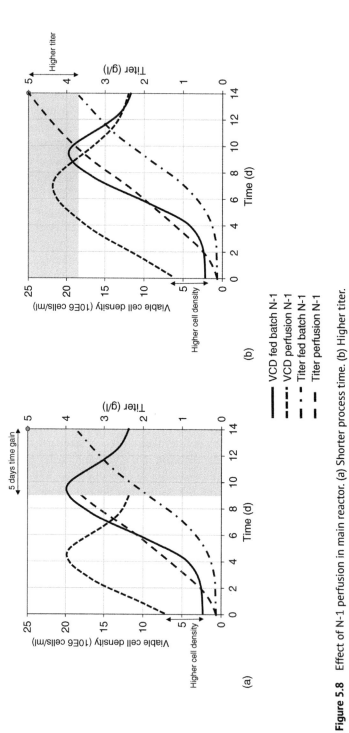

Figure 5.8 Effect of N-1 perfusion in main reactor. (a) Shorter process time. (b) Higher titer.

Table 5.2 Comparison of N-1 results from the literature.

Peak VCD (×10⁶ cells/ml)	Duration (days)	Perfusion rate	Cell retention	N-1 scale working volume (l)	References
15.8	5	1 vvd	Settler	3000	[20]
24	6	1–3 vvd	ATF	3	[16]
1600	5	n/a	Various	5 or 50	[21]
40	5	0.05–0.12 nl/cell/day	ATF	4	[17]
41–62	6	0.5–2 vvd	Filter	5	[14]
45	6	0.4–2.3 vvd	TFF	20	[18]
100	6	0.04–0.08 nl/cell/day	ATF	200	[13]

ATF, alternating tangential flow filtration; VCD, viable cell density; vvd, vessel volume exchanges per day; n/a, not available.

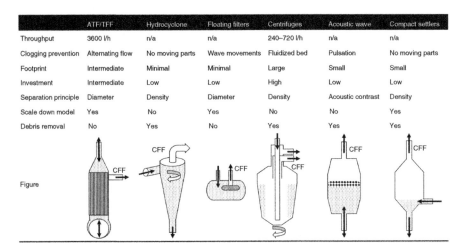

Figure 5.9 Comparison cell retention technologies.

instrumentation necessary to enable these processes. The advantages of perfusion can only be realized by the availability of reliable cell retention devices that partly could also be used for harvesting purposes to deliver CFF. Figure 5.9 gives an overview on current cell retention technologies that are available in a disposable format.

5.2.2.3.1 Tangential Flow Filtration (TFF)

When comparing four different technical solutions, such as spin filters, acoustic settlers, gravitational settlers, and centrifuge, the gravitational settler could be identified as the most appropriate device for a wide range of volumes [21]. However, the majority of these instruments do not exist as disposable versions. A widely used exception is the tangential flow filtration (TFF) or its USP-adapted version, alternating tangential flow filtration (ATF), which are both based on hollow fibers [25].

They hold back the cells while allowing the passage of the culture liquid. The tangential flow minimizes the tendency of clogging the small pores, which is even further reduced in the case of ATF where the orientation of the flow is alternating. However, active bleeding is required to remove dead cells and their debris, as the filter does only discriminate between large and small components of the liquid [26].

5.2.2.3.2 Hydrocyclones

The ability to bleed and hereby also to influence the composition and concentration of retained cells is one of the major advantages of hydrocyclones (HC) utilizing a centrifugal field for separation that is induced in form of a vortex by the tangential flow of the feed suspension. However, the lack of a physical barrier complicates operations with shear stress-sensitive cells that additionally have a low sedimentation velocity. Although its popularity decreased over the last decade, HC is still an interesting technology. This is because, due to its mechanically simple structure without any moving parts, it is feasible to design HCs also as plastic single-use devices that could be manufactured by 3D printing. The only remaining issue is a pulsation-free flow at high velocities, which is difficult to obtain by traditional peristaltic pumps. In a recent example an HC was attached to a 50 l disposable bioreactor, and antibody-producing CHO cells were successfully retained for more than 20 days at constant perfusion and VCDs of up to 50×10^6 cells/ml [27].

5.2.2.3.3 Floating Filters

Wave bags have been mentioned previously in the context of cell banking and early seed train. By implementing a floating filter with a pore size of 7 μm, which is below most of the eukaryotic cell diameters, an efficient cell retention becomes feasible. The floating on the surface of the cultivation liquid prohibits clogging, as the wave movements wash it clear. As the filter is connected to a tube, media can be replaced via a peristaltic pump in a fully closed setup.

5.2.2.3.4 Centrifuges

Most of the disposable applications to retain cells have a low level of complexity when compared with centrifuges. Historically single-use centrifugation has primarily been used in blood processing and currently gets attention as platform process for cell therapies. Interestingly, the 2000 l volumetric limit of these centrifuges nicely fits to the dimensions required for perfusion as intermediate step for process intensification. Some limitations for a wider use include the difficulties with scale-down models and the inability of centrifugation to completely remove submicron particles. The technology is based on fluidized-bed centrifugation (FBC) that suspends cells between the centrifugal force and the force of the fluid pumped into the rotor, enabling the washing and concentration of cells [28].

5.2.2.3.5 Acoustic Waves

Cell retention can also be based on acoustic waves in a disposable device. Hereby a standing ultrasound wave is created that prevents the cells from moving with the liquid stream. At certain intervals, the acoustic field is interrupted to let the cells settle and to be flushed back to the bioreactor. By fine-tuning the conditions to generate the wave barrier, it becomes possible to discriminate between dead and living cells or debris. As there is no mechanical barrier, fouling and clogging are avoided.[1]

5.2.2.3.6 Settlers

In addition to the well-known inclined settlers, round compact settlers in a disposable format were developed that have a sixfold higher settling capacity at the same footprint. They allow perfusion cultures at approximately 10×10^6 cells/ml with viability above 80% for three weeks. The settler's geometry is designed in a way to continuously remove dead cells and debris from living cells.[2]

5.2.3 Downstream Processing Unit Operations

DSP, in contrast to USP where all unit operations are focusing on cultivating cells, rather deals with isolating and purifying the product generated by cells. Therefore, the starting point of DSP for recombinant proteins is a CFF. The initial step is followed by a number of unit operations that remove all kinds of impurities to finally obtain a pure and concentrated drug substance. One focus of the recent strategies was to design a fully disposable DSP that eliminates non-value-adding activities such as cleaning [29]. A summary of all DSP related strategies can be seen in Figure 5.10.

5.2.3.1 Depth Filtration

The associated impurities can be differentiated into process or product related. At the beginning of the DSP activities, typically process-related impurities such as cell debris, host cell proteins (HCP), and DNA are dominating. If these impurities resemble particles, then depth filtration (DF) can be applied. During the passage through the filter layers, fines either are sieved according to their size, become trapped in cavities, or are adsorbed due to the surface characteristics such as charges and hydrophobicity [30]. Depth filters are typically used only once, thus fulfilling the attribute of

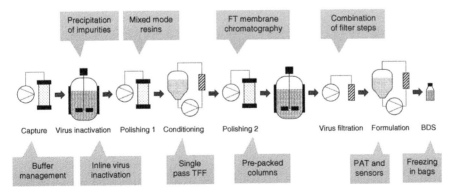

Figure 5.10 DSP process intensification strategies.

---

1 https://www.ipratech.be/cytoperf/.
2 http://www.sudhinbio.com/singleusebiosettlers.html.

single use. However, this alone would not qualify DF as element of process intensification. However, when combining steps or when creating a situation where DF could do more than just eliminate debris, DF could become a tool of intensification.

One example for multipurpose DF is adsorptive hybrid filters (AHF) that contain positively charged membranes. Depending on the ligand, these offer a 40–700% higher ionic capacity to remove anionic impurities. Typically, these filters are not applied during harvest but often directly after the capture chromatography to quickly remove impurities early in the process [31]. The same type of filters was evaluated for their virus removal capabilities that are strictly dependent on electrostatic adsorption and only minimally on physical entrapment [32].

Besides the direct application of DF to simply eliminate cells and debris at the end of the cultivation, a number of chemical treatments can be used. These approaches lead to the increased efficiency of particulate removal and simultaneously reduce the load of typical impurities early in the process. All these methods were recently reviewed in detail [33] and are further explained in the paragraphs below.

5.2.3.1.1 Filter Aids

Although DF usually delivers sufficient clarification by the abovementioned mechanism, its performance can be further improved by adding filter aids such as diatomaceous earth (DE), volcanic ash, or activated carbon. They all are inert materials forming a second layer around the primary layer, the filter fibers, and improve the flow rate by reducing compressibility while increasing permeability. The whole process can be integrated by mixing DE with the cell containing solution before passing the mixture through the DF. Lowering the pH during mixing further reduces HCP and DNA content. This method called body-feed filtration can be established entirely by single-use components, fulfilling the criteria of process intensification [34].

5.2.3.1.2 Flocculation

Another approach to eliminate insoluble particles from solutions is flocculation that can be applied together with DF. Here, these fine-dispersed colloids form larger clusters upon induction by a flocculants that are often polyelectrolytes connecting opposite charges in a pH-dependent manner. The conditions to generate large flocs that sediment due to their size must be carefully selected to avoid less stable agglomerates that are easily disrupted by shear forces. Typically, the process of mixing, flocculation, and consecutive DF is time consuming. The duration can be drastically reduced through a continuous approach. By optimizing the concentration of the polycationic flocculation agent polydiallyldimethylammonium chloride (pDADMAC) and the residence time in a static mixer, a faster process and a fourfold decrease of filter area can be achieved. Starting with an unconditioned CHO cell containing broth after a two-week cultivation, a 97% yield could be obtained in a fully disposable setup without the need of using large mixing vessels [35].

5.2.3.1.3 Precipitation

The other DF compatible method to eliminate impurities is precipitation. On the one hand, precipitates can form spontaneously; on the other hand, they can be induced by changing chemical or physical parameters. For instance, acidification below the isoelectrical point (pI) of the target protein can lead to selective precipitation of HCPs or DNA that represent the major process-related impurities. Sometimes the addition of chemicals such as caprylic acid can improve the elimination of unwanted species by neutralizing charges and inducing partial unfolding that ultimately leads to the exposure of hydrophobic patches, causing protein–protein interactions and precipitation [36]. Besides the typical elimination of process-related impurities, caprylate also supports the reduction of product-related impurities such as aggregates. Due to its additional capacity to simultaneously inactive enveloped viruses, caprylate is a true example for process intensification combining multiple functions and effects in one process step. Further intensification can be obtained by performing precipitation in a coiled flow inverter reactor (CFIR) in a continuous mode. Compared with a batch mode, DNA and HCP were cleared 33% and 16% better, respectively. The biggest impact could be seen in the increase of productivity that increased eightfold for using caprylic acid and 16-fold applying $CaCl_2$ [37].

Turning the precipitation approach around by precipitating the target protein instead of the impurities creates the additional benefit of volume reduction. Recently this idea has been applied in a continuous process intending to avoid chromatographic capture. By continuously adding $ZnCl_2$ and polyethylenglycol (PEG) via a static mixer to cell-free culture supernatant, a precipitate of the target protein is formed. This is then washed in a two-step process in hollow fiber TFF, resulting in an intermediate concentrated product with a purity and yield comparable with protein A chromatography [38]. Combining four different precipitation and dissolution techniques based on caprylate, PEG, $CaCl_2$, and cold ethanol followed by a flow through ion exchange chromatography enables full continuous processing when connected to a perfusion bioreactor [39].

Industrial application of precipitation has already been proven for a long time in plasma fractionation at large scale. The examples above utilize single-use equipment or can be adapted to disposables. Nevertheless more work needs to be done for scale-up. Currently the patenting trend shows a tendency toward approaches for a wide range of recombinant proteins [40].

5.2.3.2 In-line Virus Inactivation

Manufacturing process based on mammalian cell culture require two orthogonal virus reduction steps. One of them typically is based on the so-called virus filtration, which essentially is a nanofiltration, segregating viruses by size and being the barrier between DSP pre- and post-virus removal areas. The other method used much earlier in the process is virus inactivation by either destroying the lipid envelope or, in case of its absence, denaturing the capsid proteins. This can be done by exposure to either organic solvents, detergents, or a pH of below 4. In the case of antibodies, their capture from the cell-free supernatant is performed by binding to protein A during chromatography. The elution is triggered by lowering the pH of the mobile phase to below 3.5. Incubating the eluate at low pH for at least one hour will inactivate any present virus.

This principle of batch incubation could be replaced by a continuous flow while guaranteeing the complete mixing with the inactivating solution and a sufficient residence time under these conditions. Several different ideas have been evaluated that were all applicable either as process intensification strategy or in a full continuous mode. All examples below rely on single-use materials and do not need any vessels for incubation but might need a continuous feed stream.

One approach applied a CFI reactor where the flow is directed over a high number of 90° bends to achieve sufficient mixing over a narrow residence time distribution. The simplicity of the concept relies on a straight single-use tube in which turbulences are induced by a number of 90° bends. Here several straight helix modules are coiled over a tube and change direction by a series of bends, thus inducing vortices that enhance radial mixing. This setup resulted in the reproducible virus inactivation [41].

Instead of introducing turbulence by bends, also static mixers can be used. Ideally the first mixer is located at the beginning of the flow reactor to establish a uniform low pH. The second mixer after the incubation chamber ensures the neutralization. Comparing four different versions of incubation chambers such as U-shaped pattern, serpentine flow jig, coiled flow inverter, and jig in a box (JIB), the latter is favored owing its high reproducibility and narrow residence time distribution. Furthermore, JIB can easily be manufactured from single-use material or 3D printing in a modular way for easy scaling-up [42]. The applicability of JIB was further characterized by computational fluid dynamics (CFD) modeling. Alternating turns of 270° delivered optimal secondary flow and maintained mixing over the whole tube length. The low pressure drop allows simple integration in a continuous process flow or as process intensification solution eliminating the need for an incubation tank [43].

As alternative to the above-described tubular reactors, also packed beds of rigid nonporous beads can be used for this purpose. The bead diameter between 200 and 400 μm is large enough to prohibit any pressure buildup or fouling by accumulated particulates. In this example the inactivation is based on solvent detergent interaction not requiring an additional mixer after the incubation for neutralization. It was proven that no adsorption of the virus occurred inside the system and that the inactivation is solely dependent on the chemical additives [44]. This setup can be applied for low-pH virus inactivation as well. Here simply the second mixer for neutralization after the packed-bed reactor must be added [45]. Interestingly, this reactor type is superior to other versions based on JIB or CFI with regard to residence time distribution that is independent of bead size and reactor length. Furthermore, the reactor is fully scalable as it consists of standard chromatography column parts and could easily be premanufactured as single-use device [46].

5.2.3.3 In-line Buffer Blending and Dilution

During DSP relatively large volumes of buffer are consumed. Traditionally the different buffers are prepared in advance and stored in large tank farms until final use. Obviously there is still a room for improvement. To reduce the required quantity for storage, buffers can be prepared as either stock solutions or concentrates that only require dilution. A 10-fold concentration factor can easily reduce the storage volumes to a size that is in the range for disposable solutions. The currently available technology allows a good control by monitoring flow and following a recipe. The regulatory challenge to verify that the right diluted buffer is used can be answered through pre-point-of-use sensors for pH, conductivity, and flow [47]. The only prerequisite for that approach is accurate stock solutions. Higher levels of complexity, blending to obtain a certain pH or a certain conductivity, have to rely on the proper interplay between all sensors and cannot work on controlling flow ratio between pumps alone. Typical buffer blending relies on four different inputs; acid, base, salt, and water. By mixing these four components, a wide range of pH and conductivities values can be achieved [48].

A further simplification, enabling process intensification, is reducing the number of different buffers applied in a process. For instance, shifting from conductivity-based elution in ion-exchange chromatography to pH-based elution can eliminate the need of salt. In that case, a multicomponent buffer system must be established that allows a linear pH gradient by mixing two buffers. This concept of generic buffers has been described before, where instead of 13 different buffers for 3 unit operations, only 5 were needed when relying on generic buffers (Figure 5.11). The advantages are fewer buffer definitions with applicability in more projects and less efforts for preparation, storage, and transport.

5.2.3.4 Chromatography

With regard to recombinant therapeutic protein manufacturing, chromatography represents the most important unit operation due its unsurpassed selectivity and discrimination efficiency between the target product and other unwanted species. The significance of that step is also underlined by the many approaches to improve chromatography and intensify that step.

A chromatography step is performed by a using the distribution between a stationary and a mobile phase whose liquids are transferred by pumps. Typically pumps and sensors are combined in a system. Looking at multistep processes or multiproduct facilities, it is obvious that a lot of time and effort is spent in cleaning

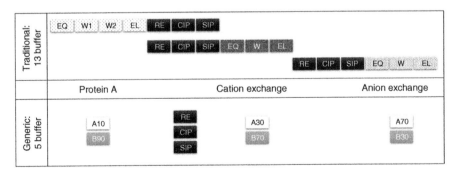

Figure 5.11 Generic buffers based on pH titration. EQ: Equilibration, W: Wash, W1: Wash 1, W2: Wash 2, EL: Elution, RE: Regeneration, CIP: Cleaning in place, SIP: sanitization in place.

Table 5.3 Comparison of stationary phases.

Characteristic	Resin	Monolith	Membrane	Fiber	Mixed-matrix membrane
Flow rate	Low	High	High	Very high	High
Pressure drop	High	Low to moderate	Low	Low	Low
Dominant transport	Diffusion	Convection	Convection	Convection	Diffusion
Binding capacity	High	Moderate	Low	Low	Moderate
Resolution	High	Moderate	Moderate	Moderate	Moderate to high
Hardware cost	Moderate	Moderate	Inexpensive	Inexpensive	Moderate
Foot print	Extensive	Extensive	Small	Small	Small

Source: Orr et al. [49].

instruments between different steps, batches, and products. These activities are not value generating and should be limited to a minimum. One way out of that dilemma is to replace the flow path of chromatography skids instead of cleaning it. There are several solutions from various vendors available. All utilize pinch valves instead of rotating valves to control the flow.

The following paragraphs describe the current range of stationary phases that are available as disposable devices. Table 5.3 summarizes their characteristics in comparison to traditional columns [49].

5.2.3.4.1 Prepacked Columns

Prepacked columns have a long history and entered R&D laboratories in the early 1990s. From there it took almost 20 years to finally reach manufacturing. Initially prepacked columns were distributed by resin manufacturers with fixed column height and a limited range of diameters. Customization and free choice of resins were not available at the beginning. However, soon this gap was closed, and nowadays a wide range of fully customizable columns can be obtained. The current maximal diameter of 80 cm is sufficient to cover the capacity needs of corresponding USP with 2000 l single-use bioreactors or can be integrated into polishing operations in large-scale commercial manufacturing.

One of the major advantages is connected to the fact that these columns can be treated as consumables, thus avoiding the requirements for equipment qualification and maintenance. Furthermore, packing efforts are outsourced, backup solutions are simplified, the risk of product carryover is absent, and packing consistency can be improved. Comparing the performance of prepacked columns with stainless steel columns, only minor differences were observed in which all were lying in the acceptable range of specifications. Even transportation did not affect the packing integrity [50].

Much more effort has been invested by the scientific community to systematically study the characteristics of lab-scale prepacked columns. For instance, a study on 25 000 ready-to-use columns prepared in a timeframe of 10 years revealed very little variations in height equivalent to theoretical plate (HETP) and asymmetry, indicating a very good control of the packing process [51]. The uniformity of packing independent of scale could also be verified by evaluating the performance of lab- and manufacturing-scale prepacked columns. In that assessment, binding capacity, HETP, and asymmetry showed no significant difference between small and large scale again demonstrating the high-quality prepacked columns [52].

5.2.3.4.2 Monoliths

A very special type of prepacked columns is monolithic columns. They are in principle a single piece of a stationary phase forming a homogeneous column. A monolith consists of a network of interconnected channels of large diameters of up to 5 µm, but no small pores. Therefore, the adsorptive surface is directly accessible but smaller than in conventional resins. Consequently, the binding capacity of monoliths is lower, but the mass transport much higher than in standard supports. However, both capacity and resolution stay almost constant independent of the flow rate. These are highly favorable prerequisites for the separation of large molecules such as DNA, viruses, or exosomes. Nevertheless, also normal antibodies can benefit from the speed of the monolithic separation, delivering a similar volumetric output in a significantly shorter time. In the last years, monoliths for large-scale applications became available. Besides the usual surface functionalities such as charges or hydrophobicity, monoliths can also be modified with affinity ligands to enable highly specific enrichment of target proteins at very high flow rates [53]. Essentially, monolithic columns can serve in process intensification approaches in two ways: as disposable, premade equipment and as element to shorten process times. Despite these advantages, monoliths still suffer from scale limitations and sometimes nonuniform pore structure. Currently the largest commercially available monoliths have a volume of 8 l.

5.2.3.4.3 Membrane Adsorbers

Membrane absorbers represent another interesting class of prefabricated chromatography devices that offer high flow rates similar to monoliths. They are currently more widespread due to their lower cost. At the beginning, charged membranes working as ion exchangers were applied in bioprocessing. As bind/elute operations on cation exchanger membranes are not feasible due to the capacity and resolution limitations, relatively early anion exchanger membranes in flow through mode were used to remove DNA and HCP [54]. Besides the rather conventional stacked-disc and radial flow membrane chromatography devices, recently a small-scale lateral flow system was evaluated. Here it could be demonstrated that in this specific configuration, the separation of a model protein mixture was even better than a comparable prepacked column. This might be one solution to escape membrane chromatography limitations [55].

Overall membrane adsorbers tolerate much higher flow rates, exhibit much shorter process times, and require less buffer volumes than conventional resins. Furthermore, scale-up is very straightforward by simply multiplying the membrane area and stacking of premade cartridges. Although there were many evaluations on the applicability of membranes for affinity chromatography [56], none of these approaches really made it to the large-scale industrial market yet [57]. Nowadays ion exchange modalities dominate the field of membrane chromatography instead. One exception for small scale is the protein A-based antibody capture device that allows a residence time of only 30 seconds while still maintaining a dynamic binding capacity of approximately 30 g/l. Even shorter residence times for membrane based antibody purification have been demanded to enable low cost and fast continuous production [58].

Membranes with quaternary amine (Q)-functionality are particularly useful for virus removal due to their high charge density. Comparing Q-membranes and Q-resins unit operations for a conventional antibody process over 10 years, cost savings of more than 20% can be achieved for disposable membranes primarily driven by a lower buffer consumption [59]. Unfortunately, Q-based membranes are sensitive to salt concentrations above 50 mM. Therefore, more than a decade ago, research started to find substitutes for that ligand. Only polyhexamethylene biguanide (PHMB) was able to maintain a significant reduction of viruses at 150 mM salt [60]. A similar effect of salt tolerance could be demonstrated with primary amines as ligands that still enabled a high virus reduction value at 22 mS/cm [61].

5.2.3.4.4 Fiber-Based Cartridges

Looking at the construction of membranes, the next step of simplification is the reduction of the stationary phase to fibers only. Here the flux in a fiber network can be extremely high. Initially, different formats like staple, aligned fibers of whole fabrics were evaluated for their performance in chromatography. Due to their macroscopic structure, high resolution of small molecules during separation is not possible. However, this disadvantage is compensated by the unparalleled convective flow capabilities that enable large-molecule separation at high speed with low back pressure. A further advantage is low material cost [62]. The resolution disadvantages of natural fibers can be circumvented by synthetic ones. Particularly electrospun polymeric nanofibers with submicron diameter offer a large surface area for adsorption. Although the binding capacity of diethylaminoethyl (DEAE)-derivatized nanofibers reached only 10% of packed beds, the 100-fold higher flow rates results overall in a 10-fold improved productivity. Due to their structure, nanofibers are less prone to fouling [63]. Further optimization and packing these nanofibers into a flow distribution device resulted in 15 times higher productivity than a resin-based process. Bind/elute cycles could be completed in 7 seconds. Small cartridges connected to a simulated moving bed chromatography skid seem to be an ideal solution for process intensification reducing both footprint and runtime of a unit operation [64]. A variation of the theme is fiber-based monolithic columns. One advantage of this approach is that these monoliths could be generated separately before pecking them into columns at scales suitable for large-scale processes [65].

5.2.3.4.5 Mixed Matrix Membranes

As the name already indicates, mixed matrix membranes (MMM) contain several materials that together contribute to versatile characteristics of these membranes. One of the early examples consisted of the incorporation of ion exchange resins into a porous ethylene and vinyl alcohol (EVAL) copolymer membrane. This setup of integrating particles into porous polymers avoids typical issues of traditional chromatography like pressure drop and flow rate limitations while maintaining high protein binding capacity and salt tolerance [66]. MMM materials have also been applied in the field of dialysis to remove toxins. In that case, activated carbon particles were embedded between the macroporous cellulose acetate [67]. However, the most advanced MMM solution relies on the combination of a flexible polymer and porous functionalized polyacrylamide hydrogel. This version of MMM with a dense body exhibits high binding capacity to large molecules at high flow rates and is commercially available as prepacked cartridge for industrial-scale protein purification purposes. Overall 40-fold faster processing speed when compared with traditional anion exchange chromatography could be achieved [68]. Just recently a novel type of MMM based on a metal–organic framework (MOF) enclosed in high-molecular-weight polyethylene network was described. The mixture of micro-channels and porous MOF results in fast permeation, rapid adsorption, and high separation power that can discriminate proteins on charge differences (Wang et al. 2019).

5.2.3.4.6 *Flow-Through and Tandem Chromatography*
The combination of two chromatography steps in a tandem stetting clearly fulfills the criteria for process intensification. Here two cases can be distinguished. On the one hand, this could mean the continued process of the first eluate without any collection or conditioning to the next column; on the other hand, even further intensification by avoiding binding and elution and selecting conditions that allow a flow-through process forms the first to the next column. Herby, the eluate of the first column must be compatible to serve as load for the second column without any collection or conditioning step in between. This requires process conditions that enable binding of impurities while passing the protein of interest directly to the next column. Flow-through approaches work without the need of intermediate hold tanks and accept smaller column sizes, as only impurities should bind, which represent just a fraction of whole sample. Therefore, the required column dimensions are in the range of prepacked columns, even for large-scale commercial processes.

In a small-scale example, the eluate of a protein A column was directly neutralized by a second pump and immediately loaded onto an anion exchange column that only bound impurities and let the antibody pass through it in a flow-through mode. The virus inactivation was moved upstream to be performed with detergent after the removal of the cells as there was no low-pH virus inactivation with a constant incubation time possible. This approach eliminated the need of an intermediate hold tank and all activities around it [69].

A similar approach evaluated the coupling of two polishing steps in a flow-through mode. Here the second and third column were coupled in a way to skip pool holding tanks by introducing in-line conditioning of the flow through. This principle is applicable to combine anion exchange with either hydrophobic interaction or mixed-mode chromatography. The pool-less strategy reduces processing time and eliminates the need of holding tanks [70].

Even product-related impurities such as aggregates have been successfully eliminated in a flow-through mode. By varying pH values, conditions could be identified that allowed unrestricted passing of the pure antibody through the column while high-molecular-weight multimers bound to the stationary phase [71].

Even flow-through coupling of three columns is possible. This was demonstrated with activated carbon, cation, and anion chromatography. The starting material for these studies was virus inactivated and neutralizes protein A eluate. At least between both ion exchangers, in-line dilution and pH adjustment were required. Besides the expected removal of HCP and DNA, also higher molecular aggregates of the target antibodies were reduced [72]. Further DoE helped to find common conditions that were amenable to all three columns to avoid any additional adjustment of the respective load and allowed the simultaneous equilibration of all columns. This enabled the direct coupling of all columns to save one third of process time and 95% of buffer consumption. The result of the output of this integrated approach with more than 95% monomer and very low residual HCP and DNA fully corresponds to expected specification of a typical antibody production process. Besides the obvious benefits that also include footprint reduction, some regulatory challenges such as monitoring, sampling, and analytical control remain to be solved [73].

5.2.3.4.7 Mixed-Mode Chromatography

Traditional bioprocesses to purify recombinant proteins contain several chromatography steps that separate the desired product from impurities by different modes. This often involves affinity, charges, or hydrophobicity. Instead of doing these unit operations sequentially in a row, several separation principles can be combined, thus fulfilling the requirements for process intensification. These approaches are collected under the term multimodal or mixed-mode chromatography (MMC) and have been seen growing interest over the last decade. An early and widely applied example of MMC is hydroxyapatite that combines the properties of electrostatic and metal affinity in one resin. Frequently this involves the integration of hydrogen bonding and electrostatic and hydrophobic interactions [74].

Ideally the different functionalities are present on a single ligand supported by a scaffold to simplify manufacturing of the stationary phase but also to allow multimodal binding. The focus on small organic ligands will also allow the inclusion of MMC into monoliths or fiber/filer-based chromatography that are amenable to higher flow rates. Besides the abovementioned adsorption mechanisms, also thiophilic interaction could be relevant in the context of antibody purification [75].

Straightforward separation by ion exchangers based on charges only can easily be optimized by varying pH and ion strength of the solution to induce binding or elution. In the case of MMC, process development becomes more complicated. As the effects are superimposing, DoE must be applied that benefits from automation and miniaturization to establish a solid basis for the purification strategy [76]. In most cases, a systematic evaluation of the influence of pH, organic compounds, ion strength, salts, and buffer types is worthwhile. It is now understood that pH affects the surface charge distribution while conductivity impacts the balance between electrostatic and hydrophobic interactions. Interestingly MMC is also susceptible to the dielectric constant of the mobile phase, which allows its exploitation for separation purposes as well. One of the major advantages of MMC is its ability to bind proteins at their pI under quasi neutral conditions, thus avoiding aggregation [77]. A further benefit is the ability to bind proteins in the presence of salt, which allows a direct coupling of chromatography steps without intermittent adjustment of conditioning steps. This approach was demonstrated when protein A eluate containing 400 mM salt and PEG to suppress aggregate formation could be directly loaded onto an MMC column [78].

Besides the aspect of process intensification, MMC is ideally suited to address challenges arising from the new complex therapeutic proteins. For instance, the class of bispecific antibodies suffers from the problem on how to eliminate product-related impurities that are difficult to differentiate from the desired target protein as their properties are highly similar. In a case study about asymmetric bispecific antibodies, typical byproducts such as half-antibodies, symmetric homodimers, or aggregates could be successfully eliminated in a single MMC step that included optimized washing procedures [79].

In addition to adsorber functions on the surface of MMC, also some kind of size exclusion can be applied to separate protein mixtures. Here the porous core is covered by a multifunctional ligand and embedded in a non-functionalized shell. The pore size exclusion limit of approximately 700 kDa keeps larger entities out while allowing binding of smaller proteins in the inner core. This is quite useful for the purification of viruses [80].

5.2.3.5 Tangential Flow Filtration
5.2.3.5.1 Single-Pass TFF

Tangential flow filtration (TFF) also known as crossflow filtration can be distinguished from direct flow filtration by the direction of the feed stream. In the case of TFF, the flow passes in parallel to the membrane face to avoid the buildup of a restrictive layer on top of the membrane. One portion passes through the filter, representing the permeate, while the remaining volume, the retentate, is recirculated back. The nature of TFF requires two tanks, the retentate tank that contains the starting and the recirculated material and the permeate vessel that is filled by the solution passing the filter. TFF can be used to concentrate or desalt solutions, and to fractionate according to size, as the pores of the membranes are categorized by molecular weight cutoff.

In most cases, normal TFF can be replaced by single-pass tangential flow filtration (SPTFF) that allows concentration of solutions without the two conventional vessels and associated piping and cleaning installations. The basic principle of SPTFF is to modulate the ratio between permeate and feed flow, thus influencing the concentration factor of the solution. This is done by increasing the residence time, which is achieved by either reducing the flow rate or extending the flow path by arranging a number of TFF modules in series. Simply, the higher the desired end concentration, the more filter cassettes needed [81].

In principle SPTFF can be applied at all stages benefiting from concentration, like shortening load times of chromatography by reducing the volume. Furthermore, it can eliminate the need for tanks or fit processes into facilities where tank sizes are not sufficient. In one example SPTFF decoupled USP from DSP by in-line concentration of the cell-free supernatant [82]. If the SPTFF-concentrated cell culture broth is stored, this can be regarded as complete segregation between USP and DSP. This setup was demonstrated by a 15- to 25-fold concentration of the harvest volume and subsequent storage at −40° or −70 °C in single-use bags [83].

Alternatively, SPTFF was employed to connect two chromatography unit operations by in-line concentrating up to 10-fold, reaching more than 100 g/l. Despite these excellent results, it must be taken into account that filtrate flux per filter area is less efficient in SPTFF. It also should be noted to perform sufficient process characterization to properly control the transition from conventional to SPTFF [84]. A similar effect was demonstrated by performing SPTFF in an antibody polishing step, reducing the volume for the consecutive anion exchange flow-through chromatography. Interestingly the eightfold concentration improved HCP clearance while maintaining virus clearance. As there was no big pressure drop, both combined operations could be driven by a single pump which that to savings in investment and facility footprint. The coupling additionally reduces column size and buffer volumes needed for equilibration [85].

5.2.3.5.2 In-line Diafiltration
Although the ideal concept of a DSP is the total absence of non-value-adding steps like conditioning and adjustment, sometimes ultrafiltration/diafiltration (UF/DF) cannot be avoided, particularly at the end to establish the proper buffer matrix for long-term storage of the bulk drug substance (BDS). Buffer is exchanged in that process by continuously adding the new buffer at the same rate as the previous buffer is removed. Typically, at least sixfold the volume of the initial buffer volume is used as DF buffer to replace the initial buffer composition and to obtain a 99.75% buffer exchange.

Recently an approach was described to integrate single-pass diafiltration (SPDF) into a fully continuous antibody manufacturing process. Here three SPTFF units were linearly coupled together via surge tanks into which the DF buffer was pumped to dilute the concentrate of the previous step to maintain the initial protein concentration. The high buffer consumption can be reduced either by increasing the number of SPTFF stages or by concentrating the solution prior to each DF as much as possible. A 4-stage SPTFF has better buffer efficiency than the original UF/DF with a retentate tank but is a more complex and costly system. The 3-stage setup is ideal for aggregation-sensitive molecules, as the number of pump passages is reduced and the exposition to stirring is limited but consumes 4 times the buffer volume of a conventional DF process [86].

The approach with 3-stage SPTFF was further refined by a countercurrent strategy to save DF buffer volumes while still achieving a 99.9% target buffer exchange. During 24 hours of operation, more than 12 kg of antibody per m^2 could be formulated [87].

Just recently a different setup was tested at small scale. The basis of this new single-pass module is a 3D-printed two-membrane construction with orthogonal flow for concentration and diafiltration streams. In a continuous mode, this design could reduce the salt concentration to less than 50% while concentrating proteins more than fourfold in a flow path of only 5 cm. However, this design causes concentration polarization, diminishing membrane permeability. Nevertheless, its simple design could be useful for small-scale applications [88].

The current commercial solution for in-line depth filtration (ILDF) consists of six SPTFF modules with sequential feed dilution and concentration, achieving a 99.9% buffer exchange with just about double the traditional buffer volume.

5.2.3.6 Drug Substance Freezing

At the end of a manufacturing process is the filling of the BDS into primary containers for long-term storage before the final drug product generation. Traditionally the protein-containing final formulation is transferred into multiple 1–5 l bottles. Usually the BDS is then frozen, which takes up to 20 hours particularly for larger bottles with wider diameters. The filling process is sequential and quite time consuming as the fill volume is individually verified on balances. Large-volume storage solutions still frequently rely on stainless steel cryovessels that have their own limitations in the form of high investment and maintenance costs, complex cleaning processes, and efforts for validation. Therefore, they are not considered in this comparison.

In the context of process intensification, it would be useful to streamline the fill and freeze operation. Firstly increasing the volume of the container and secondly minimizing the diameter of the container would save time in filling and freezing.

5.2.3.6.1 Freezing in Bottles

A straightforward approach to parallelize the bottle filling while also introducing a completely closed system is to generate single-use manifolds that can be connected to bottles. All elements are pre-assembled and y-irradiated and ready to use. Instead of a semi-manual process, this is now fully automated and improves standardization while reducing the human error risk [89]. However, bottles might not be the ideal solution anyhow, even when manual handling is eliminated as, due to the large diameter, concentration gradients occur, both during freezing and thawing. This effect could induce approximately threefold higher concentrations in the center, depending on the liquid depth [90]. Additionally, the air layer in the head space isolates the rest of bottle causing nonuniform freezing kinetics.

5.2.3.6.2 Freezing in Bags

As many unit operations both in USP and DSP are nowadays performed in bags, it is quite obvious to apply the same principle also for BDS storage. There are several advantages associated with the transition from bottle to bag fill. Bags allow a larger fill volume, and due to their flat geometry and the direct contact to large cold surfaces without any air cushion in a controlled rate plate freezer, freezing is completed faster and homogenously. Despite the low heat transfer capabilities of the bag material, this setup avoids the buildup of concentration gradients in the vessel. Even with a plate-freezing system and an optimal temperature gradient, it still will take several hours to lower the temperature by 40 °C in a 20 l bag.

Many of the usual advantages for single-use devices are true for bag fill and freeze as well. For instance, freezing in bags increases flexibility, reduces risk and validation efforts, and requires less infrastructure and peripheral equipment. However, one complication arises from the exposure of the plastic to low temperature; the material becomes less flexible and more prone to fractures and leaks. This sensitivity can partly be addressed by secondary covers such as metal shells for storage and transportation [91].

In contrast to stainless steel cryovessels, the shipment is unidirectional. Moreover, after emptying the bags, they are discarded, thus eliminating the cost for reverse logistics and cleaning. Nevertheless, the whole transport logistics chain has to be validated, including stability studies, temperature profiles, or mechanical shock and vibration resistance. Nowadays prequalified shippers for frozen single-use bags are available, reducing this effort [92].

Recently industry association performed a comprehensive analysis of prerequisites for freezing and thawing in bags. They looked at parameters for materials, container design, filling and thawing operations, storage, and transportation. Their conclusion was that these requirements are fulfilled by the currently available large-scale single-use solutions; however a close collaborative dialogue between supplier and user must be established [93].

5.2.3.6.3 Freezing in Containers

The next step beyond bags is the freezing in semirigid but still disposable plastic containers. They benefit from the inherent stability of the material that allows transportation without additional shell, simplifying handling and reducing the cost. The more than 6 mm-thick material is less prone to damage or leaks than normal bags. Due to the rectangular format, the containers are saving storage space and fit to standardized pallet formats. The so-called Cryovault[3] is available in several volumes and allows freezing of up to 75 l in one vessel and shipment of up to 300 l on one pallet.

5.3 Process Analytical Technology and Sensors

Process intensification is certainly heavily influenced by the practical operations on the shop floor. However, it should be noted that process analytical technology (PAT) can contribute a great deal in increasing the productivity. Looking at the single-use aspect in that respect, a wide range of disposable sensors has to be mentioned. In traditional batch monitoring, many in-process controls are performed by manual sampling and offline analysis that is time consuming and sometimes error prone. Intensified or continuous processes require real-time monitoring with in-line measurements [94].

The holy grail for process intensification from a PAT perspective certainly is real-time release testing (RTRT). This becomes possible by integrating measured critical quality attributes (CQA) that reflect the quality target product profile (QTPP)

---

3 https://www.meissner.com/products/cryovault-freeze-thaw-platform.

with process control. The starting point is the identification of critical control points (CCP) of the product quality attributes (PQA). Implementing these principles should allow corrective actions in real time but depends on well-established process understanding. In the end, RTRT should increase quality assurance, reduce process cycle times, minimize end product testing, shorten release times, and thus decrease inventory cost. Successful RTRT requires validated models for the different unit operations with appropriate control systems integrated into a holistic process wide control. Currently the most frequently applied approach is to conduct processing within the boundaries of the design space for critical process parameters (CPP). As this only allows feedforward control, some other strategies have to be employed to enable feedback control as well. Unfortunately, some CQA such as the absence of viral or microbial content are currently not amenable to in-line control and therefore might dictate the duration of the release procedure. Other parameters such as glycation can be steered by feedback control for glucose concentration based on Raman spectroscopy. Despite the many advantages, an RTRT strategy bears the risk that failed RTRT tests cannot be replaced by a successful end product release [95].

Ideal characteristics of single-use sensors are their ability to be sterilized, their compatibility to physical and chemical process conditions, their coverage of measuring range and accuracy, and their compliance to good manufacturing practice (GMP) guidelines. With respect to costs it might be reasonable to separate the disposable sensing element from the non-disposable measuring and transmitting part. However, this might complicate calibration and integrity testing. Pre-calibration at the vendor site is certainly desirable, as long as it can be verified at the time of use; alternatively offline calibration on-site is preferable. The ideal sensor is sterilized, calibrated, and ready to use. Some sensor types are amenable to noninvasive measurement, for instance, flow meters based on ultrasound that can simply be clamped onto tubing. Other optical devices require fiber optics connecting the emitter and detector to the standalone spectrometer, for instance, measuring UV absorption or near-infrared signals [96].

5.3.1 Sensors for USP Applications

In the context of USP, the sensors are primarily attached to the bioreactor to deliver critical process information such as temperature, pH, dissolved oxygen, carbon dioxide, and vessel pressure. These process parameters are quite standard, but new initiatives are additionally focusing on the chemical environment in the bioreactor, capturing real-time in-line date on substrate or product concentrations.

Sensors can be either integrated into the bioreactor and sterilized together with the bag or as optical sensor connected through port or a specific adaptor. Fully integrated sensors belonging to the class of electrochemical sensor or chemo- or biosensors can measure a wide range of parameters, including pH, metabolites (glucose, glutamine, or lactate), or gases such as O_2 or CO_2. Interestingly biosensors with immobilized enzymes or ligands are particularly suited to single-use reactors, as they withstand γ-irradiation, but not steaming. Radio-frequency identification (RFID) transponders are also applicable here, as their signal is not shielded by a metal wall as in the case of steel reactors. The class of optical sensors relying on specific windows in the bioreactors can measure pH, O_2, or CO_2 but suffers from a slow response time. Other concerns are signal drift or lot-to-lot variability. In the case of spectroscopic sensors, specific requirements are the absence of reflection, absorbance, or filtering. Interestingly, Raman spectroscopy that can capture a very wide range of analytes seems feasible with that approach as light scattering takes place at 180°. Impedance spectroscopy to measure biomass concentration, cell viability, and size is also applicable in a disposable format [97]. The applicability of capacitance measurements in single-use bioreactors was demonstrated during a scale-up from 50 to 2000 l. Here the critical process parameters such as viable cell concentration, viable cell volume, and wet cell weight could be measured simultaneously and continuously to compare the values in several reactors and to verify a successful scale-up. The precision of this approach was comparable with offline analysis and independent of scale [98].

In an ideal world the bioreactors are equipped with a range of sensors that allow a continuous monitoring of process and product parameters *in situ* to control process performance attributes and product quality by automatic feedback.

5.3.2 Sensors for DSP Applications

The typical sensors for DSP cover flow, pressure, temperature, turbidity, UV absorbance, pH, and conductivity. There is some overlap with sensors used in an USP context. All of them are available as disposable versions.

Removing cells and debris at the transition from USP to DSP by DF requires control of an important process parameter, the pressure. The performance of the DF is equivalent to the pressure differential across the filter. Key variables like flow rate or particle load heavily influence the efficiency of the DF step. Standard pressure gauges inserted in stainless steel piping suffer from the need to calibrate and clean them before use. Therefore, it is much easier to use pre-calibrated single-use pressure sensors. In a fully automated setting, the pressure readout could also be used for feedback control of the flow rate [99].

Besides filtration, the other significant unit operation in DSP is chromatography. Here the critical parameters are flow, pressure, pH, conductivity, UV absorption, and turbidity. Following UV profiles during elution is a critical prerequisite for peak cutting activities to select the correct fraction. Usually UV absorption measurement is an inbuilt feature of chromatography skids. However sometimes an additional UV readout at different positions in the process flow could beneficial. Here single-use UV detectors come into play. They consist of a flow-through cuvette and fiber optics that couple it to the photometer. Currently only single-wavelength detection is possible. The system can easily be integrated into tubing manifolds and can be sterilized by γ-irradiation and sanitized by NaOH [100]. Turbidity sensors that measure the particle load of a solution are quite useful for the applications that might create aggregates. Recently disposable turbidity detectors were employed in a column-less chromatography system called continuous countercurrent tangential chromatography (CCTC). Here the unpacked resin moves freely through hollow fibers to simulate the steps, load, wash, elute, and regenerate in separate but interconnected cartridges. The turbidity sensor steers closing of valves to direct the flow between circulation and forward movement in the stream [101].

As alternative to the conventional concentration determination of proteins by their UV absorbance, the same information and more details can be gathered by Raman spectroscopy. In addition to the concentration, this analysis can also differentiate between antibody isoforms. Interestingly this even works in crude samples such as cell-free bioreactor supernatant from perfusion cultivations. This could be an important addition to the arsenal of DSP in situ monitoring tools [102].

The usual last step before BDS freezing is the formulation in the final storage buffer. This is done by UF/DF that requires a clear control of multiple critical parameters. Recently multi-attribute PAT was established for that step in lab scale. The system consisting of a variable path length UV photometer, light scattering detector, and densitometer was configured to allow simultaneous in-line assessment of protein concentration, apparent molecular weight, hydrodynamic radius, and degree of buffer exchange. Unfortunately none of these sensors currently exists as single-use version, yet [103].

5.4 Conclusions

In summary, there are multiple ways to benefit from process intensification approaches that focus on single-use solutions (Table 5.4). On a high level, the abovementioned examples can be classified into several approaches that create productivity gains by different means. A special case is the targeted enhancement of phenomena that rely on chemical or physical principles. An overview on the classification of all approaches can be seen in Table 5.5. It can further be differentiated between intensification based on equipment or methods and the savings that can be generated.

Process intensification can be implemented in most processes but should be evaluated already during the development. This is particularly true where intensification implies process changes or impacts the number of cell generations. In USP the highest efficiency gain results from improving the seed/inoculation train and a step elimination strategy. Process intensification in DSP allows more options, for instance, choosing between different chromatography setups. The major benefit in DSP can be obtained by combining operations such precipitation with filtration or flow-through membrane chromatography with SPTFF. Additionally, disposables have a huge impact on intensification by eliminating labor or preparation time and enabling the uncoupling of activities.

5.4.1 Transition from Traditional to Intensified Processes

Currently we experience the coexistence of several parallel bioprocess universes. On the one hand, there are still many legacy products with old-fashioned processes; on the other hand, we can find a few modern examples adopting full continuous

Table 5.4 Classification of single-use process intensification approaches.

Intensification approach	Equipment (single use)	Method	Combination	Enhancement	Frontloading	Uncoupling	Elimination	Simplification
High density large volume cell bank	x		x		x	x		x
Wave bag expansion	x		x				x	x
N-1 perfusion	x	x			x			
ATF/TFF	x		x					
Hydrocyclone	x		x		x			x
Floating filter	x	x	x					x
Single use centrifuge	x		x				x	
Acoustic wave settler	x		x		x			
Compact round settler	x		x		x			
Depth filtration	x		x				x	
Filter aids		x	x		x			
Flocculation or Precipitation		x	x	x				
In line virus inactivation	x	x	x					x
in line buffer blending and dilution	x	x				x		
Prepacked columns	x	x					x	
Monoliths	x			x			x	
Membrane absorbers	x						x	
Filter cartridges	x						x	
Mixed matrix membranes	x		x					
Flow through chromatography		x					x	x
Mixed mode resins		x	x					
SPTFF or SPDF	x							x
Bio-capacitance or Raman sensor		x	x	x				
Freezing in bottles	x	x						x
Freezing in bags	x						x	x
Freezing in containers	x						x	x

Table 5.4 (Continued)

	Savings				
Time	Footprint	Hardware	Labor	Material	Comment
x		x	x		skip manual expansion steps
x	x	x	x	x	fewer manual steps
x			x		higher cell density to start main reactor
	x				might tend to clogging
	x		x		very small and simple device
	x			x	integrated in wave bags
x					no CIP or SIP needed
	x		x		enables debris and dead cell removal
	x		x		enables debris and dead cell removal
x			x	x	hybrid filters with charged membranes
x			x		early reduction of impurities
x			x		focus on specific impurities
	x	x			might require continuous feed stream
x	x	x	x	x	buffer dilution is easy to implement
x		x	x		elimination of non-value adding steps
x			x		no packing, high flow rates, large molecules
x			x		no packing, high flow rates, large molecules
x			x		no packing, high flow rates, large molecules
x			x		two purification steps combined
x	x	x	x	x	elimination of conditioning and washing
x	x	x	x	x	two purification steps combined
	x	x	x		elimination of vessels
				x	enables feed-back controlled feeding
x					performing filling in parallel
x	x		x		performing filling in parallel
x	x		x	x	fewer manual steps

Table 5.5 Reported time savings by intensification in USP.

USP process step	Time savings (days)	References
Large-volume high-density cell bank	25–39	[9]
Wave bag cell expansion	9	[12]
N-1 perfusion	5–6	[12]

processing. In between, the majority of processes apply to some extend the principles of process intensification representing a transition between both extremes. Particularly the conversion to more intensification offers opportunities for existing facilities, for instance, to close volumetric gaps of available and preinstalled vessels by replacing conventional TFF steps by SPTFF to eliminate the need of a retentate tank or to switch to flow-through chromatography and in-line virus inactivation. Other options to increase throughput in old plants are the introduction of buffer dilution skids that can multiply the capacity of existing buffer preparation capabilities or the utilization of wave bags in the inoculum instead of shake flasks. Furthermore, starting with a higher cell density and larger volume of a cell bank can easily be implemented.

However, the implementation of N-1 perfusion is more difficult to achieve due to the consumption of larger media volumes and the installation of cell retention devices, although the productivity gain would be substantial.

Gradual transitions should start with intensification approaches that do not require process changes, like buffer preparation and distribution or replacing self-packed by prepacked columns. In the context of chromatography, substitution of resins is certainly more complicated as this might impact product CQA and requires careful process characterization.

The transition to continuous processing as the ultimate goal for intensification can easily be achieved for the USP part as perfusion cultivation is known for decades. In DSP the situation is a bit different as full continuous processing would require a skid setup with multicolumns with more complicated technology.

When trying to integrate unit operations, three approaches have been described. The modular strategy utilizes enabler technologies such as SPDF to connect two separate unit operations into one continuous flow. The second case requires the adaptation of one or two process steps to obtain operating ranges that overlap sufficiently to combine both steps, for instance, by introducing salt tolerant mixed-mode resins that accept a high conductivity feed stream of a previous step. The third possibility is to merge multiple unit operations into one, such as charged depth filters instead of depth filtration and ion exchange [104]. An overview of combined approaches can be seen in Figure 5.12.

Facility retrofitting to increase productivity might often be limited by the capacity of the installed utilities systems. For instance, existing water for injection (WFI) supply might be too small to install additional steel bioreactors or tanks but could be sufficient when increasing the potential output by installing more single-use vessels that do not require large WFI volumes for cleaning in place (CIP). This also holds

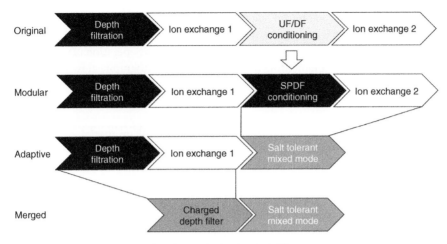

Figure 5.12 Process intensification optimization in DSP.

true for changing to buffer prep and hold in disposable tanks. Taking into account potential footprint reductions due to the process intensification, enough space could be freed to expand the BDS storage area as well [105].

A stepwise USP approach toward continuous manufacturing was described recently. Starting from a traditional 15 000 l stainless steel bioreactor with a culture duration of 12 days and a volumetric productivity of 0.2 g/l/day, the same output could be achieved by increasing the volumetric productivity to 1.5 g/l/day and reducing the working volume of the bioreactor to 2000 l, which would be suited to a single-use version. Further footprint reduction at identical output is possible by extending the cultivation time to 48 days in continuous manufacturing while diminishing the bioreactor to just 500 l [106].

5.4.2 Impact on Cost

The known cost benefit of disposables, mainly resulting from the reduction in clean utility installations, disappears at a certain batch number threshold, as then recurring consumable costs become dominating. Further cost savings beyond that become only possible if the consumable costs can be controlled. There are two options. On the one hand, the productivity of the single-use system could be increased, which at least for USP is limited as too high cell densities suffer from insufficient oxygen supply. On the other hand, the disposable elements could be utilized longer. Both the extended use and perfusion conditions that allow high cell densities are features of continuous processing. A few case studies evaluate the cost benefits of continuous processing, and process intensification is a critical stepping stone to get there.

Partial process intensification at defined steps can have a significant impact on overall costs as discussed for antibody manufacturing. For instance, introducing N-1 perfusion increased titers to fourfold. Using high-capacity resins, switching to flow-through instead of bind/elute chromatography for polishing and applying multicolumn chromatography for capture led to lower buffer consumption and smaller columns. Overall costs were diminished by a factor of 6.7–10.1 [15].

The transition from a conventional batch-based four-unit operations DSP to continuous processing by implementing multicolumn chromatography at two positions resulted in a 4 to 6-fold increase of volumetric productivity. Therefore the size of the stationary phase and consequently the buffer volume could be drastically reduced as well [107].

Another study evaluated the business impact of a full or partial continuous process instead of a conventional process. Based on economic modeling of operations and capital investment, net present values for the different platforms were established and compared. The fully integrated solution reduced costs on a ten year perspective by 55% [108]. Another study analyzing the costs for fully integrated continuous processing with a traditional bioprocess revealed a 4.6-fold productivity improvement. However, in theoretical modeling of scaling up both processes into either $4 \times 12\,500\,l$ stainless steel reactors or $5 \times 2000\,l$ disposable vessels, the cost advantage based on identical product output just reached 15% for the continuous disposable solution [109].

Cell culture media is the biggest cost contributor to intensified USP. Focusing only on this parameter, it was found out that the media costs per gram antibody are comparable for fed-batch and perfusion processes, although the bioreactor productivity was approximately fivefold higher for the perfusion mode [23].

Often manufacturers are challenged to decide between stainless steel and disposable solutions. On the one hand, the lower investment and faster installation outweigh traditional facilities. This is even more prominent if perfusion is applied. This means in many situations single-use solutions are absolutely competitive [110]. Despite all these economic benefits, there are still some technical limitations that restrict the application of disposables. Two often occurring issues are certainly the scalability and safety risks such as leaks [111].

5.4.3 Influence on Time

In the numerous examples mentioned above, very often time savings are reported. This is particularly obvious in the USP part, as most of the strategies are directed to increase volumetric productivity, in other word to generate more output in the same time. Typical time savings in the USP intensification approaches are summarized in Table 5.5. Combining all three proposed steps from that list results in a significant reduction of process time.

The approaches for USP can be summarized under "more cells faster" to reduce the overall process time. Current facilities typically have up to six bioreactors feeding into just one DSP. This is fully dependent on the different duration of the processes in USP and DSP, where USP frequently takes 5–6 times longer than DSP. The more process duration in USP can be reduced, the closer USP and DSP reach a relation that is typical for continuous processes. Even if the end result is not a continuous process, a relation below 6 : 1 would require less investments into USP hardware, clearly saving space and cost. On the other hand, the optimization of DSP processes often is not easy quantifiable and does not necessarily reduce runtime as drastically as in USP, but also contributes to decreasing hardware investments in the form of large tanks for intermediates or buffers.

Nevertheless, besides reduction in process time, the elimination of non-value-adding activities saves time. In the case of single-use materials, this is primarily related to skipping any cleaning and sanitization efforts. Unfortunately, this is not always true, as buffer preparation and distribution in mobile single-use containers add labor that does not add any value. This extra time is partly recovered if buffer blending or dilution is applied, and smaller volumes need to be prepared.

The last decade alone brought significant progress to the field of process intensification that rises expectations by the manufacturers for the years to come. This is of particular interest as the number of approved biological drugs is growing and simultaneously price sensitivity at insurance companies is increasing as well. It will be exciting to be able to contribute to the future of bioprocessing under these circumstances.

References

1 Stankiewicz, A.I. and Moulijn, J.A. (2000). Process intensification: transforming chemical engineering. *Chem Eng Prog.* 96 (1): 22–34.
2 Whitford, B. (2020). Bioprocess intensification: aspirations and achievements. *Biotechniques* 69 (2): 85–88.
3 Bielenberg, J. and Bryner, M. (2018). Realize the potential of process intensification. *Chem Eng Prog.* 114 (3): 41–45.
4 Keil, F.J. (2018). Process intensification. *Rev Chem Eng.* 34 (2): 135–200.
5 Shukla, A.A., Mostafa, S., Wilson, M., and Lange, D. (2012). Vertical integration of disposables in biopharmaceutical drug substance manufacturing. *Bioprocess Int.* 10 (6): 34–47.
6 Schmidt, S.R. (2016). The benefits and limits of disposable technologies in manufacturing protein therapeutics. *Am Pharm Rev.* 19 (5): 60–62.
7 Kiesewetter, A., Menstell, P., Peeck, L.H., and Stein, A. (2016). Development of pseudo-linear gradient elution for high-throughput resin selectivity screening in robocolumn(r) format. *Biotechnol Prog.* 32: 1503–1519.
8 Ninomiya, N., Shirahata, S., and Murakami, H. (1991). Large-scale, high-density freezing of hybridomas and its application to high-density culture. *Biotechnol Bioeng.* 38 (9): 1110–1113.
9 Heidemann, R., Mered, M., Wang, D.Q. et al. (2002). A new seed-train expansion method for recombinant mammalian cell lines. *Cytotechnology.* 38 (1–3): 99–108.
10 Seth, G., Hamilton, R.W., Stapp, T.R. et al. (2013). Development of a new bioprocess scheme using frozen seed train intermediates to initiate CHO cell culture manufacturing campaigns. *Biotechnol Bioeng.* 110 (5): 1376–1385.
11 Singh, V. (2005). *The Wave Bioreactor Story*, 1–18. Somerset: Wave Biotech LLC.

12 Tao, Y., Shih, J., Sinacore, M.S. et al. (2011). Development and implementation of a perfusion-based high cell density cell banking process. *Biotechnol Prog.* 27 (3): 824–829.

13 Xu, J., Rehmann, M.S., Xu, M. et al. (2020). Development of an intensified fed batch production platform with doubled titers using N - 1 perfusion seed for cell culture manufacturing. *Bioresour Bioprocess.* 7 (17).

14 Yongky, A., Xu, J., Tian, J. et al. (2019). Process intensification in fed-batch production bioreactors using non-perfusion seed cultures. *MAbs* 11 (8): 1502–1514.

15 Xu, J., Xu, X., Huang, C. et al. (2020). Biomanufacturing evolution from conventional to intensified processes for productivity improvement: a case study. *MAbs* 12 (1): 1–13.

16 Padawer, I., Ling, W.L.W., and Bai, Y. (2013). Case study: an accelerated 8-day monoclonal antibody production process based on high seeding densities. *Biotechnol Prog.* 29 (3): 829–832.

17 Yang, W.C., Lu, J., Kwiatkowski, C. et al. (2014). Perfusion seed cultures improve biopharmaceutical fed-batch production capacity and product quality. *Biotechnol Prog.* 30 (3): 616–625.

18 Stepper, L., Filser, F.A., Simon, F. et al. (2020). Pre-stage perfusion and ultra-high seeding cell density in CHO fed-batch culture: a case study for process intensification guided by systems biotechnology. *Bioprocess Biosyst Eng.* 43 (8): 1431–1443.

19 Wright, B., Bruninghaus, M., Vrabel, M. et al. (2015). A novel seed-train process using high-density cell banking, a disposable bioreactor, and perfusion technologies. *Bioprocess Int.* 13 (3): 16–25.

20 Pohlscheidt, M., Jacobs, M., Wolf, S. et al. (2013). Optimizing capacity utilization by large scale 3000 L perfusion in seed train bioreactors. *Biotechnol Prog.* 29: 222–229.

21 Hecht, V., Duvar, S., Ziehr, H. et al. (2014). Efficiency improvement of an antibody production process by increasing the inoculum density. *Biotechnol Prog.* 30 (3): 607–615.

22 Clincke, M.-F., Mölleryd, C., Zhang, Y. et al. (2013). Very high density of CHO cells in perfusion by ATF or TFF in WAVE bioreactor™. Part I. Effect of the cell density on the process. *Biotechnol Prog.* 29 (3): 754–767.

23 Xu, S., Gavin, J., Jiang, R., and Chen, H. (2017). Bioreactor productivity and media cost comparison for different intensified cell culture processes. *Biotechnol Prog.* 33 (4): 867–878.

24 Whitford, W.G. (2015). Single-use perfusion bioreactors support continuous biomanufacturing. *Pharm Bioprocess.* 3 (1): 75–93.

25 Clincke, M.F., Mölleryd, C., Samani, P.K. et al. (2013). Very high density of Chinese hamster ovary cells in perfusion by alternating tangential flow or tangential flow filtration in WAVE bioreactor™. Part II: applications for antibody production and cryopreservation. *Biotechnol Prog.* 29 (3): 768–777.

26 Kelly, W., Scully, J., Zhang, D. et al. (2014). Understanding and modeling alternating tangential flow filtration for perfusion cell culture. *Biotechnol Prog.* 30 (6): 1291–1300.

27 Bettinardi, I.W., Castilho, L.R., Castan, A., and Medronho, R.A. (2020). Hydrocyclones as cell retention device for CHO perfusion processes in single-use bioreactors. *Biotechnol Bioeng.* 117 (7): 1915–1928.

28 Kelly, W., Rubin, J., Scully, J. et al. (2016). Understanding and modeling retention of mammalian cells in fluidized bed centrifuges. *Biotechnol Prog.* 32 (6): 1520–1530.

29 McLeod, L. (2009). The road to a fully disposable protein purification process: single-use systems eliminate time-consuming, non-revenue-generating activities. *Bioprocess Int.* 7 (6): 4–8.

30 Schmidt, S.R., Wieschalka, S., and Wagner, R. (2017). Single-use depth filters. *Bioprocess Int.* 14 (1i): 6–11.

31 Singh, N., Arunkumar, A., Peck, M. et al. (2017). Development of adsorptive hybrid filters to enable two-step purification of biologics. *MAbs.* 9 (2): 350–363.

32 Trapp, A., Muczenski, S., Igl, L. et al. (2018). Evaluating adsorptive filtration as a unit operation for virus removal. *Bioprocess Int.* 16 (3): 58–62.

33 Singh, N., Arunkumar, A., Chollangi, S., and Tan, Z.G. (2016). Clarification technologies for monoclonal antibody manufacturing processes: current state and future perspectives. *Biotechnol Bioeng.* 113 (4): 698–716.

34 Minow, B., Egner, F., Jonas, F., and Lagrange, B. (2014). High-cell-density clarification by single-use diatomaceous earth filtration. *Bioprocess Int.* 12 (4): 36–46.

35 Burgstaller, D., Krepper, W., Haas, J. et al. (2018). Continuous cell flocculation for recombinant antibody harvesting. *J Chem Technol Biotechnol.* 93 (7): 1881–1890.

36 Trapp, A., Faude, A., Hörold, N. et al. (2018). Multiple functions of caprylic acid-induced impurity precipitation for process intensification in monoclonal antibody purification. *J Biotechnol.* 279 (May): 13–21.

37 Kateja, N., Agarwal, H., Saraswat, A. et al. (2016). Continuous precipitation of process related impurities from clarified cell culture supernatant using a novel coiled flow inversion reactor (CFIR). *Biotechnol J.* 11: 1320–1331.

38 Burgstaller, D., Jungbauer, A., and Satzer, P. (2019). Continuous integrated antibody precipitation with two-stage tangential flow microfiltration enables constant mass flow. *Biotechnol Bioeng.* 116 (5): 1053–1065.

39 Hammerschmidt, N., Tscheliessnig, A., Sommer, R. et al. (2014). Economics of recombinant antibody production processes at various scales: industry-standard compared to continuous precipitation. *Biotechnol J.* 9: 766–775.

40 Martinez, M., Spitali, M., Norrant, E.L., and Bracewell, D.G. (2019). Precipitation as an enabling technology for the intensification of biopharmaceutical manufacture. *Trends Biotechnol.* 37 (3): 237–241.

41 Klutz, S., Lobedann, M., Bramsiepe, C., and Schembecker, G. (2016). Continuous viral inactivation at low pH value in antibody manufacturing. *Chem Eng Process Process Intensif.* 102: 88–101.

42 Orozco, R., Godfrey, S., Coffman, J. et al. (2017). Design, construction, and optimization of a novel, modular, and scalable incubation chamber for continuous viral inactivation. *Biotechnol Prog.* 33 (4): 954–965.

43 Parker, S.A., Amarikwa, L., Vehar, K. et al. (2018). Design of a novel continuous flow reactor for low pH viral inactivation. *Biotechnol Bioeng.* 115 (3): 606–616.

44 Martins, D.L., Sencar, J., Hammerschmidt, N. et al. (2019). Continuous solvent/detergent virus inactivation using a packed-bed reactor. *Biotechnol J.* 14 (8): 1–9.

45 Martins, D.L., Sencar, J., Hammerschmidt, N. et al. (2020). Truly continuous low pH viral inactivation for biopharmaceutical process integration. *Biotechnol Bioeng.* 117 (5): 1406–1417.

46 Senčar, J., Hammerschmidt, N., Martins, D.L., and Jungbauer, A. (2020). A narrow residence time incubation reactor for continuous virus inactivation based on packed beds. *N Biotechnol.* 55: 98–107.

47 Malone, T. and Li, M. (2010). PAT-based in-line buffer dilution. *Bioprocess Int.* 1: 40–49.

48 Fabbrini, D., Simonini, C., Lundkvist, J. et al. (2017). Addressing the challenge of complex buffer management. *Bioprocess Int.* 15 (11): 43–46.

49 Orr, V., Zhong, L., Moo-Young, M., and Chou, C.P. (2013). Recent advances in bioprocessing application of membrane chromatography. *Biotechnol Adv.* 31 (4): 450–465.

50 Grier, S. and Yakubu, S. (2016). Prepacked chromatography columns. *Bioprocess Int.* 14 (4): 48–53.

51 Scharl, T., Jungreuthmayer, C., Dürauer, A. et al. (2016). Trend analysis of performance parameters of pre-packed columns for protein chromatography over a time span of ten years. *J Chromatogr A.* 1465: 63–70.

52 Schweiger, S., Berger, E., Chan, A. et al. (2019). Packing quality, protein binding capacity and separation efficiency of pre-packed columns ranging from 1 mL laboratory to 57 L industrial scale. *J Chromatogr A.* 1591: 79–86.

53 Lalli, E., Silva, J.S., Boi, C., and Sarti, G.C. (2020). Affinity membranes and monoliths for protein purification. *Membranes (Basel).* 10 (1): 1.12.

54 Knudsen, H.L., Fahrner, R.L., Xu, Y. et al. (2001). Membrane ion-exchange chromatography for process-scale antibody purification. *J Chromatogr A.* 907: 145–154.

55 Madadkar, P., Sadavarte, R., and Ghosh, R. (2019). Performance comparison of a laterally-fed membrane chromatography (LFMC) device with a commercial resin packed column. *Membranes (Basel).* 9 (138): 1–12.

56 Ghosh, R. (2002). Protein separation using membrane chromatography: opportunities and challenges. *J Chromatogr A.* 952: 13–27.

57 Boi, C. (2007). Membrane adsorbers as purification tools for monoclonal antibody purification. *J Chromatogr B Anal Technol Biomed Life Sci.* 848 (1): 19–27.

58 Jacquemart, R. and Stout, J.G. (2017). Membrane adsorbers, columns: single-use alternatives to resin chromatography. *Bioprocess Int.* 14 (1i): 18–19.

59 Zhou, J.X. and Tressel, T. (2006). Basic concepts in Q membrane chromatography for large-scale antibody. *Biotechnol Prog.* 22: 341–349.

60 Riordan, W.T., Heilmann, S.M., Brorson, K. et al. (2009). Salt tolerant membrane adsorbers for robust impurity clearance. *Biotechnol Prog.* 25 (6): 1695–1702.

61 Weaver, J., Husson, S.M., Murphy, L., and Wickramasinghe, S.R. (2013). Anion exchange membrane adsorbers for flow-through polishing steps: Part I. Clearance of minute virus of mice. *Biotechnol Bioeng.* 110 (2): 491–499.

62 Marcus, R.K. (2009). Use of polymer fiber stationary phases for liquid chromatography separations: Part II – applications. *J Sep Sci.* 32: 695–705.

63 Hardick, O., Dods, S., Stevens, B., and Bracewell, D.G. (2013). Nanofiber adsorbents for high productivity downstream processing. *Biotechnol Bioeng.* 110 (4): 1119–1128.

64 Hardick, O., Dods, S., Stevens, B., and Bracewell, D.G. (2015). Nanofiber adsorbents for high productivity continuous downstream processing. *J Biotechnol.* 213: 74–82.

65 Ladisch, M. and Zhang, L. (2016). Fiber-based monolithic columns for liquid chromatography. *Anal Bioanal Chem.* 408: 6871–6883.

66 Avramescu, M.-E., Borneman, Z., and Wessling, M. (2003). Dynamic behavior of adsorber membranes for protein recovery. *Biotechnol Bioeng.* 84 (5): 564–572.

67 Saiful, S., Borneman, Z., and Wessling, M. (2017). Double layer mixed matrix membrane adsorbers improving capacity and safety hemodialysis. *Mater Sci Eng.* 352: 1–6.

68 Hou, Y., Brower, M., Pollard, D. et al. (2015). Advective hydrogel membrane chromatography for monoclonal antibody purification in bioprocessing. *Biotechnol Prog.* 31 (4): 974–982.

69 Shamashkin, M., Godavarti, R., Iskra, T., and Coffman, J. (2013). A tandem laboratory scale protein purification process using Protein A affinity and anion exchange chromatography operated in a weak partitioning mode. *Biotechnol Bioeng.* 110 (10): 2655–2663.

70 Zhang, J., Conley, L., Pieracci, J., and Ghose, S. (2016). Pool-less processing to streamline downstream purification of monoclonal antibodies. *Eng Life Sci.* 17 (2): 117–124.

71 Chmielowski, R.A., Meissner, S., Roush, D. et al. (2014). Resolution of heterogeneous charged antibody aggregates via multimodal chromatography: a comparison to conventional approaches. *Biotechnol Prog.* 30 (3): 636–645.

72 Ichihara, T., Ito, T., Galipeau, K., and Gillespie, C. (2018). Integrated flow-through purification for therapeutic monoclonal antibodies processing. *MAbs.* 10 (2): 325–334.

73 Ichihara, T., Ito, T., and Gillespie, C. (2019). Polishing approach with fully connected flow-through purification for therapeutic monoclonal antibody. *Eng Life Sci.* 19: 31–36.

74 Halan, V., Maity, S., Bhambure, R., and Rathore, A.S. (2019). Multimodal chromatography for purification of biotherapeutics – a review. *Curr Protein Pept Sci.* 10 (1): 4–13.

75 Pinto, I.F., Aires-Barros, M.R., and Azevedo, A.M. (2015). Multimodal chromatography: debottlenecking the downstream processing of monoclonal antibodies. *Pharm Bioprocess.* 3 (3): 263–279.

76 Cabanne, C. and Santarelli, X. (2019). Mixed mode chromatography, complex development for large opportunities. *Curr Protein Pept Sci.* 20 (1): 22–27.

77 Pezzini, J., Cabanne, C., Gantier, R. et al. (2015). A comprehensive evaluation of mixed mode interactions of HEA and PPA HyperCel™ chromatographic media. *J Chromatogr B.* 976–977: 68–77.

78 Zhang, Y., Cai, L., Wang, Y., and Li, Y. (2019). Processing of high-salt-containing protein A eluate using mixed-mode chromatography in purifying an aggregation-prone antibody. *Protein Expr Purif.* 164: 105458.

79 Tang, J., Zhang, X., Chen, T. et al. (2020). Removal of half antibody, hole–hole homodimer and aggregates during bispecific antibody purification using MMC ImpRes mixed-mode chromatography. *Protein Expr Purif.* 167: 105529.

80 Sánchez-Trasviña, C., Fuks, P., Mushagasha, C. et al. (1621). Structure and functional properties of Capto™ Core 700 core–shell particles. *J Chromatogr A.* 2020.

81 Casey, C., Gallos, T., Alekseev, Y. et al. (2011). Protein concentration with single-pass tangential flow filtration (SPTFF). *J Memb Sci.* 384 (1–2): 82–88.

82 Arunkumar, A., Singh, N., Peck, M. et al. (2017). Investigation of single-pass tangential flow filtration (SPTFF) as an inline concentration step for cell culture harvest. *J. Memb. Sci.* 524: 20–32.

83 Brinkmann, A., Elouafiq, S., Pieracci, J., and Westoby, M. (2018). Leveraging single-pass tangential flow filtration to enable decoupling of upstream and downstream monoclonal antibody processing. *Biotechnol. Prog.* 34 (2): 405–411.

84 Dizon-Maspat, J., Bourret, J., Agostini, A.D., and Li, F. (2012). Single pass tangential flow filtration to debottleneck downstream processing for therapeutic antibody production. *Biotechnol Bioeng.* 109 (4): 962–970.

85 Elich, T., Goodrich, E., Lutz, H., and Mehta, U. (2019). Investigating the combination of single-pass tangential flow filtration and anion exchange chromatography for intensified mAb polishing. *Biotechnol Prog.* 35 (5): e2862.

86 Rucker-Pezzini, J., Arnold, L., Hill-Byrne, K. et al. (2018). Single pass diafiltration integrated into a fully continuous mAb purification process. *Biotechnol Bioeng.* 115 (8): 1949–1957.

87 Jabra, M.G., Yehl, C.J., and Zydney, A.L. (2019). Multistage continuous countercurrent diafiltration for formulation of monoclonal antibodies. *Biotechnol Prog.* 35 (4): 6–11.

88 Tan, R. and Franzreb, M. (2020). Continuous ultrafiltration/diafiltration using a 3D-printed two membrane single pass module. *Biotechnol Bioeng.* 117 (3): 654–661.

89 Hutchinson, N. (2014). Understanding and controlling sources of process variation. *Bioprocess Int.* 12 (9): 24–29.

90 Kolhe, P. and Badkar, A. (2011). Protein and solute distribution in drug substance containers during frozen storage and post-thawing: a tool to understand

and define freezing-thawing parameters in biotechnology process development. *Biotechnol Prog.* 27 (2): 494–504.

91 Goldstein, A., Pohlscheidt, M., Loesch, J. et al. (2012). Disposable freeze systems in the pharmaceutical industry. *Am Pharm Rev.* 15 (7): 53–58.

92 Gentile, C. and Marciniak, M. (2017). Storing and shipping frozen APIs in single-use containers. *BioPharm Int.* 30 (6): 36–42.

93 Goldstein, A., Bourret, J., Barbedette, L. et al. (2019). Development of large-scale bulk freezing systems. *Bioprocess Int.* 17 (1–2): 32–36.

94 Rios, M. (2020). Developing process control strategies for continuous bioprocesses. *Bioprocess Int.* 18 (5i): 12–16.

95 Jiang, M., Severson, K.A., Love, J.C. et al. (2017). Opportunities and challenges of real-time release testing in biopharmaceutical manufacturing. *Biotechnol Bioeng.* 114: 2445–2456.

96 Furey, J., Clark, K., and Card, C. (2011). Adoption of single-use sensors for bioprocess operations. *Bioprocess Int.*: 36–42.

97 Busse, C., Biechele, P., de Vries, I. et al. (2017). Sensors for disposable bioreactors. *Eng Life Sci.* 17: 940–952.

98 Metze, S., Ruhl, S., Greller, G. et al. (2020). Monitoring online biomass with a capacitance sensor during scale-up of industrially relevant CHO cell culture fed-batch processes in single-use bioreactors. *Bioprocess Biosyst Eng.* 43: 193–205.

99 Bink, L.R. and Furey, J. (2018). Using in-line disposable pressure sensors to evaluate depth filter performance. *Bioprocess Int.* 16 (10s): 2–5.

100 Renaut, P. and Annarelli, D. (2018). Evaluation of a new single-use UV sensor for Protein A capture. *Bioprocess Int.* 16 (16/10s): 11–13.

101 Fedorenko, D., Tan, J., Shinkazh, O., and Annarelli, D. (2018). In-line turbidity sensors for monitoring process streams in continuous countercurrent tangential chromatography. *Bioprocess Int.* 16 (10s): 18–21.

102 Yilmaz, D., Mehdizadeh, H., Navarro, D. et al. (2020). Application of Raman spectroscopy in monoclonal antibody producing continuous systems for downstream process intensification. *Biotechnol Prog.* 36 (3): e2947.

103 Rolinger, L., Matthias, R., Diehm, J. et al. (2020). Multi-attribute PAT for UF/DF of proteins – monitoring concentration, particle sizes, and buffer exchange. *Anal Bioanal Chem.* 412 (9): 2123–2136.

104 Rathore, A.S., Kateja, N., and Kumar, D. (2018). Process integration and control in continuous bioprocessing. *Curr Opin Chem Eng.* 22: 18–25.

105 Kelley, B. (2007). Very large scale monoclonal antibody purification: the case for conventional unit operations. *Biotechnol Prog.* 23 (5): 995–1008.

106 Chen, C., Wong, H.E., and Goudar, C.T. (2018). Upstream process intensification and continuous manufacturing. *Curr Opin Chem Eng.* 22: 191–198.

107 Gjoka, X., Gantier, R., and Schofield, M. (2017). Transfer of a three step mAb chromatography process from batch to continuous: optimizing productivity to minimize consumable requirements. *J Biotechnol [Internet].* 242: 11–18.

108 Walther, J., Godawat, R., Hwang, C. et al. (2015). The business impact of an integrated continuous biomanufacturing platform for recombinant protein production. *J Biotechnol [Internet]* 213: 3–12.

109 Arnold, L., Lee, K., Rucker-Pezzini, J., and Lee, J.H. (2019). Implementation of fully integrated continuous antibody processing: effects on productivity and COGm. *Biotechnol. J.* 14 (2): 1–10.

110 Pollard, D., Brower, M., Abe, Y. et al. (2016). Standardized economic cost modeling for next-generation MAb production. *Bioprocess Int.* 14 (8).

111 Rogge, P., Müller, D., and Schmidt, S.R. (2015). The single-use or stainless steel decision process. *Bioprocess Int.* 13 (11): 10–15.

6

Single-Use Continuous Manufacturing and Process Intensification for Production of Affordable Biological Drugs

Ashish K. Joshi and Sanjeev K. Gupta

Advanced Biotech Lab, Ipca Laboratories Ltd., Plot No. 125, Kandivli Industrial Estate, Kandivli (West), 400067, Mumbai, India

6.1 Background

There has been a continuous flow of many new innovative and biosimilar drugs into the market. Hence, new doors are starting to open up in the healthcare industry, but at the same time competition is also rising at the same rate. To address the challenges and withstand the future market, it is important to design robust, cost-effective, and flexible manufacturing processes that ensure a continuous supply of quality drugs at affordable price. This is now possible with the implementation of process intensification and advancement in the technology that can be used to launch a wide range of products in the same facility, thereby making a small facility footprint, less investment, and risk. This approach ensures that according to the need of the hour, a particular product can be made available at a fast pace in efficient manner.

The high cost of drugs many times becomes a major bottleneck in the healthcare industry since a large population cannot afford a prescribed therapy. At the molecular level, a lot of innovations is going on that include a variety of recombinant monoclonal antibody (mAB), antibody–drug conjugates (ADCs), and gene therapies. In the case of targeted gene therapy, it is crucial and challenging that the intended drug should be personalized and effective for a variety of individuals. As a result of innovations and refinement of drugs that is called biobetter, healthcare industries are targeting specific patient population pool and provide constant supply as and when required.

Also, we must be able to cope up with the dynamic environment of the product pipeline in the healthcare industry to manage the capacity uncertainty as due to the competition between the companies, capacity planning becomes very difficult when we hit the market.

To address the issues of costly manufacturing infrastructure, it is better that one can choose the option of contract manufacturing where facility buildup cost is saved since, irrespective of manufacturing of a product, facility running cost and depreciation cost will be present. However, at the same time regulatory and other challenges associated with contract manufacturing organization (CMO) should be well thought of and considered for successful manufacturing.

Therefore, it is economically favorable to manufacture the product through process intensification by which repetitive batches could be run to produce the material in large quantities efficiently using the same facility. When compared to the conventional manufacturing design even if a couple of percentage increase is there in overall yield through process intensification, it is an addition to the overall profit.

As upstream processes are refining more and more, downstream processes are also adopting newer generations of cost-effective and more efficient chromatographic media. To match with the continuous upstream, efforts are being made in downstream also to involve continuous processing that reduces intermediate hold times in the process.

Moreover, the addition of single-use (SU) technology in the process brought the revolution in the manufacturing of therapeutics that offers greater advantages over the conventional stainless steel technology as this can reduce the cleaning, sterilization in place (SIP), and maintenance cost of the systems, saving the potential time as well.

Process intensification and SU technology are more useful to mitigate the risk of capacity uncertainty when the manufacturing facility footprint is small. For example, to meet the demand and reduce the capital risks at an initial stage, scale-out is recommended instead of scale-up. In scale-out, smaller facilities will work to produce the increasing number of doses. If the manufacturer is certain about market demand and supply, then building a conventional large manufacturing facility is a good option. However, it has a disadvantage of large investment too. On the other hand, if one is targeting uncertain demands, short supply, and limited specific products, then SU technology with a smaller footprint is a better option.

To achieve success in a continuously changing market, single-use technologies give flexibility to process operations in short time with less capital investment.

Several drugs are becoming costly to many as the population is increasing. Thus, to cut down the prices, cost-effective measures are being taken by the biopharmaceutical companies.

The tools that are essential for the conversion from batch to continuous mode are automation, process integration, and digitalization as these forms the basis of future facility design while process intensification is the important building block of continuous manufacturing, which remove the difficulties and provide the advantage of saving the cost and time.

6.2 State of Upstream and Downstream Processes

To sustain in today' competitive biosimilar/biologics market, industries are bound to develop advanced technologies that can be integrated into a large segment of biosimilar products pipeline. Now, the time has come for small footprint good manufacturing practice (cGMP) facilities that are capable of processing multiple products. Nowadays, each function of a biotech industry is going through an advancement step, be it equipment, upstream process media, and downstream media/resin. Due to the continuous advancement in the equipment, technologies, and process design, the large bioreactor can be replaced with small and fully disposable technology. The productivity/titer has increased eight to tenfold, and over 5 g/l is achieved for many biosimilars and biologics, when we compare it with the productivity range of the products from 1982 to 2004 [1]. Perfusion bioreactors showed a remarkable outcome in productivity that is more than 25× higher than batch culture [2]. Moreover, to cope up with this high titer, downstream process has started including new emerging tools to replace the chromatography media with high-throughput membrane chromatography. Now even for affinity chromatography like protein A resin, there can be format available in the form of a membrane, which will have added advantages of high flow rate, high productivity, and purity over the commonly used protein A resins [3, 4]. This way an affordable bio-therapeutic can be produced with the use of novel and advanced process technologies.

Some of the operations in the downstream manufacturing process are complementary to the upstream, and it gives an illusion of being continuous, but which is not true. Since the downstream contains major and multiple steps of chromatography, an operation in continuous mode is rare and limited to only one or two steps. The actual continuous bioprocessing demands process intensification not a batch-to-batch repetitive cyclical motion. Only if the complexities are reduced and the process is at ease as compared to the previous mode, it can be called a continuous process [5].

6.2.1 Sizing Upstream Process

In upstream, when a fed-batch process is scaled up, there are two ways to increase the productivity rate. One of them is to use a larger bioreactor vessel that will demand an expansion of the facility [6], and the other way is to use multiple bioreactors that are harvested periodically week by week. However, this eventually will lead to putting high pressure on downstream. To match the pace with upstream, there will be a requirement of using multiple columns and multiple filtration approaches that can match the frequency of multiple bioreactor harvest. For example, for commercial-scale mAb production, the traditional fed-batch bioreactor usually cultures cells in 10 000–25 000 l stainless steel tanks for 7–21 days with a product yield of 2–6 g/l [7].

A new interest is growing in the field of upstream processing to get the higher titer in a short span of culture time. To achieve this target, there is a continuous advancement in the development of a high producer and stable cell line giving a high titer that can reduce the facility footprint significantly [8].

Perfusion bioreactors generally run over a long period (typically N20 days) and can achieve 10–30 times higher cell density than a fed-batch reactor [9]. At a fixed interval, part of the harvest is removed continuously, and fresh media is supplied to it. This way non-essential by-product is removed and new growth promoter substances are added periodically. In this way, a fourfold productivity (in terms of mg/l/d) can be increased using perfusion bioreactors as compared to a fed-batch unit with the same reactor volume [7]. Hence, within a limited space and capital cost, the same product quantity can be produced continuously.

Implementation of perfusion reactors has been successfully commercialized, ranging from large biopharmaceutical companies such as Pfizer, Genentech, Shire, and Genzyme/Sanofi [10–12] to small companies and innovative vaccine manufacturers such as CMC Biologics and Crucell [1, 13].

Although there are still drawbacks to the technology such as the usage of large volumes of the medium and high level of operator training required due to the complexity and intensity of the operation, the economic gain from smaller vessels and facilities has a critical impact on process considerations [2, 7].

6.2.2 Sizing Downstream Process

The downstream process is the second complex and expensive step after the upstream process. Continuous operations are now possible in purification/downstream. Major purification steps such as clarification, affinity chromatography, and intermediate and polishing steps can be combined and designed to include single-use technology for continuous operations.

The use of centrifuge in clarification is difficult to scale up and complicated to operate at a larger or commercial scale [14]. As an alternative to centrifugation, SU continuous processes have been demonstrated successfully for large-scale, commercial processing of (Humira®) with flexible and versatile disposable depth filtration systems like Stax (Pall) [15]. Two-stage depth filters from Millipore (Clarisolve, D0HC, and X0HC adsorptive depth filters) can also be used either directly from the bioreactor or after the addition of flocculent to precipitate the impurities present in the culture or cell harvest.

These depth filters maximize the loading capacity of the subsequent chromatography column by reducing host cell protein (HCP) and DNA impurities and removing the cell debris [16].

In some processes, diatomaceous earth can be added to cell culture fluid to prevent blockage in-depth filters that allow large batches to be clarified with maximum efficiency in SU formats as demonstrated by Sartoclear Dynamics (Sartorius Stedim Biotech) [17]. Various column chromatography techniques such as affinity chromatography (AFC), ion exchange chromatography (IEX), hydrophobic interaction chromatography (HIC), multimodal chromatography (MMC), etc. are employed to purify the molecule.

In general, the column sizing is done based on volumetric flow rate instead of capacity to increase productivity. This results in oversized columns that demands investment in large columns, related equipment, and space as well. Large columns can suffer from scale-related packing problems including hysteresis, edge effects, and resin compression [18].

As an alternative to this, cost-effective expanded bed adsorption (EBA), simulated moving beds, and membrane chromatography are being used. In EBA, all three steps are integrated as one, such as clarification, filtration, and capture, which are beneficial in terms of cost, time, and space requirement.

In EBA, the crude harvest is subjected to the expanded bed with upward flow. Target proteins are bound to the column while cells and other particulate contaminants pass through it. Loosely bound material, such as cells and other particulate contaminants, is washed away. Subsequently, the captured proteins are eluted by reversing the flow. The output obtained will be clarified, concentrated, and partially purified, which can be processed for further purification [19]. Though EBA offers the advantage of lower CapEx, buffer consumption, time, and space requirement, however, its major drawback is the need for recirculation and non-specific adsorption that affects bed stability and overall purification performance [20].

Second-generation EBA robust technology is an advanced EBA. According to a data presented at the BioProcess International Conference and Exposition, Rhobust MabDirect Protein A requires one third of process time, offers a better yield of 12%, and utilizes half of the buffer quantity while showing no compromise on purity along with enhanced DNA clearance, when compared with a packed bed protein A column [21]. EBA stands out to be a promising option that improves throughput even when cell density and productivity of a bioreactor increases. The use of EBA obliterates the demand for large filtration areas that are vital to deal with the fouling from crude feeds.

The simulated moving bed (SMB), another promising purification technology, bestows a fully continuous method for carrying out chromatography. The BioSMB system introduced by Pall (former Tarpon Biosystems Inc.) provides continuous feed loading along with continuous elution as multiple Pro A columns are cycled through the various stages of loading, wash, and elution at different times [22].

The accelerated seamless antibody purification (ASAP) process is a continuous and fully disposable mAb DSP that is based on the AKTA periodic countercurrent chromatography (PCC), involving protein A, anion exchange, and mixed-mode resin columns where these are cycled simultaneously [23]. An additional advantage of SMB mode is that the connection of the columns can be in series, and any breakthrough from the first column directly lead to loading onto the second subsequent column, thereby utilizing the complete capacity of the primary column without any loss of the valuable product [24]. This permits the usage of shorter bed heights column, which can function at faster residence times but may have shallower breakthrough curves, thereby increasing the overall productivity [25]. The use of SMB grants several advantages over batch resin chromatography including boosting of productivity by 30%, increase in load capacity up to 40%, and cutting down consumption of buffer by 27% [24]. Execution of procedures in SMB continuous mode will not only decrease the high risk of contamination due to human error and stoppages in process but also cut down the operation cost by reduction in the amount of resin and buffers that would save a quality amount of time required for processing [26–29].

Another substitute to resin columns are nanofiber adsorbents, namely, Puridifye's FibroSelect platform, though having a low binding capacity (10 mg/ml) can function at very high flow rates (2400 cm/h), thereby enabling high productivity along with continuous processing [30].

Monolithic platforms can also prove to be convenient for mAb purification because of their high porosity structure that provides effective mass transfer for antibody-sized target. Moreover, the manufacturing cost is also reduced due to the ease of material preparation. The monolith operation has yet to be construed from analytical scale to industrial scale [31]. These platforms are commercially available from various manufacturers including Millipore, Sepragen, BIA, and Separations [31]. Another option to resin chromatography is membranes that are now gaining popularity for their high throughput and ease of use. Mass transfer, in contrast to resin beads that relies on diffusion, is dominated by advection, allowing operation at higher flow rates. Conventional ion-exchange membranes that are well suited for flow-through (FT) application are made available by several suppliers such as Pall (Mustang) and Sartorius (Sartobind). However, the bind and elute (B&E) applications are limited due to the low ligand density on the resin beads [32]. Natrix's HD membrane technology consists of a high-density binding ligands in porous polymer hydrogels permitting significant improved flow characteristics without compromising on the binding capacity. As per the mAb process simulations, cycling smaller devices with high binding capacity leads to significant reduction in CapEx as the hardware required for larger columns is expensive and OpEx is reduced due to the high cost of media (especially protein A) [13]. Higher throughput can be achieved, with this new chromatography media, without oversizing the device, and media volume can be reduced by utilizing rapid cycling. This would lead to increase in productivity along with flexibility at much lower CapEx.

6.2.3 Continuous Process Retrofit into the Existing Facility

6.2.3.1 Upstream Process

To speed up the research and development and to produce cost-effective products, there is a need to have an improvement in different technologies including single-use bioreactor and continuous purification systems.

As compared with the fed-batch process, perfusion cell culture is run for a longer time and requires a relatively large volume of media. To address these challenges, Daisie Ogawa from Boehringer Ingelheim and her colleagues worked on seed train. Generally, Batch "N-1" seeding takes 7–10 days to speed up the production, while perfusion "N-1" seeding took only 5 days to reach the production phase with a 10-fold increase in N-stage seed density.

Daisie and the research team worked on a non-steady-state perfusion culture where they did not control cell density and added concentrated media and a diluent to supply sufficient nutrients while maintaining a low perfusion rate. Through this, the osmolality was balanced with the flushing of waste present in the spent media.

As compared with the fed-batch process, the non-steady-state bioreactor produced >10-fold higher productivity. This improved process was successfully scaled up from 2 l scale to 100 l scale bioreactor (Figure 6.1) [33].

6.2.3.2 Downstream Process

A continuous downstream process for mAb consists of several operations like multicolumn chromatography for protein A, low-pH virus inactivation in a packed column, filter train, multicolumn chromatography for continuous polishing, and

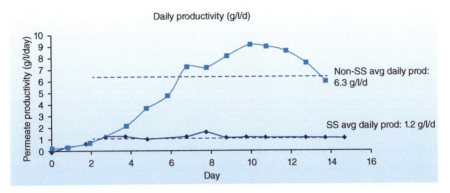

Figure 6.1 Daily productivity in a non-steady-state production phase. Source: Ogawa [33].

single-pass buffer exchange and final concentration run with continuous upstream manufacturing in a fully integrated system for two weeks, says Lindsay Arnold from MedImmune. Each of the five components of the DSP is run automated individually, but there is a need to integrate all five units.

Arnold explained that the company has developed its own approach to low-pH virus inactivation using in-line titration followed by a static mixer. The company is working with a vendor to supply the filtration-AEX membrane–virus filtration combination. One of MedImmune's challenges is the existing installed capital base low utilization. In this case, one strategy that can work best is a modular approach, wherein continuous processing can be introduced where batch processing has low productivity.

Arnold has been focusing on areas where continuous bioprocessing can reduce the cost of goods, increase process control, and adopt single-use systems.

6.2.4 Learning from Chemical Industry

Andrew Zydney from the Pennsylvania State University says countercurrent staging can also be used to develop continuous bioprocesses. Examples are diafiltration, purification using continuous chromatography, and capture via continuous precipitation.

Owing to the product recirculation in batch mode TFF, there is a requirement for a large amount of buffer. According to Zydney, a continuous countercurrent staged system with in-line dilution is a better option instead of a mixer. On one side, there is an addition of buffer to the product, and on the other side, there is a removal of the buffer by which buffer exchange is improved with multiple time buffer use. This improved operation gives constant impurity removal throughout the process, and fixed process conditions all the time eventually lead to improved product quality. By changing the number of stages and the feed flow rate, the required level of impurity removal is possible. According to him, 99.9% buffer exchange along with a 1000-fold impurity removal can be done with a three-stage system as compared to batch mode operation.

With regard to chromatography, this technique is called countercurrent staging. Here, we can use a system in which there is a series of static mixers and hollow fiber membrane modules that can replace the series of columns. Moreover, the result is a cyclical response instead of a steady-state response (Figure 6.2).

Zydney says, binding, washing, elution, and stripping solutions are passed directly on the slurry in continuous countercurrent tangential chromatography, while hollow fiber membranes and static mixers are used for separation and residence time control. As compared to the multistage batch column system where we mix the slurry with the elution buffer, countercurrent staging provides better management of buffer along with a reasonable purity level. In continuous staged precipitation, the targeted precipitating agents such as metal chelators, solvents, and affinity ligands have been proved to purify the product at low cost. Earlier, the titers in plasma fractionation industries were very low, but now it has gone up to 10 g/l.

Brower and the team have scaled a continuous chromatography system and achieved high-quality products. The company has worked on the design specification of the systems to get the know-how of automated and single-use different unit operations from bioreactor to the complete purification. These operations can be handled by a small team as compared to a large team. Process analytical technology (PAT) are involved in this system, which allows operators to have process control in system deviations from the desired parameters and data generation for multivariate data modeling and analysis.

6.3 Cell Line Development and Manufacturing Role

Cell line development traditionally consists of cell transfection, cell sorting, and clone selection based on higher productivity and growth. This selection of clones usually takes approximately 9–10 months, and to speed up the screening process, some techniques involve single-cell isolation in well plates followed by the selection of a higher producer from more than 50 plates. It is a better option to select a higher producer pool by keeping the cells in the form of mini pools instead of individual selection. This way a lot of time is saved, and only a small number of pool are screened to get the single cells. Advanced techniques like fluorescence-activated cell sorting (FACS) along with the glutamine synthetase (GS) system can play an important role to select the clone in a faster way. Clone assessment based on proliferation rate and cell-specific productivities can be done within a week with the help of microfluidics. The amount of time invested in the selection of higher producer clones can be saved, and higher producer cell lines can be developed by targeted gene insertion method as compared to random transfection events [34].

By incorporating these advanced techniques, the development of consistently producing higher titer clone can be done within a short span of three weeks, eventually reducing the time for toxicity study in the next phase of biosimilars.

Even if the product is stable for a long period in batch mode, it is not good for the genetic stability of the producer cells. The advantage of continuous bioprocessing

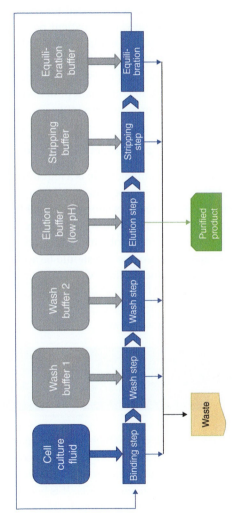

Figure 6.2 Conceptual diagram of a continuous countercurrent tangential chromatography.

increases when the product is unstable, and leaving it for a long time in the culture medium becomes a problem. Studies have been performed to understand the long-term genetic stability, mechanism, and microenvironment of cells [35, 36]. To reduce the risk associated with continuous bioprocessing during aseptic product removal, many technological advancements have been made [37, 38].

Across the industries, there is an acceptance of gene therapy and cell therapy that it should reach to production level, but the challenge is that majority of them are from the autologous process. If a patient's cells will be used for therapy, then scale-out is the only option one can choose for the various patient population, rather than scale-up. Different lots will be using separate equipment in scale-out, which is feasible only with the help of single-use technologies [39].

As compared to scale-up where only one process can be used to treat multiple patients, there is added cost in scale-out. Recombinant proteins especially MAbs are a good example of the current scale-up process train [40].

6.3.1 Speeding Up Upstream and Downstream Development

In a way to speed up the process development, there is a requirement of microbioreactors, high-throughput plates, and spin tubes, in parallel to lab-scale bioreactor from 250 ml to 10 l capacity. Small bioreactors should be capable enough in showing and predicting the futuristic conditions and results that can come at scale-up operation. The correct prediction can only be obtained after good analytical support that is integrated within the upstream process and should be automated. Important data can be obtained from cell culture media along with bioreactor, chromatography media along with purification steps, and drug substance and drug product, but how these data can be captured and stored for future evaluation is a great challenge. It can be done as described in the workflow below.

Everything becomes clear when the data which is being analyzed is evaluated along with intersystem interactions in such a way that it becomes easier for the scientist to compare the other batches and draw some meaningful conclusion and avoid any deviation that can lead to erroneous results. It is better to integrate more analytical data points in the process to have accurate real-time analytics upstream and downstream. This flow of data stream can cover the entire process including bioreactor and chromatography raw materials, purification data, and filtration data (Figure 6.3). This will give better quality and yield by giving freedom of process modification from start to end. It is called the best process when it produces the product with good quality at low cost and efficient enough in terms of flexibility and completion speed.

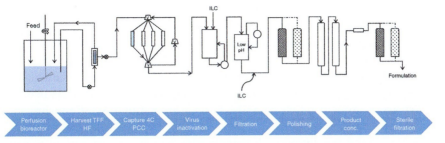

Figure 6.3 End-to-end upstream and downstream continuous processing.

6.3.2 The State of Manufacturing

Mostly all the biopharma companies are targeting the manufacturing of mAbs, antibody fragments, therapeutic peptides, and proteins for so many years now. Conventionally the production is done using stainless steel bioreactors in fed-batch mode along with the traditional QC and QA to release the batch. They need a large facility with upstream and downstream separation and huge investment where the focus is on fixed cost. However, as the demand changes, the industry needs to be changed (Table 6.1).

Biomanufacturers are making their facility flexible according to the need of the industry because of the increasing demand for complex biomolecules like bi-specific MAbs, ADCs, oncolytic viruses, CAR T cells, and RNA interference-based drugs. Flexibility in the facility is required also due to the fact that when a product leaves the facility for clinical trials or moves in the market, there should be a next molecule to take its place in the pipeline and other follow-on molecules at different stages of development.

Between 50% and 80% of the manufacturing cost is in downstream processing, and because of the multistep purification process, incorporation of continuous bioprocessing becomes difficult [41].

There are two main factors based on which most of the continuous downstream processing studies are carried out only on mAbs. One of them is due to the current

Table 6.1 Biomanufacturing now and in the future.

No.	Current state	Future state
1	Focus on mAbs	Different range of modalities
2	Fed-batch process in stainless steel	Capacity is flexible and agile
3	QC is conventional	Release in real time
4	Lead time in the supply chain is high	Quick response in supply
5	Capital cost in the facility is >600 million USD	Cost reduced to ±50 million USD
6	Focus on the fixed cost	Focus on the variable cost
7	USP and DSP are separate	USP and DSP are integrated

scenario in which the majority of the biopharma segment is commercializing only mAbs, and the other one is due to the availability of high-purity mAbs with good recovery through their platform processes that comprise protein A affinity, CEX, and AEX chromatography. However, process intensification can make some significant advancement to this [5]. Hence, in recent times, many continuous liquid chromatography systems and processes have been designed by key service providers like Pall Life Sciences (BioSMB®), Cytiva (AKTA PCC), ChromaCon® AG (Contichrom®), and Novasep (BioSC®) [42].

Recently, Cytiva has introduced nanofibers to provide greater capacities with high flow rates that are being claimed to be run for 150–200 cycles [43]. Due to this approach, we can cut down the expenses of costly protein A resin and associated processes for mAb purification. This will be considered as a fixed material cost for each cycle rather than a huge upfront cost that depreciate continuously until the particular product is in pipeline and Fibro is currently under development for manufacturing scale but available for bench scale [44, 45].

Another recent work that shows scope for future scale-up showed a process of in-line precipitation of an MAb by zinc chloride and polyethylene glycol solution [46]. Subsequently, they used tangential flow filtration to wash the precipitated antibody followed by resolubilization. With this technique, they got recovery of 97% along with 90% removal of HCPs.

The demand of the molecule that is in the market is uncertain due to the price of the drug, the need of the market, changing required dose levels as a result of clinical outcomes and recommendation, and the size of the patient population, which changes as new indications are approved. Therefore, it is required to develop the flexible drug in terms of the lowest possible cost and production size to enter into the global market to support changes in demand wherever it is required. This is possible only when the smaller facilities could be created at different locations with SU systems at lower upfront and fixed costs. There should be a scope of multiplying the facility at the same or different location to allow the production to be double whenever it is required.

6.4 Process Integration and Intensification

6.4.1 Intensification of a Multiproduct Perfusion Platform

According to Shawn Barrett (Sanofi), the company is reducing the footprint and cost of goods by converting batch mode DSP process into a single-use system processing suspension Chinese hamster ovary (CHO) culture along with continuous mode capture chromatography, thereby increasing robustness and flexibility. Their upstream team is also working on the intensification of perfusion culture to increase volumetric productivity and minimize cell-specific perfusion rate, thus minimizing media flow through. While doing so, the product quality attributes are maintained, and process robustness is checked for an extended period of runs. While doing a scale-up with this improvised process, a fixed range of cell density should be maintained.

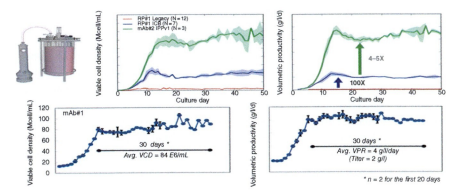

Figure 6.4 Intensified perfusion platform (Sanofi).

Sanofi's first integrated continuous biomanufacturing perfusion system (Figure 6.4) utilized chemically defined medium with a new cell line. Moreover, as compared with the old process, new process was able to increase the productivity by 100-fold, and with other biologics, productivity increased to fivefold after development at 10 l scales (Figure 6.5).

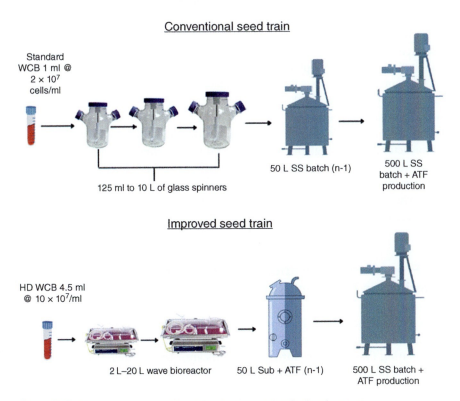

Figure 6.5 Upstream process intensification in seed train development.

6.4.2 Upstream Process Intensification Using Perfusion Process

Earlier they faced some of the challenges in terms of lower cell viability, 50% ± 10% lower productivity, and sudden hauls of growth rate while using different cell lines. The addition of a new feed concentrate in the new perfusion process solved the problem. Further, the clones that have been used in the intensified perfusion process were screened from fed-batch clone while the new process with eight different clones showed that it can give 45% ± 15% higher productivity with a titer between half and equivalent to the clones that have been screened from a fed-batch system. Intensified perfusion process of a biosimilar mAb along with the integration of continuous chromatography capture step showed reduced cost of goods (COGs) of up to 50% at 2000 l perfusion process when the plant capacity increased up to 2.5-fold.

The application of computational fluid dynamics (CFD) in bioreactors has shown a promising outcome during the mass transfer, mixing, and substrate gradient studies [47]. This will help in investigating and managing heterogeneities in the continuous bioreactors. Process development and real-time gradient analysis become easy when CFD is coupled with compartmental modeling and sensors [48–50].

Alternating tangential flow (ATF) filtration has also been developed and is used widely for external cell retention to avoid blockage observed in the membrane during the long perfusion process [51].

Moderna won the 2019 ISPE Facility of the Future award and provided the first batch of COVID-19 vaccine to the National Institute of Allergy and Infectious Diseases in less than one and a half months after sequence identification [52]. This could happen only because of continuous improvement in process intensification and digitalization of the biopharmaceutical industry from lab scale to commercial scale, for example, Xcellerex APS (perfusion bioreactor) and Fibro (nanofiber-based chromatography platform) [53].

In the perfusion process, high cell density can be achieved that can work as an N-1 bioreactor by which starting cell density at the time of inoculation in production bioreactor becomes high that eventually reduces the time required for achieving high cell density [39]. This approach is being utilized to inoculate multiple fed-batch production bioreactors [54]. So here, it is a combination of perfusion and fed-batch process where there is no need for multiple scale-up to generate the inoculum for multiple fed-batch production bioreactors.

6.5 Process Intensification and Integration in Continuous Manufacturing

Due to the advancements in the monitoring, analytics, artificial intelligence, automation, and advances in robotics, there is a kind of revolution in digital manufacturing (William Whitford, GE Healthcare). Due to this, it is possible to get real-time release testing, continuous quality determination, and better process control of product quality. There are many disciplines in digital biomanufacturing [55, 56], and only a few people have expertise in all that is a big challenge.

Other than that in digital biomanufacturing, in terms of information technology support, there are many supporters like high-end software that are suitable for FDA regulation, large and complex data storage management, and systems with artificial intelligence. According to Whitford, in digital biomanufacturing, real-time prediction, analysis, and control of critical quality attributes (CQA) and critical process parameter (CPP) are present along with process intensification control and continuous process parameter optimization.

Digital biomanufacturing also includes a plant that can be controlled remotely, self-aware, and continuously adaptive. Besides in-line or on-line, real-time, orthogonal process monitoring, incident control, management, and reporting capabilities should also be present in digital biomanufacturing. It means that here, there is a huge potential to support high-end manufacturing intelligence that is converted into optimal product harvest, monitoring support for analytics and process control, reporting QA and QC support, process development and optimization, the capability of prediction in scheduling, and supply chain optimization. In the QbD approach, digital biomanufacturing can accommodate successfully that results in multi-attribute methods of analysis, for example, quadrupole Dalton mass spectroscopy, by which the impact of the process on multiple CQAs can be identified and multiple old assays can be replaced. According to Whitford, multi-attribute methods of analysis are regulator friendly and cost-effective, support advanced process control, and support reporting on multiple product attributes in the near real time.

Commercial automated monitoring technology support includes aseptic at-line sampling from bioreactors and several downstream processes for multiplexed analysis and at-line analytics. Along with this, there is no problem in getting different monitoring solutions for real-time, continuous, and specific analysis.

In situ Raman probes are available that are single-use and adaptable, which can simultaneously measure glutamate, lactate, glucose, glutamine, ammonium ion, osmolality, viability, and total cell density. With the advancement of new instrumentation, real-time glycan analysis is also possible within 30 minutes (Figure 6.6).

In digital biomanufacturing, advanced process monitoring will lead to process development change and control, using new analytical techniques and new data sources. For example, rather than using a representative value, such as pH or glucose levels, measuring glycoform in near real time provides a better approach to using changes in product attributes for process control. Along with this, there is a possibility that the available mass of data can be used to control the process in a better way by monitoring the various process parameters in upstream processing (USP) and DSP.

Potential deviations can be handled easily when there is a proper protocol for execution, control, and precise analysis in a fully integrated manner. This can be possible with the help of introducing an automation philosophy and integration of upstream and downstream bioprocesses to work as a continuous process [5]. Application of automation philosophy in process integration leads to improved and cost-effective process design, higher productivity product quality, and safety.

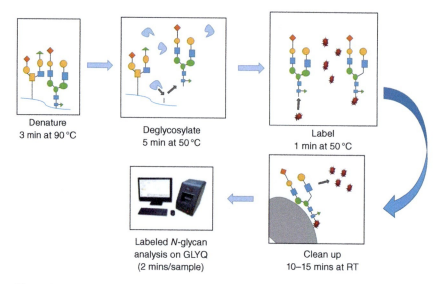

Figure 6.6 Glycan analysis using an advanced instrument (GlyQ).

When compared with batch process, in continuous bioprocessing, there is improved facility and raw material utilization along with cost- and energy-saving benefits.

Smart manufacturing concepts are called Industry 4.0 by which there is better control over the process, and continuous bioprocessing is a part of the same. As compared with batch processing where there is a need for continuous manual interventions to run and control the process at a defined time interval, continuous bioprocessing is automated, steady, and sustainable with the ability to overcome the limitations of the batch process.

During single-use scale-up in continuous bioprocessing, waste disposal is an alarming situation [57, 58]. Though it is obvious that continuous bioprocessing provides better flexibility with single-use systems, it also raises the concern related to environmental safety as in the current scenario the world is trying to avoid plastics as much as possible.

6.6 Single-Use Manufacturing to Maximize Efficiency

As the biopharma market is increasing day by day that counts for ~25% of the total pharma market, industries are adopting biopharma products in their portfolio. However, owing to their complexities, their production is difficult, and to overcome these difficulties, they require advanced technologies. In addition to that, one needs to have the approval of both, the process and the product too from regulators, and due to the risk of high cost and change in production process, some industries are not willing to adopt the new advanced technologies like single use technology (SUT) while some are adopting to be in the market with new products and future benefits.

To design a facility, one should consider each specific unit operation to be in balance with decision-making factors. Each unit operation has a defined purpose and each step is different. The equipment used in the unit operations may or may not be the same. For example, all chromatography steps may use the same instrument where TFF needs another while VI needs a simpler utensil. One must think about the questions listed as follows:

(1) Is this a start-up or a large, established biopharma?
(2) What type of product is being manufactured?
(3) What is the scale of operations?
(4) What is the in-house knowledge and capability?
(5) What level of demand is expected and, therefore, how much capacity is needed?

Their answers will tell, what should be the design and capacity of the facility. At the upstream side, stringent CIP and SIP requirement is present due to the higher risk of bioburden. On the other hand, in downstream processing, the drug is purified in terms of increasing purity and quality. Due to this, there will requirements of high-end analytical instruments that lead to high SUT consumable cost as compared with bioreactor consumables.

There are continuous and innovative improvements in single-use systems by which the issues associated with the perfusion process have been resolved in recent years, for example, SU bioreactors, filters, columns, and cell retention devices [5].

Table 6.2 compares the benefits and challenges of the two manufacturing platforms – continuous stainless steel vs. single-use systems.

In some cases where there is a requirement to get into the market soon, it is advisable to choose the SUT for faster results because the cost of SUT may be bearable and cost-effective than the missing opportunity of early market capture. We have to decide either to go with SUT only in USP or DSP or both based on strategies like switching between multiple products, preclinical batches, or clinical batches.

6.6.1 The Benefits of SUT in the New Era of Biomanufacturing

The cost benefits with SUT are there if the titer is high and working volume is less. For example, if the requirement is not much then, one cannot build a facility. Then wait and watch for demand to come, and SUT can be time saving as it takes ~5 years to build an SS facility.

No or less cleaning validation, small footprint, and flexible process design are the other benefit of SUT.

6.6.2 Managing an SUT Cost Profile

It is good to have SUT at an early stage of development and then shift to SS facility to shorten the time. SUT can be integrated between different unit operations such as single-use chromatography followed by a filtration step.

The limitations of SUT include writing SOP based on SU items that is critical, inventory maintenance of SU items, SUT's flow kit restrictions on maximum flow and pressure rating, and less sensitive SU sensors as compared to traditional sensor extractable, leachables, and integrity testing. With SUT, vendor is important who is ready to supply the consumables constantly (Figure 6.7).

Table 6.2 Benefits and challenges that single-use systems can offer compared with stainless steel-based manufacturing platforms.

S. No.	Multiuse facility		Single-use facility	
	Advantages	Disadvantages	Advantages	Disadvantages
1	Well-known systems with standardization	Inflexible infrastructure	Capacity flexibility	Lack of mechanical strength, i.e. difficult scale-up
2	Less disposable waste	Challenging in scale-up	Can be easily scaled out	Waste generation
3	Available in large capacities	Higher maintenance requirements, utilities	Easy product/facility changeover	Risk of extractable and leachable
4	More advanced sampling and process control	Cleaning validation required	Reduced cleaning validation	Lack of standardization
5	—	Higher risk of contamination	Lower energy demand and water use	The requirement of sustainable supply of consumables
6	—	Higher capital expenditure (CAPEX) required to start	Lower risk of contamination	More dependence on automation and a very skilled operator needed

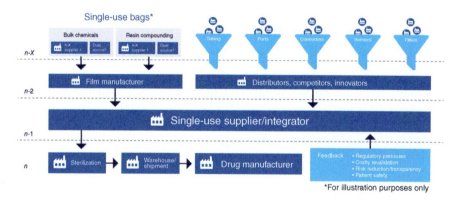

Figure 6.7 Single-use supply chain.

6.6.3 In-Line Conditioning (ILC)

In-line conditioning is nothing but an approach by which a buffer of desired strength is prepared from the stocks through an automated operation in the system without any manual intervention. This can assure you of the right buffer that is within the specification. If the system detects any deviation in the specification, it corrects the preparations according to the set specifications [59].

By the use of SUT in USP and DSP, there is an improvement in mAb production. However, still to be benefited in terms of cost-effective production, there is a requirement and scope of some additional advanced approaches. In SS bioreactor wherein there are several steps required in seed generation from shake flask to small bioreactor, wave bioreactors can be used up to 500 l volume [60].

6.6.4 Impact of Single-Use Strategy on Manufacturing Cost of Goods

Conventional processes are running with the old process and not operating to the full capacity. At the same time, equipment are also not flexible to accommodate new advancement in the technology and meet the expectation of new FDA or EMA guidelines, so the COGs has been increased to follow the guidelines by some other ways. On the other hand, SUT is getting more popular owing to less or no CIP validation requirements, less CapEx, low risk, low cost, less time consuming, and less or no contamination [61].

Biosolve and SuperPro Designer are the software by which one can simulate and develop an economic model that helps in the evaluation of the best possible manufacturing process design with SU strategy. Since most of the manufacturers hesitate to share the information of their process and facility, there is little information available in the field of mAb with SUT. However, it is possible to have SUT in the existing facility that gives higher throughput and handles the regulatory hurdles more efficiently, leading to the affordable processing.

Through Biosolve analysis, it has been shown that since there is less requirement of maintenance, utilities, labor, and waste management with SUT, there is a reduction of 22% OpEx cost per gram of mAb as compared to stainless steel facility [62]. In PepTalk conference organized at Cambridge Healthtech Institute, a case study showed that in a facility where disposable bulk freeze containers and buffer hold bags have been accommodated, $250 000 have been saved annually in water for injection (WFI) generation, and $60 000 in labor time for setting up and CIP of SS facility [63].

Biosolve Process model for different titers of mAb for a 1000 and 5000 l scales evaluated that in production facility, replacing the polishing resin with SU membrane format has significantly reduced the buffer volume by up to 55% and the cost by 19–33% lower [64].

By accepting SUT to build a new facility, there is no worry of space crunch, scheduling constraints, difficult installations, and nonoptimized layouts [65]. Though the initial cost will be present, it is less as compared to building a SS facility with their skids and hardware for multiple products. Since the instruments will be smaller, the footprint will also be smaller.

Biopharm Services Limited has published a white paper in which they have mentioned that by using modeling software, it has been shown that a process with 30×2000 l disposable bioreactor facility costs $250 million while with SS facility of the same capacity has a cost of $352 million [66].

According to one more model case study, as compared to a new MU 2×1000 l facility, the SU facility as CapEx annually saves €11, and €1 million extra for OpEx is the only cost [63, 67]. The reason for this reduction in CapEx comes from decreased engineering and equipment costs that is by 83% and 37% respectively.

However, 51% increase in running cost is due to the higher consumption of consumables. [67].

According to the FlexFactory concept from Xcellerex, with the incorporation of SU operations in the facilities, the area of a cleanroom can be designed to accommodate individual units with their SU tubing sets and less risk of operator contamination [68].

In International Society for Pharmaceutical Engineering (ISPE) Strasbourg conference, it has been shown that through FlexFactory facility design for a 2000 l mAb facility, as compared to SS facility, the cost is reduced by 67% as water usage is reduced by 87%, and space requirement is reduced by 45% eventually, leading to a 32% reduction ($104.8 per gram of mAb) [69].

In the future, flexible facilities with a modular approach can be designed and built off-site to be installed in any unclassified areas in the world. In this way, the whole process can be designed with this approach by which speedy production is possible having an option of scale-out [70].

This type of model is the same as Xcellerex FlexFactory for SUT. Furthermore, the CoG for an mAb is ~$85/g, while the cost saving for a new facility is expected to be $25 million as CapEx.

For preferred lower risk of contamination, exposure to biohazardous substances, and real-time process conditions, SU PAT such as disposable real-time monitoring systems is gaining popularity. For example, SciLog Bioprocessing Systems offers disposable SU sensors like SciTemp SciPress and SciCon sensors for in-line temperature, pressure, and conductivity monitoring.

For applications of media, feed, and buffer preparation, Sartorius Stedim Biotech offers SU bioprocessing devices with built-in pH and temperature disposable sensor, for example, magnetic mixer LevMixer®.

For low-pH virus inactivation, pH adjustments, dilutions, and product formulation, Sartorius Stedim Biotech also offers the Flexsafe bioreactors with welded BioPAT ViaMass sensor discs developed by ABER Instruments, and through RF impedance spectroscopy, it measures cell density level in cell cultures.

6.6.5 Limitations of SUT

The concern of extractable and leachable compounds cannot be ignored as SU tools are often manufactured from derivatives of plastics. This includes plasticizers, curing agents, and antioxidants, which may affect the quality of the final product. Also, by the use of plastic SU products, a debate has been started over their potential environmental hazards as there is very limited possibility of recycling due to the mixed plastic content. Thus, the only options left are incineration and landfill [71].

Some other group says that this drawback can be compensated with SUT by overall reduced carbon footprint in the form of energy in water and cleaning agents as compared to multiuse stainless steel facilities [72, 73].

Continuous supply of the SU same quality items is the other issue as this material is required for further manufacturing due to the validation and QA regulations.

Other limitations include less economic when it comes to large scales (multiton per year) as seen in the case study above (maximum size for SU bioreactors is 2000 l, prepacked resin columns are limited to 45 cm and recently 60 cm inner diameter) [74]. However, in the future, these limitations can be overcome by the advancement in the SUT.

6.7 Process Economy

With regard to the selection of continuous biomanufacturing for CT and commercial batches, the management includes the suggestions of their partners for its success. The main challenge to adapt to a new process in the plant includes strategic planning, educating the users, and handling of the new issues [5].

As stated earlier in upstream, the perfusion-related issues have been resolved, and the process has become robust, easy, and scalable with a low risk of contamination. In downstream, process intensification has gained the advantage of handling an increased rate of producing high titers. As compared to batch mode, the continuous operations have shown sixfold to tenfold cost reduction [75]. The remarks made against the use of continuous bioprocessing upstream and downstream are only because there is limited availability of successfully implemented case studies and less exposure to the process [76].

In today's mAb production, to achieve success, the market is continuously getting inclined toward the single-use technology change that gives some flexibility in operations. With this, one can get huge profit at a lower capital cost within a short period (Figure 6.8). The benefits include the following:

(1) *Less labor requirement*: Since the CIP and SIP are not required, hence also there is no need for cleaning validation too.
(2) *Shorter changeover time*: Single-use technologies enable the operator to make the systems ready for the next batch within some hours only as compared to the SS facility where a week is required.
(3) *Reduced footprint*: The facility is much smaller that eventually reduces the cost of other utilities required to run the facility including cooling and heating unit, air conditioning, etc.

Every biopharma company wishes to have in their pipeline the molecules that are innovative and/or drugs that are blockbuster in terms of patient care. So cost-effectiveness becomes major support for an industry in this competitive world. To sustain and survive in the competition, the industry must be able to produce drugs at a lower manufacturing cost by accepting the changes according to the time and requirements like process intensification and flexible multiproduct facility design.

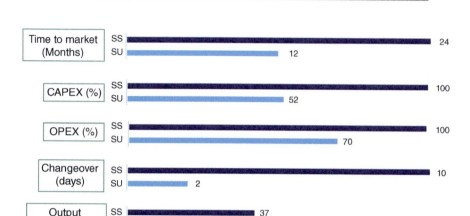

Figure 6.8 Process economy single-use vs. stainless steel facility.

6.7.1 Biopharma Market Dynamics

Together with plasma products, biopharmaceutical molecules specifically mAbs have increased the market value from $11 billion to $230 billion in the last 20 years [77]. In the next one or two decades, mAbs biosimilars are going to be the main fuel to the engine of market growth.

Though the biosimilars are in the market for a decade in the form of erythropoietin (EPO) and growth hormone, biosimilar mAbs after getting approval from the US market has become the highest revenue-generating molecules for almost every industry, and some mAbs are so popular that more than 20 companies are working on it at the same time. This number is constantly increasing for other upcoming biosimilars, showing in the future how tight the competition is going to be between the industries.

The new biosimilars are giving a tough fight to the innovator and existing biosimilars in such a way that the existing drugs in the market are losing their revenue. However, due to the strict regulatory guidelines, biosimilar approval has become difficult for the industries.

As compared to innovator molecule wherein the manufacturing process is old, new biosimilars are coming up with an improved version of the instruments and process that help them lower the price of the drug and sustain in the market. The pace of therapeutic and molecular innovation is also paving a new path for the future of biopharmaceuticals.

As the new anticancer drug like Keytruda™ and Opdivo™ are attracting attention to the world, there are so many biosimilars in the developmental stages for these molecules, but the main concern is their high price that may not be affordable for all patients.

Other than biosimilars, innovations are also coming into the market such as gene and CAR T cell therapies, ADCs, and bi-specific antibodies. The challenge in cell and gene therapy is that they are person specific, followed by the way they will be manufactured in low quantity as personalized medicines for the small patient population.

6.7.2 Management of the Key Risks of a Budding Market

How and where the manufacturing facility should be built and up to what capacity is another question for the industries since there is no surety of the dynamics of the market. There are three factors on which the decision is based:

(1) *Market fragmentation*: As explained earlier, some drugs are for a small patient pool, and you don't know that in the future until you go to the market and find how many competitors will be there. Thus, deciding production scale based on demand only becomes very difficult.
(2) *Globalization*: Emerging markets are the place where good revenue can be generated for many biosimilars, so many industries would like to manufacture their drugs in those markets because it gives a somewhat healthcare security along with a sign of economic growth.
(3) *Demand*: The demanded quantity for the particular drug is uncertain. In some cases, according to the requirements, the process will result in <500 kg, while in some cases it can be <100 kg as per requirement. A large facility built and low demand will lead to high COG. In addition to that if the facility is small and demand is high, an opportunity can be missed. The facility generally is built before regulatory approval, and unfortunately, if the product is not approved, the facility may be unused.

So to build an optimally sized facility, manufacturer must keep in mind the above mentioned factors and ensure the continuous supply of raw materials to fulfill the demand. There are some more challenges to it. For example, if some changes have to be made after regulators' suggestion and if by doing so the process becomes costly, raw materials or processing time increases significantly, and then COG becomes high. That is why a robust process is necessary that can yield the same quality product even after the significant changes in the process.

Not only process costs like chromatography resins, cell culture media, and consumables add up to the COGs for a product, but also a large cost is a facility running cost that includes electricity, security, and GMP qualification. This fixed cost is 60–70% of the cost and it continues whether or not the process is running. The only solution to this problem is to improve throughput by process intensification in a controlled way as to not waste even one percent of extra material that is of high value in the market.

As the upstream is increasing its efficiency in terms of high titer, the downstream is also reducing the cost with new resins with high binding capacity and better impurity removal with a small amount of resin. Now, the processes are getting designed in such a continuous and single-use mode that it reduces or minimizes the hold times in between the steps.

Today, the new concept of the modular facility is advancing where it can be designed for multiple products according to their requirements. To reduce the risk of the unused or underused large facility at the early stage of production, scaling out is a better choice instead of scaling up within a single large facility. Scale-out is where there may be multiple small facilities that produce the same or a large amount of product as produced by a single large volume facility.

This can be understood based on the example set by the two big biopharma companies, Amgen and Samsung. In 2015, Samsung announced that it will build a large facility in Songdo, Incheon, South Korea. This will produce 4500 kg product/year with 180 000 l capacity, and the facility cost will be $750 million. On the other end, Amgen has built a 120 000 sq. ft. modular small facility in Singapore, which gives an option of easy switching between different product operations.

According to the annual report for 2020 submitted by BioPlan's, more CMOs will be evaluating process intensification and adopting the continuous bioprocessing over the next year (53% will be testing DSP vs. 38% biomanufacturing facilities) while 40% CMO vs. 28% biomanufacturers would like vendors to focus on advanced continuous upstream technologies for them [78].

Based on the preliminary data, this year, approximately 20 new technologies have been identified that include on top single-use bioreactors (noted by 45.9% of respondents) and cell culture media optimization, followed by upstream and downstream continuous bioprocessing [79].

6.8 Future Perspective

One of the main reasons for inclination toward new continuous bioprocess technologies is less than 1 g/l upstream product titers in batch mode [80]. This limitation was overcome by the perfusion process where there is a constant balance of cell count and regular exchange of cell culture supernatant with fresh media. However, there is still a scope of improvement in this technology as some of the issues are yet to be addressed like maintaining long-term aseptic conditions, increase in process time and cost, non-homogeneity in bioreactor vessel, and log hour calculations as a regulatory viewpoint.

Superior control strategies are required to monitor the complexities and control the continuous process, ensuring better product quality, as suggested by FDA [81]. To cope up with the future expectations of monitoring control in continuous bioprocessing, PAT and automation tools [78] are essential that significantly reduce the time required for process monitoring.

The development toward industry is supported by the application of new and advanced sensing tools like spectroscopic and soft sensors, chemometrics, wireless sensors, and advanced image analysis by which informative process monitoring control becomes easy in continuous bioprocessing [5]. These devices are different from the traditionally available sensors like pH, pressure, flow, and temperature, in such a way that they generate the data associated with the state of process development, for example, wireless sensors with noninvasive instrumented particle technology [82], by which even in an agitated condition the process data can be accessed [83].

Other than the use of the wireless sensor for process monitoring, a new innovative tool available is advanced image analysis that, when coupled with chemometrics along with machine learning algorithms, proved to be an innovative tool for continuous process monitoring in fermentation, according to a study [84]. Spectroscopic sensors when tested could sense and record multiple compounds at a time with no time delay after processing the spectra with chemometrics [85]. In the demanding manufacturing environment, new studies like infrared spectrometers [86] and Raman spectroscopy [87] are being modified for being applied.

Meticulous Research, leading global market research company, published a research report titled "Continuous Bioprocessing Market by Product (Filtration, Chromatography, Bioreactor, Centrifuge), Application [Commercial (Monoclonal Antibodies, Vaccine Production), Research], and End User (Pharma & Biotech, CDMOs)-Global Forecasts to 2027." According to this latest publication, the global continuous bioprocessing market is expected to grow at a CAGR of 8.8% from 2019 reaching $7.33 billion by 2027.

References

1 Wurm, F.M. (2004). Production of recombinant protein therapeutics in cultivated mammalian cells. *Nat. Biotechnol.* 22 (11): 1393–1398.
2 Wang, L., Hu, H., Yang, J. et al. (2012). High yield of human monoclonal antibody produced by stably transfected drosophila Schneider 2 cells in perfusion culture using wave bioreactor. *Mol. Biotechnol.* 52 (2): 170–179.
3 Hou, Y., Brower, M., Pollard, D. et al. (2015). Advective hydrogel membrane chromatography for monoclonal antibody purification in bioprocessing. *Biotechnol. Progr.* 31 (4): 974–982.
4 Kang, Y.(K.)., Ambat, R., Hall, T. et al. (2015). Development of an acidic/neutral antibody flow-through polishing step using salt-tolerant anion exchange chromatography. *Pharm. Bioprocess* 3 (8): 477–487.
5 Kumar, A., Gargalo, C.L., Udugama, I.A., and Gernaey, K.V. (2020). Why is batch processing still dominating the biologics landscape? Towards an integrated continuous bioprocessing alternative. *Processes* 8: 1641.
6 Shimoni, Y., Moehrle, V., and Srinivasan, V. (2013). Process improvements increase production capacity of a legacy product. *Bioprocess Int.* 11 (10): 26–31.
7 Rose, S., Black, T., and Ramakrishnan, D. (2003). Mammalian cell culture: process development considerations. In: *Handbook of Industrial Cell Culture* (eds. V.A. Vinci and S.R. Parekh). Totowa, NJ: Humana Press.
8 Low, D., O'Leary, R., and Pujar, N.S. (2007). Future of antibody purification. *J. Chromatogr. B* 848: 48–63.
9 Lim, J., Sinclair, A., Shevitz, J., and Carter, J.B. (2011). An economic comparison of three cell culture techniques: fed-batch, concentrated fed-batch, and concentrated perfusion. *BioPharm. Int.* 24 (2) [Online]. http://www.biopharminternational.com/economiccomparison-three-cell-culture-techniques.
10 Shukla, A.A. and Thommes, J. (2010). Recent advances in large-scale production of monoclonal antibodies and related proteins. *Trends Biotechnol.* 28 (5): 253–261.

11 Butler, M. and Meneses-Acosta, A. (2012). Recent advances in technology supporting biopharmaceutical production from mammalian cells. *Appl. Microbiol. Biotechnol.* 96 (4): 885–894.

12 Kelley, B. (2009). Industrialization of MAb production technology. *MAbs* 1 (5): 443–452.

13 Franzreb, M., Müller, E., and Vajda, J. (2014). Cost estimation for protein A chromatography. *Bioprocess Int.* 12 (9): 44–52.

14 Jungbauer, A. (2013). Continuous downstream processing of biopharmaceuticals. *Trends Biotechnol.* 31 (8): 479–492.

15 Muhl, M. and Sievers, D. (2010). Cell harvesting of biotechnological process by depth filtration. *Bioprocess Int.*: 86–88.

16 Schreffler, J., Bailley, M., Klimek, T. et al. (2015). Characterization of postcapture impurity removal across an adsorptive depth filter. *Bioprocess Int.* 13 (3): 36–45.

17 Minow, B., Egner, F., Jonas, F., and Lagrange, B. (2014). High-cell-density clarification by single-use diatomaceous earth filtration. *Bioprocess Int.* 12 (4): 2–9.

18 Gottschalk, U. (2008). Bioseparation in antibody manufacturing: the good, the bad and the ugly. *Biotechnol. Progr.* 24 (3): 496–503.

19 Shukla, A.A., Etzel, M.R., and Gadam, S. (eds.) (2006). *Process Scale Bioseparations for the Biopharmaceutical Industry*. CRC Press.

20 Poulin, F., Jacquemart, R., De Crescenzo, G. et al. (2008). A study of the interaction of HEK-293 cells with streamline chelating adsorbent in expanded bed operation. *Biotechnol. Progr.* 24 (1): 279–282.

21 Hofmann, M. and Hall, P. (2014). Design and manufacture of cGMP, scalable expanded bed adsorption chromatography columns. *Bioprocess Int.* [Poster] https://bioprocessintl.com/sponsored-content/design-manufacture-cgmp-scalable-expanded-bed-adsorption-chromatography-columns/.

22 Klutz, S., Magnus, J., Lobedann, M. et al. (2015). Developing the biofacility of the future based on continuous processing and single-use technology. *J. Biotechnol.* 213: 120–130.

23 Mothes, B., Pezzini, J., Schroeder-Tittmann, K., and Villain, L. (2016). Accelerated, seamless antibody purification. *Bioprocess Int.* 14 (5): 34–58.

24 Angarita, M., Muller-Spath, T., Baur, D. et al. (2015). Twincolumn CaptureSMB: a novel cyclic process for protein A affinity chromatography. *J. Chromatogr. A* 1389: 85–95.

25 Kaltenbrunner, O., Diaz, L., Hu, A., and Shearer, M. (2016). Continuous bind-and-elute protein a capture chromatography: optimization under process scale column constraints and comparison to batch operation. *Biotechnol. Progr.*: 1–11.

26 Müller-Späth, T., Aumann, L., Melter, L. et al. (2008). Chromatographic separation of three monoclonal antibody variants using multicolumn countercurrent solvent gradient purification (MCSGP). *Biotechnol. Bioeng.* 100 (6): 1166–1177.

27 Baur, D., Angarita, M., Muller-Spath, T. et al. (2016). Comparison of batch and continuous multi-column protein A capture processes by optimal design. *Biotechnol. J.* 11 (7): 920–931. https://onlinelibrary.wiley.com/doi/epdf/10.1002/biot.201500481.

28 Napadensky, B., Shinkazh, O., Teella, A., and Zydney, A.L. (2013). Continuous countercurrent tangential chromatography formonoclonal antibody purification. *Sep. Sci. Technol.* 28: 1289–1297.

29 Farid, S.S., Pollock, J., and Ho, S.V. (2015). Evaluating the economic and operational feasibility of continuous processes for monoclonal antibodies. In: *Continuous Processing in Pharmaceutical Manufacturing* (ed. G. Subramanian), 433–454. Weinheim: Wiley-VCH.

30 Hardick, O., Dods, S., Stevens, B., and Bracewell, D.G. (2015). Nanofiber adsorbents for high productivity continuous downstream processing. *J. Biotechnol.* 213: 74–82.

31 Barroso, T., Hussain, A., Roque, A.C.A., and Aguiar-Ricardo, A. (2013). Functional monolithic platforms: chromatographic tools for antibody purification. *Biotechnol. J.* 8 (6): 671–681.

32 Jorjorian, P. and Sears, G. (2015). The use of membrane chromatography throughout a product's life cycle. In: *Report and Survey of Biopharmaceutical Manufacturing Capacity and Production*, 12e (ed. E.S. Langer), 72–76. BioPlan Associates, Inc.

33 Ogawa, D. (2019). Continuous Manufacturing for the Modernization of Pharmaceutical Production: Proceedings of a Workshop. National Academies of Sciences, Engineering, and Medicine; Division on Earth and Life Studies; Board on Chemical Sciences and Technology. Washington (DC): National Academies Press (US).

34 Cytiva, B. Optimizing process efficiency in upstream manufacturing by Dr. Andreas CastanStaff Scientist, GE Healthcare Life Sciences https://www.cytivalifesciences.com/en/us/news-center/optimizing-process-efficiency-in-upstream-manufacturing-10001 (accessed 20 June 2021).

35 Jones, S., Castillo, F., and Levine, H. (2007). Advances in the development of therapeutic monoclonal antibodies. *BioPharm. Int.* 20: 96–114.

36 Agrawal, V. and Bal, M. (2012). Strategies for rapid production of therapeutic proteins in mammalian cells. *Bioprocess Int.* 10: 32–48.

37 Ebeler, M., Lind, O., Norrman, N. et al. (2018). One-step integrated clarification and purification of a monoclonal antibody using Protein A Mag Sepharose beads and a cGMP-compliant high-gradient magnetic separator. *New Biotechnol.* 42: 48–55.

38 Käppler, T., Cerff, M., Ottow, K.E. et al. (2009). In situ magnetic separation for extracellular protein production. *Biotechnol. Bioeng.* 102: 535–545.

39 Ultee, M.E. (2020). New directions in bioprocess development and manufacturing. *BioProcess Int.*

40 Gennari A, Donohue J. (2019). Cracking the code in cell and gene therapy manufacturing. Presented at BioNJ's Manufacturing Briefing, Newark, NJ.

41 Huter, M. and Strube, J. (2019). Model-based design and process optimization of continuous single pass tangential flow filtration focusing on continuous bioprocessing. *Processes* 7: 317.

42 Carvalho, R.J., Castilho, L.R., and Subramanian, G. (2017). Tools enabling continuous and integrated upstream and downstream processes in the manufacturing

of biologicals. In: *Continuous Biomanufacturing-Innovative Technologies and Methods*, 31–68. Hoboken, NJ: Wiley.

43 Scanlon, I., Hardick, O., Morris, C. et al. (2018). *High-Throughput Downstream Processing Using Cellulose Fiber Chromatography and ÄKTA™ Chromatography Systems*. Uppsala, Sweden: GE Healthcare Life Sciences https://cdn.gelifesciences.com/dmm3bwsv3/AssetStream.aspx?mediaformatid=10061&destinationid=10016&assetid=28844.

44 Schreffler J. (2019) Comparison of protein-A fiber and intensified protein A resin processes. Presented at BioProcess International Conference, Boston, MA.

45 Mehta K. (2019) Industrialization aspects of membrane affinity (ProA) technology. Presented at BioProcess International Conference, Boston, MA.

46 Li, Z., Gu, Q., Coffman, J.L. et al. (2019). Continuous precipitation for monoclonal antibody capture using countercurrent washing by microfiltration. *Biotechnol. Prog.* 35 (6) https://doi.org/10.1002/btpr.2886.

47 Haringa, C., Mudde, R.F., and Noorman, H.J. (2018). From industrial fermentor to CFD-guided downscaling: What have we learned? *Biochem. Eng. J.* 140: 57–71.

48 Tajsoleiman, T., Spann, R., Bach, C. et al. (2019). A CFD based automatic method for compartment model development. *Comput. Chem. Eng.* 123: 236–245.

49 Gargalo, C.L.; Heras, S.C.; de Las Jones, M.N.; Mansouri, S.S.; Krühne, U.; Gernaey, K.V. Towards the development of digital twins for the bio-manufacturing industry. https://doi.org/10.1007/ 10_2020_142.

50 Busse, C., Biechele, P., de Vries, I. et al. (2017). Sensors for disposable bioreactors. *Eng. Life Sci.* 17: 940–952.

51 Farid, S.S., Thompson, B., and Davidson, A. (2013). Continuous bioprocessing: The real thing this time? In: *Proceedings of the10th Annual bioProcessUK Conference*. London, UK.

52 Adams B. (2020). Moderna wins bragging rights as it kick-starts first experimental coronavirus clinical trial. http://www.fiercebiotech.com/biotech/moderna-wins-bragging-rights-as-itkickstarts-first-experimental-coronavirus-clinical-trial.

53 May M. (2019). GEN. Modular bioprocessing makes adaptability a snap. http://www.genengnews.com/insights/modular-bioprocessing-makes-adaptability-a-snap/ (accessed 20 June 2021).

54 Gagnon, M., Nagre, S., Wang, W. et al. (2019). Novel, linked bioreactor system for continuous production of biologics. *Biotechnol. Bioeng.* 116: 1946–1958. https://doi.org/10.1002/bit.26985.

55 Whitford, W. (2017). The era of digital biomanufacturing. *BioProcess Int.* 15 (3): 12–18.

56 Whitford, W. and Julien, C. (2007). Analytical technology and PAT. *BioProcess Int.* 5 (1 Suppl): 32–41.

57 Scott, C. Sustainability in bioprocessing. *BioProcess Int.* 2011, 9 https://bioprocessintl.com/manufacturing/monoclonal-antibodies/sustainability-in-bioprocessing-323438/ (accessed 20 June 2021).

58 Lonza Innovations in Pharma Biotech & Nutrition. Available online: https://annualreport.lonza.com/2019/segments/pharma-biotech-nutrition/innovations.html

59 Raghunathan, M. Striving for bioprocessing excellence: balancing modern approaches to manufacturing. Product Strategy Leader GE Healthcare Life Sciences. U.S. News, Biologics: The Drugs Transforming Medicine. https://www.usnews.com/news/healthcare-of-tomorrow/articles/2017-07-25/biologics-the-drugs-that-are-transforming-medicine.

60 Eibl, R., Kaiser, S., Lombriser, R., and Eibl, D. (2010). Disposable bioreactors: the current state-of-the-art and recommended applications in biotechnology. *Appl. Microbiol. Biotechnol.* 86 (1): 41–49.

61 Langer, E. (2015). Reasons for increasing use of disposables & single-use systems. In: *Report and Survey of Biopharmaceutical Manufacturing Capacity and Production*, 12e, 285. BioPlan Associates, Inc.

62 Levine, H.L., Stock, R., Lilja, J.E. et al. (2013). Single-use technology and modular construction: enabling biopharmaceutical facilities of the future. *Bioprocess Int.* 11 (4): 40–45.

63 Goldstein A, Molina O. (2016). Implementation strategies and challenges: single use technologies. PepTalk Presentation.

64 Xenopoulos, A. (2015). A new, integrated, continuous purification process template for monoclonal antibodies: process modeling and cost of goods studies. *J. Biotechnol.* 213: 42–53.

65 Higham, S.R. (2006). Capital costs for biopharmaceutical process retrofit projects. *Pharm. Eng.* 26 (3): 1–6.

66 Biopharm Services (2016). *Mab Manufacturing Today and Tomorrow*, 309–318. Elsevier B.V.

67 Eibl, R. and Eibl, D. (eds.) (2011). *Single-use Technology in Biopharmaceutical Manufacture*. New Jersey: John Wiley & Sons.

68 Rios, M. (2010). Flexible manufacturing: evolving technologies combine to enable a new generation of processes and products. *BioProcess Int.* 8: 34–46.

69 Hodge G. (2009). FlexFactory: the economic and strategic value of flexible manufacturing capacity. ISPE Conference Presentation.

70 Jornitz, M.W. (2015). Cleanroom and facility planning: getting beyond cost per square foot. *Pharm. Technol.* [Online]. http://www.pharmtech.com/cleanroom-and-facility-planning-getting-beyond-cost-square-foot.

71 Hammond, M., Marghitoiu, L., Lee, H. et al. (2014). A cytotoxic leachable compound from single-use bioprocess equipment that causes poor cell growth performance. *Biotechnol. Progr.* 30 (2): 332–337.

72 Colton, R. (2008). Recommendations for extractables and leachables testing. *Bioprocess Int.* 5 (11).

73 Sinclair, A., Leveen, L., Monge, M. et al. (2008). The environmental impact of disposable technologies. *BioPharm. Int.*: 1–9.

74 Jorjorian, P. (2014). Seeking the next generation of single-use technologies. *Bioprocess Int.* 12 (4): 54–56.

75 Schaber, S.D., Gerogiorgis, D.I., Ramachandran, R. et al. (2011). Economic analysis of integrated continuous and batch pharmaceutical manufacturing: a case study. *Ind. Eng. Chem. Res.* 50: 10083–10092.

76 Langer, E. (2020). Biomanufacturing: demand for continuous bioprocessing increasing. *BioPharm. Int.* 33: 5.

77 Jagschies, G., Lindskog, E., and Lacki, K. (2017). et al. *Biopharmaceutical Processing: Development, Design, and Implementation of Manufacturing Processes.* Elsevier.

78 Randek, J. and Mandenius, C.-F. (2017). On-line soft sensing in upstream bioprocessing. *Crit. Rev. Biotechnol.* 38: 106–121.

79 BioPlan Associates. (2019). 16th Annual Report and Survey on Biopharmaceutical Manufacturing Capacity and Production. Rockville, MD.

80 Xu, J., Xu, X., Huang, C. et al. (2020). Biomanufacturing evolution from conventional to intensified processes for productivity improvement: a case study. *mAbs* 12: 1770669.

81 FDA (2004). *PAT Guidance for Industry – A Framework for Innovative Pharmaceutical Development; Manufacturing and Quality Assurance.* Rockville, MD.

82 Zimmermann, R., Fiabane, L., Gasteuil, Y. et al. (2013). Measuring Lagrangian accelerations using an instrumented particle. *Phys. Scr.* 2013: 014063.

83 Freesense. https://www.freesense.dk (accessed on 11 December 2020).

84 Pontius, K., Junicke, H., Gernaey, K.V., and Bevilacqua, M. (2020). Monitoring yeast fermentations by nonlinear infrared technology and chemometrics – understanding process correlations and indirect predictions. *Appl. Microbiol. Biotechnol.* 104: 5315–5335.

85 Landgrebe, D., Haake, C., Höpfner, T. et al. (2010). On-line infrared spectroscopy for bioprocess monitoring. *Appl. Microbiol. Biotechnol.* 88: 11–22.

86 Gargalo, C.L., Udugama, I.A., Pontius, K. et al. (2020). Towards smart manufacturing: a perspective on recent developments in industrial measurement and monitoring technologies for bio-based production processes. *J. Ind. Microbiol. Biotechnol.* 47: 947–964.

87 Oh, S.-K., Yoo, S.J., Jeong, D.H., and Lee, J.M. (2013). Real-time estimation of glucose concentration in algae cultivation system using Raman spectroscopy. *Bioresour. Technol.* 142: 131–137.

Part III

Digital Biomanufacturing

7

Process Intensification and Industry 4.0: Mutually Enabling Trends

Marc Bisschops[1] and Loe Cameron[2]

[1] Pall Corporation, Nijverheidsweg 1, 1671 GC, Medemblik, The Netherlands
[2] Pall Corporation, 20 Walkup Drive, Westborough, MA 01581, USA

7.1 Introduction

Over the past two decades, the biopharmaceutical manufacturing industry has matured significantly. McLaughlin identified four stages in the evolution of biopharmaceutical manufacturing [1].

In the 1980s, most efforts were directed toward "plan for success." The target was to navigate candidate products through the clinical stages and obtain regulatory approval. As a consequence, within biopharmaceutical manufacturing, the focus was on ensuring safety and efficacy. Identification and removal of impurities became the highest priority in the process development. To ensure availability of material for clinical studies and product launch, companies established partnerships with contract manufacturing organizations (CMOs).

The second wave in the industrialization of biopharmaceutical manufacturing in the 1990s was focused on "titer and yield." Significant advancements in cell culture sciences resulted in the increasing titers for recombinant proteins and in particular for monoclonal antibodies. In addition to this, the yields of downstream processing operations were improved, resulting in an overall higher output of the overall drug substance manufacturing. The manufacturing facilities that came online were often still based on the output assumptions from the earlier generation manufacturing platforms that occasionally lead to the excess capacity. This was exacerbated by some drug failures in (late)-clinical phases and caused the industry to start the adoption of single-use technologies to enhance the flexibility in manufacturing and allowing the industry to respond better to the dynamics.

McLaughlin refers to the 2000s as the third wave in industrialization, focused on "first in human." Scientists started to realize that the long process development timelines became a bottleneck with appreciable financial impact. As a consequence, the industry started adopting technologies that significantly enhanced efficiency of process development and manufacturing. This resulted in accelerated cell line selection and screening and high-throughput experimentation techniques. For

Process Control, Intensification, and Digitalisation in Continuous Biomanufacturing, First Edition.
Edited by Ganapathy Subramanian.
© 2022 WILEY-VCH GmbH. Published 2022 by WILEY-VCH GmbH.

manufacturing, there was a marked drive toward platform processes that would accommodate a larger variety of biopharmaceutical products with moderate customization or adaptation to product specific needs. This also resulted into further adoption of single-use technologies to provide flexibility in manufacturing, reduce timelines in establishing capacity, and reduce downtime between manufacturing campaigns. With these efforts, most constraints with respect to clinical manufacturing were addressed.

The fourth and last wave that McLaughlin describes is referred to "continuous vision" or "integrated and intensified." In the 2010s, further advancement in biopharmaceutical manufacturing sciences shifted the focus toward manufacturing efficiency and costs. This coincided with the launch of the first technologies that enabled continuous bioprocessing, such as multicolumn chromatography. Together these developments enabled manufacturing platforms that focused on process efficiency without compromising on efficacy and safety. In 2019, the first monoclonal antibody produced in a fully integrated (end-to-end) continuous bioprocessing platform was produced for clinical studies [2]. In 2020, successful completion of the first clinical study was reported [3].

The biopharmaceutical industry is relatively young compared with other manufacturing industries. As a consequence, it skipped first stage of industrialization of manufacturing based on power and steam (often referred to as "Industry 1.0"). The second wave targeted mass production, which introduced concepts such as the conveyor belt and assembly lines. The first biological manufacturing plants aligned quite well with this "Industry 2.0" where manpower and automated manufacturing were combined to establish a more effective manufacturing platform. The current state of the biopharmaceutical manufacturing aligns best with what is covered under "Industry 3.0" where computers and equipment were brought together into automated systems that result in more consistent and efficient manufacturing processes.

The current digital era and its opportunities for further enhancing efficiency and robustness of biopharmaceutical manufacturing are most likely the next wave in the industry's maturation. The impact of digitization on manufacturing is often referred to as "Industry 4.0." This term has become a label that involves any and all use of data, advanced automation, digital, cloud computing, networks and many other opportunities that arise from the use of modern digital technologies.

The alignment of the evolution of the biopharmaceutical manufacturing as described above with the industrial maturation from "Industry 1.0" toward "Industry 4.0" is illustrated in Figure 7.1.

In this chapter, the opportunities and recent advancements in the field of Industry 4.0 and process intensification are reviewed. For this purpose, we will review a biomanufacturing process along various dimensions. We will start with the impact of process intensification on the primary process flow that involves the series of unit operations in the biomanufacturing process. The second axis along which the manufacturing process is viewed is the flow of materials and buffers. The third dimension that is considered in this chapter is the flow of information and data. Finally, some comments will be made on the evolution of a manufacturing from process development to commercial manufacturing. This dimension with

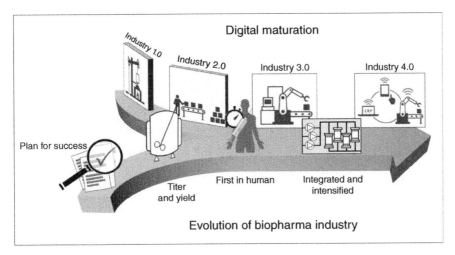

Figure 7.1 Alignment of the evolution of the biopharmaceutical industry and the global digital maturation of manufacturing toward Industry 4.0.

technology transfer remains hidden in perspectives on process intensification. Also, the challenges and opportunities related to this area are often overlooked.

7.2 Enabling Technologies for Process Intensification

7.2.1 Process Intensification in Biomanufacturing

Most initiatives related to process intensification targets increasing facility throughput, reducing the footprint in manufacturing, and shortening timelines.

An important contributor to process intensification comes from enhanced scientific understanding and material sciences. This has been the main driver for the tremendous improvements in cell culture and for the increased capacities in some chromatography adsorbents and enhanced throughputs of membranes and filters. The second movement in the industry that targets process intensification is based on engineering approaches and principles of lean manufacturing. This movement focuses on minimizing and/or eliminating the classical areas for waste as defined in lean manufacturing approaches, such as transportation, inventory, movement, waiting, overprocessing, overproduction, and defects [4].

This resulted in the establishment of continuous unit operations for bioprocessing and connected manufacturing platforms. With the increased specific (volumetric) productivities and reduced interstage hold tank volumes, the trend toward further adoption of single-use technologies in downstream processing platforms becomes a technically and economically viable option. In the case of further implementation of single-use technologies, flexibility in multiproduct facilities is enhanced (reduced overprocessing), turnaround times between batches is reduced (reduced waiting), and contamination risks are addressed by providing a processing platform in a functionally closed single-use platform (reduced defects).

7.2.2 Process Intensification in Cell Culture

The tremendous advancements in expression levels in cell culture processes over the last decades have probably been the biggest contributor to the maturation of the biopharmaceutical industry's manufacturing capabilities. Cell titers have gone up from below 1 g/l in the early stage to 3–5 g/l routinely nowadays. Some companies even report 5–10 g/l in fed-batch cell culture. This is the result of enhanced clone selection techniques and improved growth medium for cell culture processes.

The tremendous titer increase has been a key enabler in the adoption of single-use bioreactors. This has allowed producing relevant amounts of monoclonal antibody in significantly smaller volumes. Nowadays, a 2000 l bioreactor with an expression level of 5 g/l can produce the same product mass as 20 000 l bioreactor with an expression level of 0.5 g/l two decades ago.

Perfusion cell culture is often considered the preferred cell culture technology for intensified monoclonal antibody manufacturing. The volumetric productivity of perfusion cell culture is significantly higher than traditional fed-batch cell culture. Continuous supply of growth media, including temporary local storage of significant volumes of liquid in the immediate vicinity of the bioreactor, however, may moderate some of the benefits of perfusion cell culture. In addition to this, the longer campaign lengths may affect the overall facility flexibility, and the increased risks and impact of bioburden contamination need to be carefully considered. This may be one of the reasons why in a recent industry survey on biomanufacturing, respondents indicated to expect expansion to be more related to fed batch than perfusion cell culture [5].

A more recent advancement in process intensification for cell culture is focused on the seed train. During this, the animal cells are expanded from one or a few vials from the working cell bank to create enough cells to efficiently produce the protein of interest. This involves a series of cell culture steps that are normally performed in spinner flasks or T-flasks. Transferring cells from one to another spinner flask typically involves open manipulations and is prone to operator error and bacterial contamination. The number of steps in the pre-culture steps can be minimized by combining a high-density working cell bank and the use of a first pre-culture bioreactor with a larger turndown ratio. Further enhancements can be obtained by using a high cell density perfusion process in the N-1 bioreactor (the bioreactor that precedes the production bioreactor).

7.2.3 Process Intensification in Downstream Processing

Probably the most significant enabling technology for continuous bioprocessing has been the introduction of multicolumn chromatography solutions. Chromatography has been (and will probably remain) the most important workhorse in the recovery and purification of biopharmaceutical products. It is traditionally performed in a batch column that is subsequently loaded, washed, eluted, cleaned, and equilibrated. This sequence of discrete steps has long been the most important roadblock for implementing fully connected and/or continuous downstream processing concepts.

In the early 2010s, various solutions for continuous chromatography were launched [6]. The impact of these solutions on the (volumetric) specific productivity of the chromatography process may range depending on the attributes such as titer and process conditions, but specific productivities between 2 and 10 times higher than in batch chromatography have been reported [7, 8]. Especially for chromatography steps with relatively high-cost affinity chromatography adsorbents, continuous multicolumn chromatography has been shown very effective. Protein A capture step therefore has been the most relevant area for applying continuous chromatography [8–12].

Process intensification on the virus inactivation step would not typically focus on the volumetric specific productivity. The reason for that is that the contact time or the incubation time of the intermediate product solution under low-pH conditions determines the volume of the contacting zone, irrespective whether it is an incubation chamber (or mixer) or a flow through device. Instead, the focus in process intensification is more directed toward minimizing or eliminating the time during which product is held prior to the inactivation step. For this reason, the fully automated repetitive batch approach offers an appropriate solution combining efficiency and simplicity. This approach is, for instance, the basis for the Cadence VI system developed by Pall Biotech (Port Washington, NY). The system allows a first single-use mixer to be filled with the eluate of a (continuous) Protein A chromatography step while a second single-use mixer performs the incubation at low pH. Once the incubation and subsequent neutralization have been completed, the second mixer is emptied and becomes available for collecting new eluate. At that point, the product solution in the first mixer will be acidified, and the incubation can start [13]. This concept is illustrated in Figure 7.2. Downstream of the virus inactivation system, a break tank will be required to manage the intermittent release of inactivated process solution.

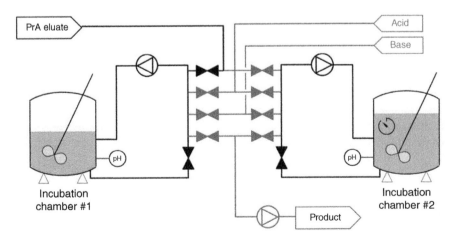

Figure 7.2 The repetitive batch approach for virus inactivation relies on two incubation chambers that alternate between collecting the eluate from the Protein A chromatography step (left: chamber #1) and the incubation/inactivation phase (right: chamber #2).

Tangential flow filtration (TFF) is a very common unit operation in the downstream processing of biopharmaceutical products. Traditionally, these processes require recirculation of the process fluid across the TFF module or cassette, and in most cases, this involves a recirculation vessel. Single-pass tangential flow filtration (SP-TFF) is an approach that eliminates the need of the recirculation vessel. It involves a cascade of TFF stages. As the fluid passes through these stages, the permeation process happens until the concentrated liquid reaches the outlet. The technology is available in a single-use format as the Cadence® in-line concentrator (ILC). For diafiltration processes, a special architecture has been developed, which is commercialized as the Cadence in-line diafiltration (ILDF) technology.

Both ILC and ILDF technologies are relatively straightforward modules and can essentially be implemented as a plug-and-play device. Yet, in a well-controlled, fully automated platform, the operation would involve pumps with flow controls and pressure sensors to adequately monitor the process. In addition to this, one should consider adding sensors to measure the critical process parameters (CPPs) (flow rate, pressures) and critical quality attributes (CQAs) (product concentration).

7.2.4 Process Integration: Manufacturing Platforms

Continuous bioprocessing has become one of the most important enabling contributors to process intensification in the biopharmaceutical industry. As a consequence, most platforms that are being developed and/or implemented involve at least some continuous bioprocessing solutions.

Some companies have started implementing fully integrated (end-to-end) continuous manufacturing platforms. Examples and proof of concept at process development scale were presented by Genzyme [10], Bayer [14, 15], and Merck [8]. BiosanaPharma was the first company to actually bring a monoclonal antibody into a clinical study that was produced on a fully integrated continuous bioprocessing platform [2]. Conceptually, all these platforms rely on the same scheme, as shown in Figure 7.3 relying on a fed-batch bioreactor. However, quite a few companies are exploring the use of perfusion cell culture for integrated continuous bioprocessing platforms as well. In all situations, the workhorse in the downstream processing platform is the continuous multicolumn chromatography technology.

Fully integrated continuous bioprocessing platforms cannot be operated without an adequate level of automation to ensure continuous flow of the product from one unit operation to the next. This requires, as a bare minimum, adaptive control of the flow rates in each unit operation.

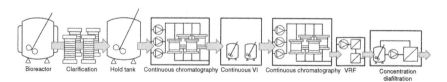

Figure 7.3 Simplified process map for a fully integrated continuous bioprocessing platform (based on a fed-batch bioreactor).

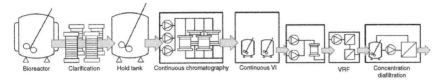

Figure 7.4 Simplified process map for a hybrid platform based on a combination of batch and continuous bioprocessing solutions (based on a fed-batch bioreactor).

While integrated (end-to-end) continuous bioprocessing has been considered as the holy grail by many, some companies chose a more pragmatic approach and implemented hybrid platforms that consist of a combination of continuous bioprocessing solutions and more traditional batch unit operations. A conceptual diagram for a hybrid platform is shown in Figure 7.4.

Proof of concept for a hybrid manufacturing platform for monoclonal antibody production has been published by various end users [9, 16]. An example of the successful implementation of a hybrid platform in a cGMP facility at significant scale was published by Sanofi Aventis (Frankfurt, Germany) [17].

7.2.5 The Two Elephants in the (Clean) Room

In process intensification, most companies primarily focus on the primary flow of the product, which follows the unit operations as also displayed in the process maps in Figures 7.3 and 7.4. This leaves the two other dimensions displayed in Figure 7.5 (being (a) the flow of materials and buffers and (b) the flow of data and information) and the evolution of the platform from process development to commercial manufacturing unaddressed.

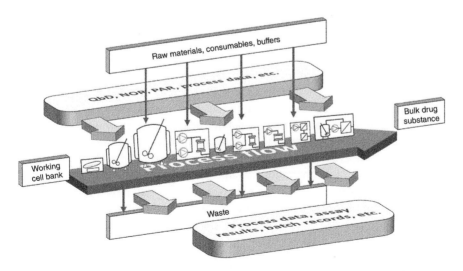

Figure 7.5 Three dimensions through which the opportunities and challenges for process intensification will be reviewed. The fourth dimension (scalability and/or development stage) is not shown.

In a typical monoclonal antibody process, the total buffer consumption takes up a much larger footprint than the process unit operations themselves. A monoclonal antibody process based on a 2000 l fed-batch cell culture can – depending on the expression level – involve as much as 10 000–20 000 l of buffers. All these buffers need to be prepared, stored, released, and transported to the point of use (POU), where they are typically held for another few days until they are needed. This involves significant logistical challenges. From the perspective of lean manufacturing and waste (muda) reduction, there are huge opportunities in this area. This is one of the reason why more and more companies are exploring options for on-demand buffer preparation at POU.

The two leading approaches for this are in-line buffer dilution from concentrates and in-line buffer conditioning from principal solutions. In both situations, the volumes of concentrates or stock solutions are significantly reduced, which has a tremendous impact on inventory and transportation. For both approaches, the system is fully automated and can be integrated into the plant automation. This would allow the buffer release parameters to be automatically included in the batch manufacturing report. In many process steps, these are CPPs, and hence they need to be well controlled and are subject to review to release the bulk drug substance.

The process intensification schemes shown in Figures 7.3 and 7.4 rely on automated systems that should work together in harmony to ensure an uninterrupted flow of the process solutions between unit operations. This creates challenges in both common automation paradigms: distributed control systems (DCS) and programmable logic controllers (PLC) paired with supervisory control and data acquisition (SCADA) systems. In DCS management of set points, control and data collection are centralized, and all unit operation actions are coordinated. This has advantages to mixed unit operation-intensified processes but can negatively impact the speed of control for the fast-moving processes like chromatography. In PLC/SCADA systems, control is typically localized to the unit operation, while set points and data collection are centralized. This has positive impacts on speed of control but makes coordination between unit operations more difficult. Typically a hybrid approach is required to overcome these challenges, and proof of concept of such control systems has been established by Bayer [14, 15], Merck [18], and BiosanaPharma [3, 19] for their integrated continuous platforms. These control systems do not only control the flow rates but also collect all relevant process data from the individual skids and store the data in a central data repository.

The collective data that is being produced during experimentation or manufacturing represents a significant volume. With 10 skids, each carrying an average of 20 instruments (sensors, valves, and sensors) and a sampling interval of 2 seconds, the amount of data adds up to 360 000 data points per hour. For a perfusion campaign of 30 days, this amounts to 10 million data points. This makes it fair to label the amount of raw data as the second elephant in the clean room, and although it may not be as visible as the volumes of buffers, the burden of handling large amounts of data becomes quite clear during the batch manufacturing report review.

This challenge is commonly referred to as "big data." Most definitions of the "big data" problem go back to the definition provided by Gartner [20] that refers to three "V's": high-volume, high-velocity, and high-variety information assets.

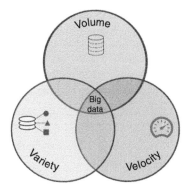 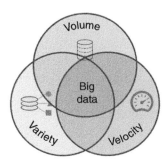

Figure 7.6 Conceptual illustration of the big data challenge. As the three aspects, volume, velocity and variety, exhibit a larger overlap, the big data challenge becomes more prominent.

In the world of process intensification and continuous bioprocessing, the convergence of the three V's increases, resulting in a significant big data challenge (illustrated in Figure 7.6). This convergence is the result of the following trends:

- *Volume*: A typical biopharmaceutical manufacturing process involves multiple unit operations, each carrying multiple sensors and instruments that produce or need data. In process intensification and continuous bioprocessing in particular, the volume of data is typically (at least) 1 order of magnitude larger than in traditional batch processes because it involves many repetitive cycles on the manufacturing systems. Rather than running three to five cycles on a batch Protein A chromatography column, process intensification may result in 20–30 cycles on 3–6 chromatography columns. To cope with continuous flow processing, the process systems also often have a slightly more complex architecture. A typical continuous chromatography system has at least twice as many sensors, pumps, and valves as its batch equivalent.
- *Variety*: In biopharmaceutical manufacturing, the data sources are quite diverse. In traditional processes, most data come in as a one-dimensional flow of structured data vs. time. As unit operations start to run in parallel, the one-dimensional time-based nature of the data flow becomes less trivial. In addition to this, offline in-process control (IPC) samples may be recorded that have to be related to events in the process rather than to specific time points. The data variety becomes even more complicated with analyzers that involve complex spectra such as Raman or near-infrared spectroscopy. The variety of data sources is further augmented if unstructured data is included from, for instance, electronic batch records and information on traceability of raw materials and consumables.
- *Velocity*: In traditional batch processes, the rate at which data comes in may be reasonably high. Yet, in continuous bioprocessing and process intensification, the pace at which the process is operated is faster. In addition to this, the intermediate product solution is not held in surge containers between subsequent unit operations. Instead, the product is continuously flowing from one unit operation to the next with very limited interruptions. As a consequence, there is a need to process the data faster to turn it into meaningful information that allow decisions to be

made (i.e. to use it to control the process). Therefore, the speed of data processing in a continuous process is typically an order more critical than in batch processes.

It is important to note that the definition of "big data" provided by Gartner does not limit the challenge to only collecting the data, but expands it to cost-effective ways of "information processing data that enable enhanced insight, decision making, and process automation."

7.3 Digital Opportunities in Process Development

Downstream process development has been revolutionized in the 2010s with the rise of high-throughput experimentation techniques. This allowed exploring a wider range of process conditions within reasonable timeframes and costs. This has been one of the cornerstones for successful process characterization studies following the quality-by-design (QbD) principles proposed by the regulatory authorities. Exploring a multidimensional design space using a robust design of experiments (DoE) approach became possible with the introduction of robotics and miniaturization in process development and associated analytics.

For continuous bioprocessing, similar high-throughput experimentation techniques do not yet exist, and it is doubtful whether they are realistic. The time scale associated with reaching a steady state in continuous processes is just one barrier to high-throughput experimentation. Another significant challenge in studying the design space using a DoE approach in continuous bioprocessing relates to the high specific productivity of these technologies. The consequence is that significant amounts of (intermediate) product solution is required for each experiment.

The solution to this challenge is based on a model-assisted design approach for continuous downstream processing technologies. For instance, the basic process design parameters for a continuous chromatography process are normally derived from straightforward batch experiments. Adequate and relatively simple design models have been developed to translate such process conditions to a continuous chromatography process [21, 22]. These models have mainly been used for capture processes, such as the Protein A affinity chromatography step in monoclonal antibody processes, but can easily be applied to other chromatography processes as well.

This model-assisted design is most likely the strongest foundation for creating a platform for QbD for continuous downstream processing. In the framework of QbD, CPPs are identified based on how they impact the CQAs of the product. Also, the CPPs for the batch process can be translated to the equivalent continuous unit operation using the model-assisted design approach [23, 24]. This may be easy to be accepted for continuous Protein A affinity chromatography as the separation by itself is extremely robust due to the very high selectivity of the affinity ligand. For polishing of a monoclonal antibody using mixed-mode chromatography, an example of this approach was provided by Utturkar et al. [25]. The data of the mixed-mode

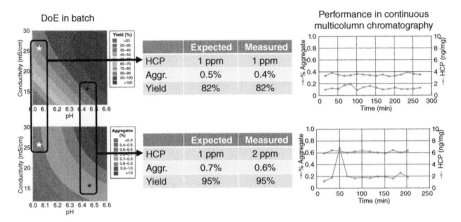

Figure 7.7 Translating a design space for mixed-mode chromatography based on batch DoE data into a continuous multicolumn chromatography process step. Source: Data and graphs based on work by Utturkar et al. [25]. Source: Based on Utturkar et al. [25].

chromatography step is shown in Figure 7.7. The design of experiments performed in batch chromatography were used to generate a design space for two of the CPPs. These process conditions were then translated into a multicolumn chromatography process. For two selected points in the design space, the performance of the multicolumn chromatography process was verified experimentally, yielding exactly the expected performance in terms of yield an impurity removal (host cell proteins and aggregates).

The work described above relies on relatively straightforward design models that often involve some simplifications to make the models more practical in their use. More advanced strategies rely on the use of rigorous numerical simulations using first-principles process models for describing process performance. An example for the use of such models was described for Protein A chromatography, suggesting that numerical simulations can be a powerful aid in increasing process understanding with limited experimentation [26]. In essence, such numerical simulation tool like described is a very simplified equivalent to a "digital twin" for the multicolumn chromatography system.

In one of the first definitions of a digital twin, it was described as a digital or virtual representation of a physical system created to enhance the understanding and information of the physical system. The physical system and its digital twin are connected in the sense that they exchange information to train the digital twin and to provide insight and understanding to the real-life world. In an ideal scenario, the digital twin and the physical system run in parallel and continue to exchange information as to continuously optimize the physical system. This would allow the digital twin to continue to adapt its models to the real world and provide more accurate process information. In such environment, this can reduce the workload during bioprocess development significantly by complementing experimental data with *in silico* results [27].

7.4 Digital Opportunities in Manufacturing

In cell culture processes, dynamic control of CPPs, such as dissolved oxygen (DO) and pH, has been common since its inception. For downstream processing, unit operations have historically been controlled with a more static control philosophy. The architecture of some continuous unit operations, however, allows for fairly elegant solutions to implement dynamic control strategies and move more toward the philosophy of process analytical technologies (PAT).

One example is the use of the signal of a titer measurement in the load zone of a multicolumn chromatography system to ensure consistent loading of the chromatography adsorbent, irrespective of titer and/or capacity decay. Such strategy has been developed for the periodic countercurrent chromatography (PCC) technology by Cytiva (Uppsala, SE) based on the UV absorbance after the primary load column [28]. Combining multiple sensors, such as conductivity and UV spectrophotometric methods, provides an alternative to enhance the resolution of the concentration measurement in the load step of the chromatography process [29]. The use of other more selective, analytical measurements may further enhance the control of (continuous) chromatography process in intensified platforms and allow a fully dynamic response of the chromatography process to variations in the upstream process. With the consideration of enhanced (more selective) analyzers, the data variety increases that makes the control strategy subject to a "big data" challenge as described in Section 7.2.5.

Manufacturing platforms that are based on process intensification and/or end-to-end continuous bioprocessing solutions, automation is a critical element to ensure robust and stable operation. Reducing or eliminating interstage hold containers, for instance, requires that flow rates between unit operations are harmonized. In addition to this, start-up and shutdown procedures become more complex than in traditional platforms without interconnected unit operations. Proof of concept for fully automated platforms based on hybrid or fully continuous bioprocessing platforms has been presented by various companies [3, 8, 15, 18, 19]. This also presents an opportunity to reduce the risk of operator error. In a recent industry survey, operator error was flagged as the most important reason for batch rejection in commercial manufacturing and the second most important reason in clinical manufacturing [30]. Some companies have shared the vision of centralized control of the entire manufacturing platform. This enables the use of a more holistic PAT strategy, where the CPPs across the entire manufacturing platform are controlled based on process information. This opens the door toward enhanced process control strategies such as model predictive control [31] and control strategies based on machine learning and artificial intelligence. Such strategies, however, will benefit from uninterrupted access to any and all relevant process information, coming from a variety of different sources (analyzers, process skids, batch records, etc.).

One area where digital solutions could have a huge impact is the accumulation and sharing of process-related information. One of the biggest challenges during technology transfer is incomplete process understanding, causing oversight of relevant issues during the initial risk assessment and gap analysis. Digital twins can help in consolidating knowledge, experience, and process information and can help in minimizing risks during process portability and technology transfer.

7.5 Digital Opportunities in Quality Assurance

An absolute necessity for (critical) raw data management in biopharmaceutical manufacturing is that it is stored in accordance with the regulatory requirements laid down in 21 CFR part 11 and Annex 11 and the multiple associated guidance documentation. One should expect that having all data stored in one central data repository makes the validation of data management less complex as it would only involve one electronic system. Yet, considering the complexity of the various data structures, the validation effort should not be underestimated. However, having all critical manufacturing data in one repository eliminates the need of transporting data across different databases. With this, such strategy aligns remarkably well with the principles of lean manufacturing.

Material traceability is one of the key elements of batch and lot definitions. In manufacturing platforms that involve elements of continuous bioprocessing, the continuous flow part of the manufacturing platform may present a challenge as the discrete amounts of product produced may no longer be segregated in time so clearly. For batch definition and material traceability in a fully integrated (end-to-end) continuous bioprocessing platform, the use of residence time distribution models has been proposed. Validation of this approach through tracer experiments has been presented [18]. A model-based approach for characterizing the residence time distribution in a fully connected downstream processing platform was presented by [32]. It seems to make sense to combine these two approaches. This is one of the areas where digital twins can also save time and reduce risks during scale-up and technology transfer. Their use allows generating meaningful insights in much shorter time than performing a full experimental characterization of the process dynamics.

With biopharmaceutical manufacturing becoming more complex and regulations more demanding, batch manufacturing records become more complex and time consuming. It has been reported that reviewing a batch manufacturing report for a traditional biopharmaceutical manufacturing process may take anywhere between 48 and 500 hours [33]. The larger amounts of data that are typically generated in intensified and continuous bioprocessing platforms can potentially augment this challenge. The repetitive nature and/or (cyclic) steady state of the process, however, also provide an opportunity to determine whether or not the process has been running consistently and whether trends or deviations have occurred. For this purpose, multivariate data analysis (MVDA) has been proposed to review data derived from continuous multicolumn chromatography processes [34]. The concept of principal component analysis (PCA) for evaluating the performance of multicolumn chromatography processes is conceptually illustrated in Figure 7.8. Such approach has shown to be effective in highlighting small deviations, column-to-column variabilities, drift in performance (of a single column), and even small differences in feed compositions.

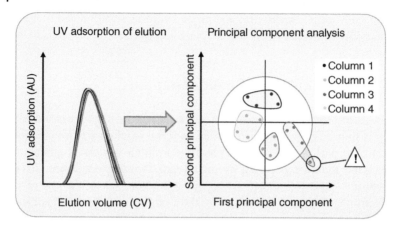

Figure 7.8 Illustration of the use of principal component analysis (PCA) for the UV absorbance of the eluate in a multicolumn chromatography experiment. AU; absorption units. Source: Manser and Glenz [35].

The use of PCA for reviewing large amounts of data in a repetitive or continuous process aligns well with the approach of batch review by exception. PCA by the very nature of it highlights specifically the primary source for deviations and excursions from the average data set. The use of various tools related to MVDA actually turns the large amounts of data generated by intensified manufacturing processes into an advantage, allowing faster and more accurate and more robust data review.

7.6 Considerations

7.6.1 Challenges

Adequate automation and control are critical for manufacturing platforms that are based on continuous bioprocessing technologies or process intensification to work robustly. Such automated and control strategies rely heavily on process information that is sufficiently accurate and timely delivered. With process intensification, the use of single-use technologies across the entire manufacturing platform becomes a viable option and requires sensors and online analytics that are compatible. However, technologies with adequate characteristics and performance have been slow to reach the market. Sensors and online analytics face different challenges that must be addressed to close this gap.

Traditional sensors in stainless steel facilities are already online that overcomes the first major hurdle. To transition these technologies to single use, they must also meet new complex specifications that can be summarized as either separation from the process or advanced stability across several factors. Ideally this is all achieved without performance impacts.

One of the options is to design single-use sensors that are not in direct contact with the process fluid. Very elegant clamp-on sensors exist for measuring the flow rate in a single-use tubing, but these are not yet as accurate and reliable as stainless steel

Coriolis or mass flow meters. Since flow rates are CPPs in many unit operations in a continuous downstream processing platform, this may become an issue.

Some other single-use sensors are in direct fluid contact. This brings in some new challenges for the sensor design. First of all, the part that is in fluid contact has to be manufactured at acceptable costs to justify it as a single-use consumable. The materials of construction need to be selected such that the sensor can be γ-irradiated to the point of sterility without losing functionality. In addition to this, the sensor needs to be pre-calibrated and retains its calibrated status through integration into the single-use assembly and γ-irradiation and finally during the shelf life (typically one or two years). These requirements make it very challenging to design and manufacture single-use sensors with the same reliability and accuracy as their traditional (stainless steel) equivalents.

An example is the single-use pre-calibrated pH probes that are being used for monitoring purposes in some systems. One promising approach has been the use of optical probes based on fluorescence principles that display many of the desired characteristics for single use. However, their response time and range have remained limited for use in critical and fast downstream processing applications. Additionally they are not compatible with all process fluids, and the appreciable drift resulting from photobleaching of the fluorescent dye makes them unsuitable for use across an entire continuous process. As a consequence, in many unit operations, and particularly in process intensification and continuous bioprocessing applications, the main workhorse for pH remains the classical glass probes. Nowadays, pre-calibrated single-use pH probes exist that employ the traditional principles of traditional glass pH probes. These seem to better meet the demands of single-use bioprocessing.

An example of such application is the control of the pH as a CPP in virus inactivation. It is used to control the addition of acid and base into the process fluid to allow consistent operation within the validated design space. In continuous downstream processing platforms, the virus inactivation process is often designed in a single-use format, and hence, a single-use pH probe is used to monitor and control the process.

As pH probes are prone to drift, the data that is generated needs to be carefully reviewed, and the associated process risk needs to be mitigated. Recalibration of the pH probe is not possible without breaching the functionally closed environment of the process, and hence, alternative mitigation strategies need to be established to address the risk of the inactivation process to operate outside its design space. One example of such mitigation strategy based on operating procedures for the virus inactivation unit operation in a fully continuous manufacturing platform has been presented in the public domain [19].

Analytics present even larger challenges than sensors because many are not yet online. Traditional manufacturing approaches have centralized analytical support in quality control (QC) labs. The equipment designed for these labs is large, complex, delicate, and expensive. The process samples are typically manually loaded into sample vials, and the systems are designed for high-throughput analysis. This approach has allowed manufacturers to spread the cost of this equipment across multiple manufacturing suites and consolidate the expertise needed to properly operate this equipment. Because intensified processes cannot tolerate the delays inherent to this approach, a new analytical strategy is needed. Autosampling technologies address these delays by automatically transferring process fluid to the analytical equipment or a robotic sample handler. This removes much of the manual interaction decreasing the time from sample to result, but it does not address the cost or design of the analytical equipment. Longer-term development is needed to eliminate the need for autosampling. Bringing equipment designed for a different environment and use case into the manufacturing suite cannot be seen as a permanent solution, and autosampling itself introduces robustness concerns into the process.

Eliminating autosampling through purpose-built online analytics faces many of the same challenges of single-use sensors. However, there are successful examples. Spectroscopic techniques can be separated from the process by integrating optical windows into the single-use systems. Techniques requiring direct access to the process fluid are particularly challenging, especially those where additional sample preparation is needed (e.g. flow cytometry). Viable solutions have not yet reached the market, but new analytical techniques, miniaturization of traditional techniques through microfluidics, and consolidation of multiple analyses into single devices have all shown a great promise.

7.6.2 Gene Therapy

The trends and examples described in this chapter focus on manufacturing of monoclonal antibodies. This has been the most relevant class of biopharmaceuticals over the past two decades for biomanufacturing. The rise of platform technologies for monoclonal antibody manufacturing in the 2000s has enabled an accelerated industrialization of the biopharmaceutical industry.

Gene therapy manufacturing is lagging approximately 15–20 years behind on monoclonal antibodies. The first gene therapy product was approved in 2003, approximately 15 years after the first approval of an antibody for human use in 1986 (Muromab, a murine IgG2a anti-CD3 to prevent transplant rejection).

Being in the same industry and relying on very similar manufacturing technologies, there is a good reason to assume that gene therapy manufacturing will have an accelerated progression through the industrialization waves described by McLaughlin [1]. The majority of gene therapy products for *in vivo* genetic modification are based on adeno-associated virus platforms as the vector. This allows establishing platform approaches for manufacturing not too dissimilar to what has been observed in the third wave of the industrialization described in Section 7.1.

Another important reason why gene therapy manufacturing is lagging a bit behind in embracing process intensification and digital opportunities is that the analytics for IPCs are still elaborate and sometimes have limited accuracy and reproducibility. This challenge immediately affects the data that is available for making meaningful decisions on further process optimization, developing and training digital twins, and establishing a robust design space.

It is important to acknowledge, however, that the regulatory expectations and requirements for gene therapy products are still evolving. In the recent response of FDA to the Biologics License Application (BLA) on the valrox (valoctocogene roxaparvovec) gene therapy treatment of severe hemophilia A, they suggested that two-year safety and efficacy data would be needed for the approval [36]. This was an unexpected setback for BioMarin and indicated that the regulatory expectations are not fully clear.

7.7 Conclusions

Biopharmaceutical manufacturing is a relatively new discipline, and from its inception, it has always benefited from a strong scientific foundation. With that, it was able to rapidly respond to the needs of the regulated environment that characterizes this industry. Within less than 40 years, the biopharmaceutical manufacturing industry was able to adopt principles like lean manufacturing and evolved relatively fast from the concepts that are typical for "Industry 2.0" toward more automated manufacturing that are characteristic for "Industry 3.0." Along this journey, the biopharmaceutical industry has started embracing concepts such as single-use technologies, continuous bioprocessing solutions, and process intensification.

The next logical stage for biopharmaceutical manufacturing is to also embrace the rapidly emerging opportunities of the digital era. The current trend toward process intensification and continuous bioprocessing will likely accelerate the impact of Industry 4.0 on biopharmaceutical manufacturing.

Process intensification and continuous bioprocessing have all characteristics to generate a "big data" problem. Intensified manufacturing platforms generate large amount of data ("volume") from different sources and data structures ("variety") and requires a very fast processing rate ("velocity") to allow enhanced process control. Yet, if managed well, this also opens the door to opportunities like model predictive control, machine learning, artificial intelligence, and enhanced statistical visualization to enable faster review of manufacturing documentation.

Intensified bioprocessing platforms and continuous bioprocessing are fully aligned with Industry 4.0 and can benefit tremendously from the various opportunities of further digitization of the industry. Digital integration of various systems is involved in manufacturing, from inventory management to quality control and from process development to commercial manufacturing at scale. Furthermore, the data management is one of the "elephants in the clean room" and can gain from the transformation that comes with Industry 4.0.

References

1 McLaughlin, J. (2016). *Integrated continuous and batch operations for efficient initial clinical manufacturing of biopharmaceuticals*. Paper presented at Bioproduction Summit (12–13 Dec 2016).
2 GEN. (2019). First mab produced via fully continuous biomanufacturing. *Genetic Engineering News* (24 Sep 2019).
3 BiosanaPharma. (2020). BiosanaPharma announces successful outcome of comparative phase I study of BP001, a biosimilar candidate to Xolair® (omalizumab). *Press Release* (30 Mar 2020).

4 Smart, N.J. (2013). *Lean Biomanufacturing*. Woodhead Publishing.

5 BioPlan Associates. (2019). Sixteenth annual report and survey of biopharmaceutical manufacturing capacity and production. *BioPlan Associates Inc.* (April 2019).

6 Bisschops, M. (2017). The evolution of continuous chromatography: from bulk chemicals to biopharma. In: *Process Scale Purification of Antibodies*, 2e (ed. U. Gottschalk), 409–429. Wiley-VCH.

7 Pagkaliwangan, M., Hummel, J., Gjoka, X. et al. (2019). Optimized continuous multicolumn chromatography enables increased productivities and cost savings by employing more columns. *Biotechnol. J.* 14 (2): 1800179.

8 Brower, M., Hou, Y., and Pollard, D. (2015). Monoclonal antibody continuous processing enabled by single-use. In: *Continuous Processing in Pharmaceutical Manufacturing* (ed. G. Subramanian), 255–296. Wiley-VCH.

9 Mothes, B., Pezzini, J., Schroeder, K., and Villain, L. (2016). Accelerated, seamless antibody purification: process intensification with continuous disposable technology. *Bioprocess Int.* 14 (5): 34–58.

10 Warikoo, V., Godawat, R., Brower, K. et al. (2012). Integrated continuous production of recombinant therapeutic proteins. *Biotechnol. Bioeng.* 109: 3018–3029.

11 Godawat, R., Konstantinov, K., Rohani, M., and Warikoo, V. (2015). End-to-end fully integrated continuous production of recombinant monoclonal antibodies. *J. Biotechnol.* 213: 13–19.

12 Angelo, J., Pagano, J., Müller-Späth, T. et al. (2018). Scale-up of twin-column periodic countercurrent chromatography for MAb purification. *Bioprocess Int.* 16 (4): 28–37.

13 Pall Life Sciences (2018). *Cadence™ Virus Inactivation System*. Application Note USD3268.

14 Klutz, S., Magnus, J., Lobedann, M. et al. (2015). Developing the biofacility of the future based on continuous processing and single-use technology. *J. Biotechnol.* 213: 120–130.

15 David, L., Schwan, P., Lobedann, M. et al. (2020). Side-by-side comparability of batch and continuous downstream for the production of monoclonal antibodies. *Biotechnol. Bioeng.* 117 (4): 1024–1036.

16 Ötes, O., Flato, H., Winderl, J. et al. (2017). Feasibility of using continuous chromatography in downstream processing: comparison of costs and product quality for a hybrid process vs. a conventional batch process. *J. Biotechnol.* 259: 213–220.

17 Ötes, O., Bernhardt, C., Brandt, K. et al. (2020). Moving to CoPACaPAnA: implementation of a continuous protein A capture process for antibody applications within an end-to-end single-use GMP manufacturing downstream process. *Biotechnol. Rep.* 26.

18 Pinto, N. (2020). Automated material traceability in end-to-end continuous biomanufacturing for batch disposition. In: *Bioprocess International Conference and Exhibition*. Boston, MA.

19 Pennings, M. (2019). *Viral clearance validation for a fully continuous manufacturing process for phase 1 studies*. Paper presented at Integrated Continuous Biomanufacturing V, Brewster MA (7 Oct 2019).

20 Gartner Glossary. (2012). "Big Data." https://www.gartner.com/en/information-technology/glossary/big-data (accessed 22 June 2021).

21 Bisschops, M. and Brower, M. (2013). The impact of continuous multicolumn chromatography on biomanufacturing efficiency. *Pharm. Bioprocess* 1 (4): 361–372.

22 Gjoka, X., Rogler, K., Martino, R.A. et al. (2016). A straightforward methodology for designing continuous monoclonal antibody capture multi-column chromatography processes. *J. Chromatogr. A* 1416: 38–46.

23 M. Bisschops. (2019). *Quality by design for continuous downstream processing*. Paper presented at Driving Value Through Intensified Bioprocessing, Oxford, UK (27 Jun 2019).

24 M. Bisschops. (2019). *Quality by design in a continuous environment – What's required?*. Paper presented at Bioprocess International Conference and Exhibition, Boston MA (11 Sep 2019).

25 Utturkar, A., Gilette, K., Sun, C.Y. et al. (2019). A direct approach for process development using single column experiments results in predictable streamlined multi-column chromatography bioprocesses. *Biotechnol. J.* 14 (4).

26 Bisschops, M. and Brower, M. (2017). Dynamic simulations as a predictive model for a multicolumn chromatography separation. In: *Preparative Chromatography for Separation of Proteins* (eds. A. Staby et al.), 457–477. Wiley & Sons.

27 C. Taylor and C. Herwig (2019) Integrated process modelling – very useful bioprocess digital twin. *Pharmaceutical Engineering*, iSpeak Blog (16 July 2019).

28 Chmielowski, R.A., Mathiasson, L., Blom, H. et al. (2017). Definition and dynamic control of a continuous chromatography process independent of cell culture titer and impurities. *J. Chrom. A.* 1526: 58–69.

29 Rolinger, L., Rüdt, M., and Hubbuch, J. (2020). A multisensory approach for improved protein A load phase monitoring by conductivity-based background subtraction of UV spectra. *Biotechnol. Bioeng.*: 1–13.

30 Bioplan Associates. (2020). *Seventeenth Annual Report and Survey of Biopharmaceutical Manufacturing Capacity and Production*, Bioplan Associates.

31 Rios, M., Herwig, C., Graham, L. et al. (2020). Developing process control strategies for continuous bioprocess. *Bioprocess Int.* 18 (5): 12–16.

32 Sencar, J., Hammerschmidt, N., and Jungbauer, A. (2020). Modeling the residence time distribution of integrated continuous bioprocesses. *Biotechnol. J.* 15 (8): 2000008.

33 Zebib, J. (2019). Review by exception: connecting the dots for faster batch release. *Biopharm. Int.*

34 M. Bisschops, M. Strawn, B. To and J. Coffman (2014) *Multivariate data analysis: managing large amounts of data coming from continuous biomanufacturing processes*. Paper presented at Recovery of Biological Products XVI, Rostock, Germany, 28 July 2014.

35 Manser, B. and Glenz, M. Quality and regulatory considerations for continuous bioprocessing. In: (ed. G. Subramanian). (this book).

36 Tarry, M. (2020). FDA turns down bioMarin's hemophilia A gene therapy. *Biospace*, online (19 Aug 2020).

8

Consistent Value Creation from Bioprocess Data with Customized Algorithms: Opportunities Beyond Multivariate Analysis

Harini Narayanan[1], Moritz von Stosch[2], Martin F. Luna[1], M.N. Cruz Bournazou[2], Alessandro Buttè[2], and Michael Sokolov[2]

[1] Institute of Chemical and Bioengineering, Department of Chemistry and Applied Biosciences, ETH Zurich, Switzerland
[2] DataHow AG, Zürichstrasse 137, Dübendorf 8600, Switzerland

8.1 Motivation

Biopharmaceutical drugs generate yearly revenues exceeding €200 billion, representing about 20% of the pharma market and its largest growing sector [1]. However, their manufacturing requires a complex and regulated journey to transfer a patented drug from lab scale (from 1 ml) to production scale (up to 50 000 l). During the development of the underlying manufacturing process, a great number of potential parameters must be considered starting from the selection of the production organism through the operating parameters of the bioreactor to the subsequent purification steps. Due to the high capital expenditure until market entry and increasing competition, biopharma companies are facing pressure for cheaper process development, faster time to market (13 years on average spent of 20 years patent lifetime, where about 18 months are dedicated to process development), and more consistent production, i.e. reducing failure runs (currently up to 5% resulting in the tens of millions euro in revenue losses per failed batch). There are indications that one-day delay in market entry could cost €0.5 million in net present value per drug [2]. In Europe, more than €3 billion are yearly spent on process development, and more than €66 billion in drug manufacturing [3]. The challenge of fast process development and robust operation is partially due to a low degree of data digitalization, standardization, and central storage and a level of automated and adaptive operation procedures including many sources for human errors [4]. In recent years, top management of large pharma companies has therefore attributed digitalization as one of the main priorities [5].

To tackle the various bottlenecks, the biopharmaceutical industry is looking for digital solutions within their process intensification initiatives toward process analytical technology (PAT), continuous bioprocessing, and robotic experimental systems as a means of resources, capacity, cost, and risk reduction. We have an aspiring vision of the future where smart data analytics and model-based

Process Control, Intensification, and Digitalisation in Continuous Biomanufacturing, First Edition.
Edited by Ganapathy Subramanian.
© 2022 WILEY-VCH GmbH. Published 2022 by WILEY-VCH GmbH.

process digitalization and automation are at the heart of operational excellence in biopharma [6].

However, the current state-of-the-art *in silico* tools are often not sufficient to support the *in vitro* requirements of process intensification [7]. In this chapter, we present advanced modeling concepts that significantly outperform the current industrial benchmarks for different applications, namely, process variable forecasting, product quality prediction, monitoring and control, scale-up and process optimization.

With the proposed methods, combining bioengineering know-how and customized machine learning techniques, the goal is to consistently support decision making across the complicated process development and tech transfer activities, to enable for the corresponding professionals to consecutively advance toward the standards of industry 4.0.

8.2 Modeling of Process Dynamics

In bioprocessing, the upstream process (USP) constitutes the bioreactor that produces the product, followed by the downstream process (DSP) encompassing the purification and polishing steps. This chapter focuses on USP, where the state of the bioreactor is characterized by its design represented by controlled operating parameters such as chosen cells, media, pH, and temperature set points and dynamically measuring key process variables, including viable cell density (VCD) (or biomass), viability and concentration of metabolites (e.g. glucose, glutamine, lactate), and product fingerprint. Process dynamics are usually quantified once per day for mammalian cell cultures and once every couple of hours for microbial cell cultures. Tracking the evolution of these variables is critical to ensure that the process is running within specifications.

Figure 8.1 demonstrates the central role of this information for different applications, all of which are covered by different sections of this chapter. For the process design and optimization, it is essential to forecast the evolution of the process a priori based on solely the process parameters (or process conditions) and initial conditions. Given that bioprocesses are very time consuming and resource intensive, it is pragmatic to simulate different plausible experimental conditions and optimize the process as much as possible *in silico* to finally perform only the promising experiments leading to the high productivity, quality requirement, or process robustness. The connection between process and product information modeling is covered in Section 8.3. The potential of an iterative optimization approach to ensure effective utilization of resources is described in detail in Section 8.4. Additionally, such models can also be coupled with orthogonal modeling techniques for other applications such as process monitoring and control, which is covered in Section 8.5, and for scale-down and scale-up modeling presented in Section 8.6. Finally, Section 8.7 puts the potential of the demonstrated modeling strategies into the perspective of continuous bioprocessing.

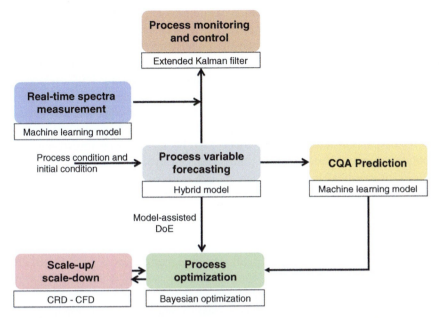

Figure 8.1 Schematic representation of the interaction of models used to forecast process variables with major applications of bioprocesses, namely, quality prediction and characterization, process optimization, process monitoring and control, and scale-up/scale-down.

For USP, various approaches to model the dynamics of the bioreactor have been attempted ranging from pure first-principles-based models to purely data-driven (also called black-box) approaches. For the former, macroscopic kinetic modeling approaches [8, 9] and more detailed models based on cybernetic approaches [10–12] and metabolic flux analysis [13] have been proposed. However, for the latter, the application of several machine learning algorithms spanning from a simple historical partial least squares (PLS) [14] to more sophisticated approaches such as random forests [15], support vector regression (SVR) [16], Gaussian processes [17], and artificial neural networks [18] have been reported. It is noteworthy that given the available commercial packages, the industrial benchmark approach remains the well-known and simple-to-interpret PLS.

Data-driven models capture relevant process behavior based on statistical correlation between input and output variables, and thus causation cannot be inferred from such models. These models can typically be generated quite efficiently and often without considerable expertise. However, a substantial amount of good quality data is required to train a reliable black-box model. Additionally, such models are usually not valid in regions that were not explored during the training. While interpretations of variable importance and correlations and sources of noise and abnormalities can be extracted, the generation of new process knowledge is limited.

On the contrary, first-principles models (FPMs) are based on physical, chemical, and biological principles typically formulated as a mixed system of differential (ordinary or partial) algebraic equations. As long as the underlying principles hold true, FPMs show very strong extrapolation capabilities. On the one hand, generating and validating the underlying physical mechanisms is often very time consuming. It is not possible to create FPMs when the underlying phenomena are not fully understood. Additionally, experiments must be properly designed to estimate the parameters independently and avoid parameter uncertainty [19], making it impossible to use the historical data available in the biopharmaceutical industry.

Hybrid modeling is emerging as a pragmatic solution that combines the advantages of the two modeling paradigms and has been promoted as a PAT to realize quality by design (QbD) [20–22]. The concept has been extensively investigated for microbial cultures over three decades [21, 23–25]. Few hybrid models have also been developed for mammalian cell cultures, though initially in more explorative academic setting [26].

8.2.1 Hybrid Models

Mainly two architectures, serial and parallel arrangements, have been proposed for hybrid modeling [23]. The arrangement is referred to as serial when the black-box model is used to estimate the unknown terms in a system of mechanistic equations and parallel when the black-box model is used to reduce the errors made by the mechanistic model. A simple serial hybrid model for the bioreactors can be formulated as follows:

$$\frac{dC_i}{dt} = \mu_i X_v \tag{8.1}$$

where C_i denotes the concentration of different process variables, μ_i indicates the lumped pseudo-specific rates of consumption/production, and X_v is the VCD. μ_i is estimated using a machine learning algorithm such as a neural network as a function of the concentration and process conditions (cf. Figure 8.2). C_i for mammalian cell culture typically include VCD; concentration of metabolites such as glucose, lactate, ammonia, glutamine, and glutamate; and concentration of monoclonal

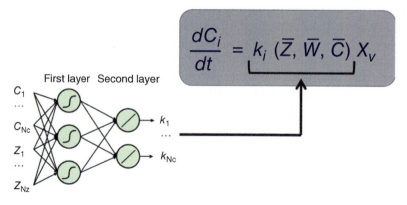

Figure 8.2 A schematic representation of the serial architecture hybrid model for bioprocesses with the artificial neural networks used as the data-driven model to learn the lumped specific rates.

antibody. For microbial cell culture, C_i typically consists of biomass and concentration of substrate (such as glucose) and byproducts (such as ethanol). Depending on the process, if or not the microbial cell culture produces a product, concentration of the product can additionally be included in the system of equations.

Further, the mass balance depends on the mode of feeding. For fed-batch cell culture with daily bolus feeding, the equation reads as follows:

$$\frac{dC_i}{dt} = \mu_i(t) X_v(t) \quad T^+ < t < T+1 \tag{8.2}$$

where T varies from 0 to t_{batch} (last day of experiment) and T^+ indicates the time just after the bolus feed is added.

For a fed-batch bioreactor, the system of equations is adapted as follows:

$$\frac{dC_i}{dt} = \mu_i X_v + \left(\frac{F}{V}\right)(C_{i,\text{feed}} - C_i) \tag{8.3}$$

$$\frac{dV}{dt} = F \tag{8.4}$$

where F is the volumetric flow rate of the feed stream, V is the volume of the bioreactor, and $C_{i,\text{feed}}$ is the concentration of the species i in the feed stream.

Additionally, for continuous feeding (or perfusion reactor), the equation is formulated as follows:

$$\frac{d(C_i V)}{dt} = \mu_i X_v V + F C_{i,\text{feed}} - F_p C_i - F_b C_i \tag{8.5}$$

$$\frac{dV}{dt} = F - F_p - F_b \tag{8.6}$$

where additionally F_p is the volumetric flow rate of the perfusion stream and F_b is the bleed rate of the perfusion bioreactor.

A schematic representation of the hybrid model is represented in Figure 8.2. $\overline{Z}$, $\overline{W}$, and $\overline{C}$ indicate the different information sources being the process condition, the controlled variables, and the process variables, respectively. A detailed description of the classification of information sources can be found in [26, 27].

As a common procedure, the dataset used is split into training and test set, where the models are calibrated on the training dataset and model performance is evaluated on the completely unknown test set. To train a hybrid model, ordinary differential equation (ODE) integration is performed inside an optimization, and the iteration is repeated until convergence. Using an optimization algorithm, the weights of the neural network (or another machine learning algorithm [28, 29]) are obtained that minimize the squared error between model prediction and measurement. Regularized objective functions with L2 norm is preferred to avoid overfitting. The optimal number of nodes (and if required also the number of layers) and regularization parameter is chosen using K-fold cross-validation [30]. Cross-validation is a commonly used technique to assess the model performance when predicting a left-out data subset during training. The performance of the model is evaluated using a root mean squared error in prediction (RMSEP) per variable by averaging across different test experiments and time points.

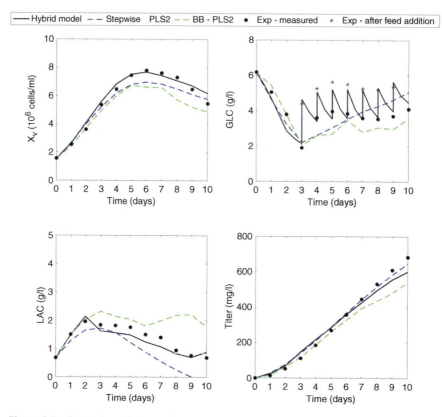

Figure 8.3 Dynamic evolution of four key process variables, namely, viable cell density, glucose, lactate, and titer as predicted by the hybrid model compared against two statistical models based on PLS.

Figure 8.3 shows profiles of four key variables namely VCD (X_v), glucose (GLC), lactate (LAC), and titer predicted by the hybrid model developed using a benchtop scale (3.5 l) bioreactor data of mammalian cell culture operated in fed-batch mode with bolus feed of glucose. The evolutions predicted by hybrid models are compared with predictions of two statistical models based on industrial benchmark method, PLS (detailed description in [26]).

It can be observed that hybrid models are more accurate in predicting the evolution of the key process variables when compared with the industrial benchmark approaches. The comparison of the absolute errors made by the different models in predicting these variables is tabulated in Table 8.1. Additionally, it is highlighted from the glucose profile that the hybrid models capture relevant physical behavior due to the imposed mass balances. In other words, hybrid models predict the degradation of glucose in daily intervals while the statistical models fail to capture this pattern.

Figure 8.4 further showcases the prediction of hybrid model applied to a perfusion bioreactor case study that uses data simulated from a sophisticated mechanistic model of the bioreactor system available in the literature. These demonstrations are

Table 8.1 Comparison of the absolute root mean squared error in prediction (RMSEP) (averaged over all times and experiments) made by the different models, namely, the hybrid model, pure black-box PLS model (BB – PLS2), and stepwise PLS2 model.

	Hybrid	BB – PLS2	Stepwise – PLS2
X_v (10^6 cells/ml)	0.61	0.82	0.80
GLC (g/l)	0.78	0.83	0.87
LAC (g/l)	0.68	0.92	0.95
Titer (mg/l)	45.51	64.30	55.39

Note: Refer [26] for detailed description of the statistical models.
Source: Narayanan et al. [26].

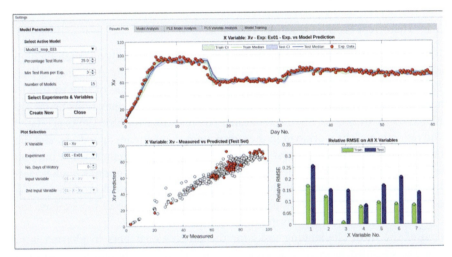

Figure 8.4 Demonstration of the performance of hybrid model developed for a perfusion bioreactor using DataHowLab software. The top panel shows the dynamic evolution of selected process variable as predicted by the model with confidence bands and actual experimental data. The bottom panels show the observed vs. predicted plot of selected process variable and relative RMSE of all process variables during model training and test phases.

created in the DataHowLab software. Again, it is observed here that predictions of the hybrid models are well in agreement with the measurements. This can be inferred from the low relative RMSEP for the different process variables. Additionally, the observed vs. predicted plot for both training and test dataset for X_v lie close to the $y = x$ line without any deviation, indicating that the model has a low variance (thus more generalizable) and does not have systematic bias, respectively. Similar analysis can be performed for other process variables (by simply choosing from the dropdown menu in the DataHowLab). A final analysis that is provided includes the experiment-wise inspection of the profiles of different process variables by comparing the original measurements with the different model predictions (also aided with the confidence interval).

8.2.2 Conclusion

Given the increased accuracy in the prediction of the process variables including titer (or product quantity), hybrid models would be superior in process optimization, for instance, to maximize titer production or minimize certain harmful metabolites in the culture. Additionally, the mechanistic backbone enforced in the form of mass balances makes such models robust in extrapolation, in contrast to purely data-driven models that greatly suffer in predicting beyond the design space where they are trained. Such extrapolation capability is essential for the process optimization.

8.3 Predictive Models for Critical Quality Attributes

8.3.1 Historical Product Quality Prediction

While, during early USP development, the focus is predominantly on reaching high performance, i.e. high titers, thereafter also other critical quality attributes (CQAs) such as aggregates, fragments, charge variants, and glycans are of utmost importance. Many of the decisions of the process development and operations teams are centered on product quality considerations. The relevance of these CQAs is related to the reduction of the workload in the subsequent downstream operations and the establishment of a robust process with a characteristic and reproducible product quality fingerprint. While, for new biopharmaceutical product entities, such fingerprint can be defined strategically based on the promising material produced in the preclinical studies by the developing organization, in the case of biosimilars, the previously defined specifications must be rigorously followed.

One of the most challenging tasks for the process engineers in pharma is to define the so-called process design space, which is the clearly outlined operation boundaries and selections of all process parameters to generate a product within the defined fingerprint specifications. This activity is highly motivated by the regulatory authorities in their QbD initiative with the goal to reach a better understanding between process and product. A second difficult mission is the establishment of a robust process monitoring and control procedure ensuring transparency, trackability, and efficient reactivity of the process so to fulfill the desired CQA characteristics. Also, this objective is stimulated by the health authorities through the PAT initiative with the purpose of empowering technologies, supporting robust process monitoring and control.

Given the large dimensionalities of the potentially important process parameters, the at-line and online process measurements and the different CQAs, process engineers require supporting data analysis tools for concise decision making. Thereby, it must be pointed out that particularly the limited process understanding of the upstream bioprocesses results in decisions being made under uncertainty. Therefore, in addition to the dimensionality, noise in the process and in its analytics and (human) errors in the process documentation and operation must be taken into account. Another important constraint in the pharmaceutical industry is the necessity to provide such decisions under time pressure, trying to reduce process development costs and accelerating time to market.

To provide quantitative decision support on process design and monitoring, two data analyses have been established as common benchmarks in the pharmaceutical industry. Design of experiments (DoEs) coupled with response surface method regression are used to understand the relationship of process parameters to CQAs. Multivariate visualization with principal component analysis (PCA) supports understanding important correlations in either the process or product variables, while multivariate regression with PLS links the process history to the final product outcome. Both methods rely on purely linearly parameterized models, which however if coupled with an adequate data pretreatment and regression stabilization have shown to yield an effective possibility to provide an important insight into the nonlinear nature of upstream bioprocesses.

As an example, let us consider a published case study of historical process modeling using several designed experimental campaigns at ambr15 scale [31]. Figure 8.5a,b demonstrates the scaled cross-validation error for the prediction of two different CQAs, fragments (lower molecular weights [LMW]) and charge variant (C3). The corresponding errors are often standardized with the standard deviation of the respective CQA to enable a simpler comparison across different

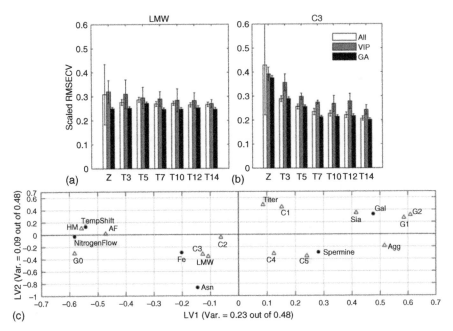

Figure 8.5 Cross-validation error of historical PLS1 models for the prediction of two CQAs, LMW (a) and the charge variant C3 (b) based on different amounts of process history (from only using designed set point and initial conditions Z to using all process history until day $T = 14$) and different feature selection methods (using all variables vs. selections with VIP and GA methods). (b) PLS2 model-based assessment of the interrelationship of process variables (black circles) and CQAs (triangles). Source: Sokolov et al. [31].

CQAs. Along the horizontal axis, different models are distinguished based on different length of process history incorporated, ranging from only the designed process parameters Z to the complete process history until day $T = 14$. First, this work demonstrated that historical process models, i.e. models based on the information on the process design coupled with dynamic process measurements until a certain point of time, can decently predict most of the final CQAs. Moreover, three variants of the corresponding models were generated, using all (in white) and selected process information with the variable importance in projection (VIP) (in gray) or the genetic algorithm (GA) (in black) methods. One can observe in Figure 8.5a that LMW can be decently predicted already at process start, while C3 requires important information until day $T = 5$, where, for instance, a pH shift was performed. Moreover, the GA feature selection enables in both cases to improve not only the accuracy but also the robustness of the models; as compared with the other methods, it tends to reduce noise and redundancy most rigorously.

In addition to the predictive ability of such models, their capability to support process understanding is of central importance. Given the high correlation in the process and product information spaces, the so-called PLS2 approach enables to link the entire process information to the complete product quality profile in a single model. This has the immense advantage of interpreting only one model for process understanding. Figure 8.5c presents the example of superimposed loadings of process (black circles) and CQA variables (triangles). Such visualization provides the possibility to directly evaluate the process variables that are likely to have a potential effect on CQAs within a single plot. For instance, the level of temperature shift seems to affect the high mannose (HM) and afucosylated (AF) glycoforms, while iron (Fe) addition into the feed tends to affect C3 and LMW and asparagine additions tend to negatively affect the final titer.

This approach can also be used in even larger spaces of correlated variables. In bioprocessing, these encompass online and spectral data, which will be considered in the following section, as well as omics data. The following use case applied PLS2 models to evaluate which metabolomic information could improve the prediction of the overall glycan pattern and hence the understanding of this fingerprint from the dynamics of the process. This work demonstrated that an improvement of the predictive ability could be achieved based on additional extracellular information. Moreover, this approach helped to identify potentially important biomarkers (shown in green in Figure 8.6), which were selected based on a variable importance assessment in the predictive models. With complex process analytics becoming more diverse, assessable, and affordable, such identification can indeed support more robust real-time process monitoring with the goal of predicting the CQAs in real time and the ultimate goal of quality by acceptance and quality by release.

The aforementioned examples were motivating the utilization of multivariate techniques to create and better understand the interrelationship between process and product information. It is important to point out that such link can also be established by methods often applied by the machine learning community. Two examples presented in Table 8.2 are support vector regression with a radial-basis kernel function (SVR-rbf) and multitask elastic nets (MEN). The former has the advantage of being nonlinear compared with PLS and hence being able to learn

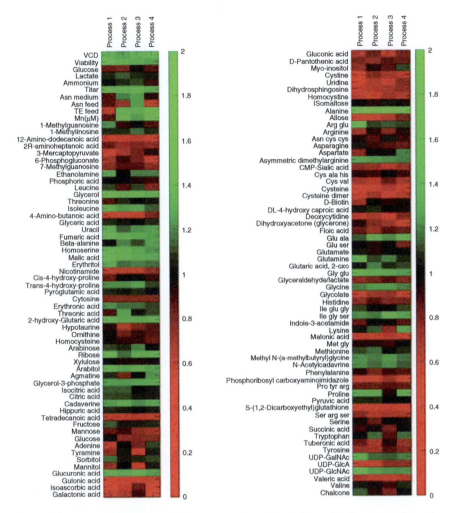

Figure 8.6 Heat map of extracellular metabolites selected based as important (green) from regression analysis. Source: Zürcher et al. [32].

also more complex interrelationships. The latter has the advantage in its intrinsic regularization, centering the model on only important variables.

Table 8.2 compares the performance of these models on the same industrial use case from Figure 8.5 based on their absolute prediction errors (RMSEP) of the CQAs in an external test set. First, one can observe that there is not a single model, which always outperforms the others so that the choice of the optimal predictive model is likely to be case specific (here to CQAs but usually also to dataset). In many cases, the multivariate models show promising performance, while for titer, G0 to G2, and C1, at least one of the machine learning models features better predictive power. This shall motivate to consider such approaches, especially if an optimal predictive ability is desired. When comparing and utilizing such models in process optimization and control, it is important to not only compare the predictive accuracy but also model

Table 8.2 Comparison of absolute RMSEP of four different models for final CQAs of a cell culture process.

	PLS2-vip	PLS1-vip	SVR-rbf	MEN
Titer (g/l)	394.68	397.28	349.89	339.72
Agg (%)	0.29	0.31	0.33	0.32
LMW (%)	0.26	0.18	0.20	0.17
C1 (%)	1.56	1.68	1.23	1.39
C2 (%)	1.20	1.63	1.74	1.21
C3 (%)	1.57	1.71	1.66	1.65
C4 (%)	0.31	0.30	0.31	0.31
C5 (%)	0.07	0.08	0.07	0.06
G0 (%)	4.06	3.84	3.10	3.78
G1 (%)	3.43	3.25	2.54	3.26
G2 (%)	0.83	0.83	0.65	0.73
HM (%)	0.41	0.36	0.39	0.39
AF (%)	0.55	0.49	0.53	0.51
Sia (%)	0.27	0.25	0.27	0.29

robustness. However, for applications focused on exploration and process understanding, linear multivariate methods can be superior due to their more attractive interpretability.

8.3.2 Synergistic Prediction of Process and Product Quality

Finally, in the context of process design and optimization, it is important to highlight the joint relevance with the methods for process prediction presented in Section 8.2. Figure 8.7 presents schematically that to build a solid predictive connection between process design Z and final product quality Y, the dynamic information X is first represented with the help of a hybrid model. In the next step, process design and (predicted) process dynamics are linked with the final CQAs using a historical model. This two-step procedure enables to simulate, for a certain process design Z, the evolution of the process and product quality, and to stabilize such prediction compared to pure black-box approaches that neglect the valuable dynamic information and simply link Z to Y. Such suggested, knowledge-centered approach enables predictive methods with stronger extrapolation power, which provide a much more robust basis for process optimization towards the desired specifications as described in the following section.

8.4 Extrapolation and Process Optimization

Systematic process optimization requires an understanding on how process parameters impact the process performance. Though platform processes based on human

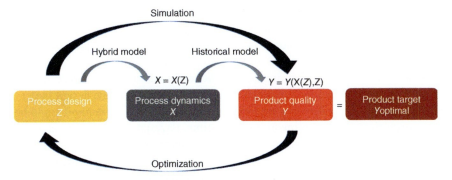

Figure 8.7 Two-step methodology to connect process design Z to final product quality Y based on process dynamics (X) prediction with hybrid models and subsequent historical models. This procedure enables process simulation for not tested process design and optimization toward product quality specifications. Source: Sokolov [33].

embryonic kidney (HEK), Chinese hamster ovary (CHO), *Escherichia coli*, etc. dominate the production of biologicals (limiting the need to redesign the entire process train), the optimization of a large number of process parameters (selecting tens of optimal conditions including cell line, medium, and operating parameters) is still required to render the production of a new biological economically attractive.

In the absence of fundamental process understanding, data-based process understanding can be established by combining DoE methods for the generation of information-rich datasets with data-driven models trained on the generated data. These approaches are very systematic, provide planning security, and can directly exploit the ever-increasing experimental capacity provided by (miniaturized) high-throughput platforms. However, the number of experiments increases (at best) polynomially with the number of process parameters under investigation, aiming at deciphering the main effects, interactions, and nonlinear behavior. To reduce the overall number of experiments, an intensified design of experiment (iDoE) method could be adopted if the temporal domain can be exploited, i.e. if intra-experiment step changes in the process parameters can be performed. For upstream microbial bioprocesses, this approach has been able to reduce the number of experiment up to threefold [34–36].

In the case that fundamental process understanding in the form of a mathematical model is already available in the beginning of the process optimization exercise (that can at least roughly describe the process behavior), then this model can be exploited to design experiments in such a way that the behavior of the process is investigated where it interests the most, e.g. close to the optimum (exploitation) and/or where the process is understood the least (exploration). This approach has been referred to as iterative learning control and iterative run-to-run (batch-to-batch) optimization or model-based design of experiment and optimal experimental design, depending whether the focus was more on exploitation or on exploration, respectively. Classically, optimal experimental design focused on improving the model parameter estimates (capitalizing on the Fisher information matrix) [37–39] and/or on discriminating between structures of different models

[40–43], the models being of mechanistic nature. However, with the advent of data-driven methods, whose structure and parameters are determined from data, the focus has shifted to the exploration of the feature space, i.e. the space of inputs to the data-driven models. Brendel and Marquardt have, for instance, shown that model-based space exploration approach is better in exploring the space than a classical DoE when dependent variables should be considered in the design [44].

Rather than choosing between exploration and exploitation, a trade-off could be chosen, exploring and exploiting at the same time. Bayesian optimization is based on this principle where different acquisition functions can be chosen that differently handle the trade-off when exploring the space [45]. For the iterative design of four parallel experiments, von Stosch [46] has used a similar idea, dedicating one experiment to full exploitation while assigning the remaining three to explore the space such that the experimental conditions are as different as possible while yielding still 80% of the performance of the exploitation experiment.

Here, a similar approach that simultaneously seeks to exploit and explore (cf. Figure 8.8) has been followed. An ensemble model is used to describe our process, as it can provide us with a prediction (here the median of all predicted values is used) and an associated prediction interval (the area between the 0.1 and 0.9 quantile). Subsequently, a given m number of experiments are sequentially designed that can be executed in parallel (e.g. on a miniaturized high-throughput platform). Once the optimal conditions for all parallel experiments are determined, the experimental run is executed (note: an emulator is used to simulate the experiments, allowing to objectively compare the obtained results and compare them with the theoretical optimum). The objective function is slightly adapted for the design of each of the parallel experiments by weighing exploration against exploitation differently. Two different strategies for updating the objective functions, as listed in Table 8.3, have been investigated.

In the case of strategy 1, it can be seen that design 1 is focused on exploitation, maximizing the target. In designs 2–4, the exploitation is increasingly constrained by the size of the prediction interval, i.e. the larger the prediction interval, the greater the penalization. Typically, for exploration the aim would be to maximize the prediction

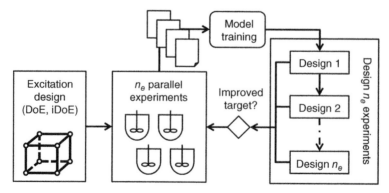

Figure 8.8 Schematic representation of the adopted iterative run-to-run optimization of the sequential model-based design of ne parallel experiments.

Table 8.3 Objective functions used in the iterative parallel optimization.

	Strategy 1	Strategy 2
Design 1	$\max_{x} \{\widetilde{T}(x)\}$	$\max_{x} \left\{ \widetilde{T}(x) \cdot \dfrac{e^{0.5 \cdot \widetilde{T}(x) - T_I(x)}}{1 + e^{0.5 \cdot \widetilde{T}(x) - T_I(x)}} \right\}$
Designs 2–4 ($j = 1..3$)	$\max_{x} \left\{ \widetilde{T}(x) \cdot \dfrac{e^{c_j \cdot \widetilde{T}(x) - T_I(x)}}{1 + e^{c_j \cdot \widetilde{T}(x) - T_I(x)}} \right\}$	$\max_{x} \left\{ T_I(x) \cdot \dfrac{e^{\widetilde{T}(x) - p \cdot \widetilde{T}_1}}{1 + e^{\widetilde{T}(x) - p \cdot \widetilde{T}_1}} \cdot \prod_{i=1..j} \sum \dfrac{|x - x_i|}{lb - ub} \right\}$

Notes : $\widetilde{T}$ is the median predicted titer value obtained for the vector of process parameters, x, i.e. the process condition vector. T_I is the width of the prediction interval. The index on the $\widetilde{T}$ and x refers to the design. lb and ub are the lower and upper bounds used for scaling. p is a performance threshold, 85% in this study. c is a vector of prediction interval limits, $c = [0.3, 0.5, 0.7]$.

interval, as areas in which the prediction interval is large are those where the model can learn. However, in particular during the first iterations, the prediction interval at the optimal exploitation condition will be large as the optimal conditions are typically very different to those composed in the original dataset, wherefore the model is extrapolating. By constraining the extrapolation with the risk of misprediction, i.e. the prediction interval, the obtained optimal conditions yield typically better experimental performance than when unconstrained [47–50]. In the case of strategy 2, the idea of strategy 1 for the first design has been followed. In contrast, designs 2–4 aim at maximizing the prediction interval, penalized only by the distance of the investigated conditions to those identified in prior designs and a minimum performance threshold.

An emulation model is used to compare the performance of the two strategies. The emulation model consists of the material balances for VCD, the classic substrates (glucose, glutamine), byproducts (lactate, ammonia), and titer. In total, 14 continuous factors can be optimized, such as temperatures, pHs, feeding profiles for different phases, and duration and timings of the phases. The data generated with the emulation model are corrupted with random noise. The initial values also contain typical experimental variations. The optimization problems of strategies 1 and 2 are solved using a Bayesian optimizer, followed by a gradient optimization using the Matlab functions bayesopt and fmincon, respectively.

The results obtained for the iterative parallel design of four parallel reactors using an emulated cell culture model are shown in Figure 8.9 for eight iterations. Both strategies seem efficient in maximizing titer over the number of iterations. For the full exploitation design, design 1 of each strategy, the trend of the titers is to successively increase, maybe a bit faster with strategy 2. In the case of strategy 1, the obtained experimental titers for iterations >3 are more consistent among each other and also more consistent with the predicted values than for strategy 2, which would be expected given that strategy 2 is exploring the parameter space where the model is "weak." This can also be seen, looking at the width of the prediction interval as a measure for learning potential, i.e. the width in the case of strategy 1 is generally much lower than for strategy 2. It is interesting to note that in the case of

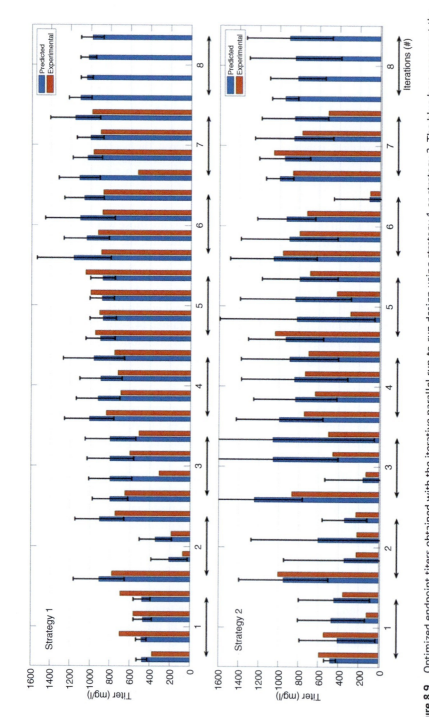

Figure 8.9 Optimized endpoint titers obtained with the iterative parallel run-to-run design using strategy 1 or strategy 2. The blue bars represent the median ensemble model predicted titer values, the black lines represent the prediction intervals, and the orange bars represent the experimental value obtained with the respective optimal conditions.

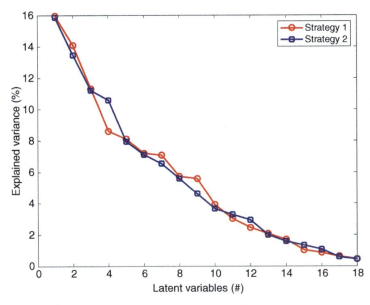

Figure 8.10 Explained variance over number of latent variables obtained for a principal component analysis of the matrix of optimal conditions $X = [x_{1,\ldots,n_e,1\ldots n_{\text{iterations}}}]$ for strategy 1 and strategy 2.

strategy 1, though the predicted titers across an iteration are many times similar, the respective process conditions are fairly different, yielding also varying experimental results. Hence, also strategy 1 is effective in exploring the parameter space. This becomes even more apparent looking at Figure 8.10, where the results of a principal component analysis are displayed, which was performed on the matrices of optimal process conditions of each strategy. It can be seen that in the case of both strategies, the explained variance decreases in a similar way for an increasing number of latent variables, suggesting that the process parameter dimensions are explored similarly well with both strategies. All in all, both strategies are efficient in exploring the parameter space in the direction of the process optimum, which with both strategies is attained during the first 24 experiments. In comparison, performing a classical design of experiment study on the 14 continuous factors typically requires a greater amount of experiments.

8.5 Bioprocess Monitoring Using Soft Sensors

Apart from process optimization, another important application is process monitoring and control, where modeling tools play a key role. System monitoring is crucial in the operation of industrial bioprocesses. Any decision or control activity, as well as the evaluation of the bioprocess performance, requires measurements of the critical process parameters (CPP) that are both fast and reliable. In the last decades, soft sensors have arisen as efficient and robust options within the PAT paradigm [40, 51–53]. A soft sensor is a measuring tool that combines an algorithm (software) with measurements or signals from the system (one or more traditional sensors) [54]. The algorithm used is considerably more complex than a linear transformation of a single signal and can process multiple data inputs to return variables that are difficult, expensive, or time consuming to measure.

The constant ramping in computing power together with the rising of digitalization and connectivity of digital devices makes soft sensors a logical step in the evolution of process monitoring and control. Signals from traditional sensors are already digitalized, making it very easy to be either processed in situ or transmitted to be used somewhere else, either online or offline. Since the processed signal is also digital, it can be directly used in other engineering task such as model-based control or online optimization. This is very important to close the loop in the Industry 4.0 practice, where the connection between the real system and the digital representation has to be bidirectional to achieve real automation.

An example has been presented already in Section 8.3, where offline measurements of extracellular variables (glucose, lactate, etc.) and process parameters (medium composition, feeding profiles) are used in combination with machine learning methods to predict the CQAs of mammalian cell bioreactors. The measurement of CQAs like the glycosylation patterns is much more expensive and time consuming than the analytics used in this case. Thus, once the sensor has been properly trained and validated, it can drastically reduce the analytical effort by partially replacing the more expensive and time-consuming assays.

There are different ways of classifying soft sensors: according to the algorithm (mechanistic or data-driven models), to the availability of the measurement (online or offline), or to the dynamics of the sensor (dynamic or static soft sensors). Here, the last type of classification is chosen since the complexity of the method and its implementation varies greatly between both classes [6].

In a static soft sensor, the algorithm uses measurements of one or more variables to make inference about the state of the system at the time of measurement. No process history or dynamics are involved in the calculations, which greatly simplifies the training and implementation of the sensors. These soft sensors resemble traditional ones, since they only depend on the current state of the system and can be applied at any given time without making assumptions of previous states of the system. On the other hand, dynamic sensors use information not only from the current state but also from the process history and dynamics. These sensors demand more work to be developed and present a higher level of complexity. However, they are usually more robust since they include more and dynamic data and first-principles process knowledge.

8.5.1 Static Soft Sensor

Examples of static sensors used in the biotech industry include systems of electrodes, such as electronic noses and electronic tongues [55], but probably the most interesting technology is spectral sensors based on near-infrared (NIR) and Raman spectroscopies [56, 57]. Both techniques are spectroscopic method. However, while NIR spectroscopy is based on absorbance on the infrared region, Raman spectroscopy is based on the inelastic scattering of monochromatic light. Raman signal is usually weaker than NIR, but it has the advantage of very low interference from water molecules.

Spectral data is multivariate in nature, since it consists of one signal for each wavelength of the spectrum. For pure chemical compounds, the peaks in the spectrogram are characteristic of each molecule, and the intensity of the signal at a given wavelength can be used to estimate the compound concentration. However, for biological process, this is not the case, since the samples consist on a mixture of many complex biological macromolecules, like proteins and sugars. Thus, the different characteristic peaks overlap, creating a rather messy spectrogram. To cope with this issue, multivariate data analysis methods are required to combine all the information from the different wavelength and correlate them with the concentration of the species of interest. Several methods have been used in the past, with PLS being the first and most established one. However, the nonlinear behavior of the system may be too complex for PLS models, and other machine learning methods are becoming more popular, like supported vector machines (SVR) and Gaussian process (GP) or combinations of both (stacking methods). Which method is better depends on the particular system and dataset, and the soft sensors must be validated with new experimental data after training.

Due to the great complexity and high noise-to-signal ratio, preprocessing of the signal is key in the development of reliable models [58]. Besides traditional techniques like mean centering and scaling, smoothing of the spectra (using method like the Savitzky–Golay algorithm) is very important to reduce noise. Outlier removal methods (using criteria like Hotelling's T2 distance) help to avoid unreliable data points that would hinder the training of the model. Wavelength selection can be very useful to avoid regions of the spectrum where the solvent masks any other species rendering the information from these wavelengths useless or just to use the most informative wavelengths to decrease noise.

In the case of bioreactors, Raman spectroscopy is a very promising tool to create soft sensors capable of measuring different species with a single device. With only one submerged probe, the spectral signal can be used in combination with the aforementioned models to measure very important process variables such as VCD, glucose, lactate, ammonia, and protein titer, among others. The simple hardware requirements of using only one probe, the relatively fast sampling frequency (second to minutes) and the online availability of the measurements, make these soft sensors very suitable for monitoring and control of bioreactors. Figure 8.11 shows how spectral data can be used for monitoring a fed-batch cell culture bioreactor. First, a dataset containing six runs with both spectral and analytical (high performance liquid chromatography [HPLC]) data of several species is used to train GP models that correlate both sets of measurement. The spectral data was preprocessed using smoothing and feature selection techniques. All the preprocessing and training were done using in SpectraHow. After the model was trained with six experiments, the soft sensor is tested in an independent experiment using only spectral data. Measurement from the soft sensor for VCD and glucose concentrations are presented in Figure 8.11, together with the offline analytic used as references. As can be seen,

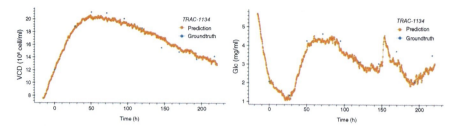

Figure 8.11 A soft sensor trained in SpectraHow is used to predict VCD and glucose concentrations for a cell culture bioreactor.

the soft sensor estimations closely follow the reference values while also displaying a much higher sampling frequency.

8.5.2 Dynamic Soft Sensors

Dynamic soft sensors presented here rely not only on instantaneous measurement but also on historical data and knowledge of the process dynamic. Besides the example presented in Section 8.3, where a historical model uses data from different time points to predict unobservable variables, soft sensors that uses dynamic models together with system measurements present a powerful tool for bioprocess monitoring and control.

These dynamic sensors are also called state observers [59] and are already very popular in - diverse engineering fields, like aeronautics or chemical engineering. Perhaps the most famous state observer is the Kalman filter, with its different versions: linear, extended, unscented, etc. [60, 61]. The idea behind this approach is that the sampled data together with the process model can be fused considering the uncertainty of both information sources. This uncertainty is contained in the covariance matrix of the system, which is also dynamical, and its evolution should be tracked by the method. The system measurements are usually unbiased but can be noisy and subject to oscillation (especially when dealing with weak signals like in the case of high dilutions). The dynamic model has no noise in its predictions, but the deviations from the real behavior of the system due to the unavoidable plant/model mismatch or the occurrence of process deviations (pump malfunctioning, temperature oscillations) may slowly create a drift away from the real system. The soft sensor would take the best from both worlds, and the fused estimation provided by the filter should have a smaller error than both the measurement and the model alone.

The main equations of the Kalman filter are as follows:

$$x_n = f(x_{n-1}, z_{n-1}) + q \tag{8.7}$$

$$\hat{y}_n = h(x_n) + r \tag{8.8}$$

$$\hat{x}_n = f(x_{n-1}, z_{n-1}) + K(\hat{y}_n - h(x_n)) \tag{8.9}$$

where x_n and z_n are the states and process variables at time n, f is the state transition function (the dynamic model), h is the measurement function that relates the model with the sensor measurement $\hat{y}_n$, K is the Kalman filter gain, and $\hat{x}_n$ is the improved estimation. The process and measurement noise q and r and their respective covariance matrices Q and R are used to calculate the state covariance P. This matrix should also be integrated and updated dynamically (the equations are not shown), and it describes the uncertainty of the system. It is fundamental in the critical step of the filter: the calculation and tuning of the gain K. The different variations of the Kalman filter differ on how this step is done.

While soft sensors based on the Kalman filter have been successfully applied to other engineering fields for years, their implementation in the biotech industry is relatively recent [62]. Bioprocesses are very complex systems, requiring a nonlinear version of the filter. In USP, the lack of reliable first-principles models hindered the implementation of these methods. The development of more complex models in the last decades (using either mechanistic or hybrid approaches) together with the increase in computing power to execute them has prompted a rise in Kalman filter sensors. As an example, let us consider the mammalian cell bioreactor presented by Narayanan et al. [27]. An *in silico* dataset is generated for a fed-batch bioreactor for the production of a monoclonal antibody. Glucose and glutamine are supplied to the reactor in pulse injections, while the principal species concentration (VCD, glucose, glutamine, lactate, ammonia, protein titer) are measured using an online probe. One hundred experiments, lasting 14 hours each, are generated and divided into a training and a test set. Most of the training set is used to develop a hybrid model like the one presented in Section 8.2, while a handful of experiments are used to tune an extended Kalman filter that combines the model with the probe measurements. Figure 8.12 presents the comparison between the model, the probe, and the soft sensor for VCD and glucose along with the reference value for an experiment from the test set. It can be seen that the Kalman filter sensor is unbiased compared with the hybrid model and has a smaller noise level than the probe signal. Overall, the error of the Kalman filter sensor is smaller than both information sources alone for both species.

8.5.3 Concluding Remarks

Soft sensors present powerful tools for the biotech industry. The improved accuracy, online availability and the possibility of doing indirect measurements can help to reduce costs and increase robustness of the production processes by enabling advanced process control with potentially less manual operations and human errors. In combination with other control and optimization techniques, they would become a central part of the Industry 4.0 paradigm.

8.6 Scale-Up and Scale-Down

Another important aspect worth considering here is that different stages of the bioprocess are performed at different scales. Screening of the various process conditions such as pH, temperature, media, etc. are performed at small scales in deep

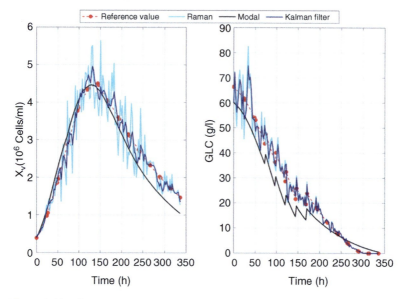

Figure 8.12 Comparison of a Kalman filter soft sensor with hybrid model predictions and probe measurements for an *in silico* fed-batch bioreactor experiment, along with the noise-free reference values.

well plates, ambr systems, or parallel microbioreactors to understand the complex dynamics. This understanding can be formalized into models used to optimize the process toward certain goals such as maximizing titer or meeting quality profile. Subsequently benchtop scale experiments are used to demonstrate robust process dynamics and quality profile. Finally, after confirmation (at pilot scale), the process is transferred to production.

8.6.1 Differences Between Lab and Manufacturing Scales

However, it is well known and documented that stress conditions at industrial scale induce changes in the performance of cell factories compared with more homogeneous lab-scale processes [63]. Not only mammalian processes [64, 65] but also robust microbial processes struggle under heterogeneous conditions that impose sudden shifts in pH, temperature, and substrate concentration, among others [66–70]. Unfortunately, it is very difficult to predict the effect of these stress conditions beforehand. This leads to major difficulties in bioprocess development and regulation compliance at production scale [71].

While energy input and material costs are of relatively low concern at lab scale, the energy required to stir an industrial bioreactor of $10\,m^3$ or more can be a major cost; aeration costs represent one of the major costs in wastewater treatment, for example, or substrate quality used in specific experiments (e.g. defined media) is not profitable in production conditions. A well-known example of further scale effects is the speed of the stirrer mainly at its tip, which increases with the radius of the stirrer at constant revolutions per minute. Additionally, the dimensions of the vessels pose relevant physical constrains, e.g. transport phenomena that are neglectable in benchtop bioreactor become relevant as are heat removal, diffusion, and hydrostatic pressure. Trivial tasks in the lab as powder mixing, storage, or handling hazardous material may become big hurdles at industrial scale. Also, sterility at large scale can become a significant percentage of the total costs depending on the process and organisms selected. Finally, especially in continuous processes, the interaction of upstream with the downstream equipment should be considered from early stages since volumes are significantly larger than buffer tank capacity.

8.6.2 Scale-Up

A proper understanding of this large-scale phenomena and its prediction at the initial stages of development becomes essential as production scales increase and new technologies with even lower power inputs (e.g. single-use systems) gain popularity. In bioprocesses, scale-up needs to consider both the changes due to the transport phenomena and physical limitations and the variations in the cellular response, making it arguably the major challenge in the bioprocess development.

The interaction between mixing and bioreactions has been largely investigated by modeling methods. In the early 1970s, Tsai and coworkers investigated this phenomenon using the concepts of complete segregation and maximum mixedness [72–74]. Bajpai and Reuss introduced some refinements (e.g. circulation time distribution) [75] but using an unstructured kinetic model for bioreaction.

Compartment models can be used to describe the heterogeneous concentration profiles inside large bioreactors [76–78], considering the spatial distribution of species concentration. Still, the major issue is that specific rates are known to depend on the metabolic state and history that led to further extensions as are probability density functions for continuous population balance equations (PBE) [78–80] on the one hand and cell-based models along with Monte Carlo techniques [81, 82] on the other. Most recently, simple interaction by exchange with the mean (IEM) mixing models has been shown to describe the glucose concentration segregations in industrial scale and laboratory-scale bioreactors properly [83]. Finally, Lagrangian particle tracking can produce a temporal signal that is used as the boundary condition for a biological [84, 85] to enable a lifeline analysis.

The most recent advances in the development of scale-down concepts include the coupling of computational fluid dynamics (CFD) models of bioreactors with cellular growth kinetics (cellular reaction dynamics [CRD]) and the mechanistic description of population groups in heterogeneous environments [86–90]. The CFD–CRD models have been used to define specific stress exposure times that are assumed to occur at the larger scale, based on mixing characteristics (CFD simulations) and dynamics of cellular responses. However, the evaluation of the detailed physiological adaptation to oscillations and their incorporation into CRD models can be an enormous amount of work, as is obvious from the works of Vanrolleghem and Canelas [91, 92].

For the prediction of gradients in a bioreactor, an alternative to the experimental approach relies on computational models that integrate both physical and biochemical phenomena [93]. In recent years, CFD simulation has become an effective tool to model multiphase reactors rapidly and accurately. Based on the numerical solution of the conservation equations for mass, momentum, and energy, the applicability of CFD simulations has been demonstrated for single-phase and two-phase flows in various chemical engineering applications. The same approach can be successfully used for bioreactors by implementing suitable biochemical models for the reaction term. However, in the very case where large concentration gradients are formed, the results are less satisfactory. Several attempts have been made to simulate a bioreactor by coupling fluid dynamics and microbial kinetics obtained from laboratory-scale experiments [67] using either a simplified compartment model [76, 94] or a full multiphase CFD approach. Very often, lower substrate consumption than expected or an overprediction of biomass yield were observed. In consequence, the parameters of the biological model (either kinetic or metabolic) had to be suited to the reactor scale. The authors of these studies agree that the key issue for successful integration of fluid dynamics and microbial kinetics is a better description of both micromixing and microbial dynamics [76, 93]. Lapin et al. [95, 96] considered this question through a CFD Euler–Lagrange model. Morchain et al. propose a single variable structured model and a population balance model dealing with the dynamic adaptation of microorganisms exposed to concentration fluctuations [89, 97].

8.6.3 Scale-Down

The main strategy to tackle scale-up difficulties is the design of scale-down experiments. This is the experimental setup such that the conditions of the industrial scale are emulated as close as possible in the lab scale. Scale-down methods range from changes in the feeding strategy, through imposing shifts in process parameters (e.g. pH or temperature) up to complex bioreactor setups as are the multiple reactor compartment systems.

Different methods have been developed over the recent years to mimic industrial scale conditions in the laboratory. The most used method is multiple bioreactors connected to each other. By this, different sections of a large-scale bioreactor are represented by different small bioreactors. Examples are different substrate concentrations, pH, and temperature, among others. In multi-compartment scale-down bioreactors, a perfectly mixed stirred tank reactor (STR) is connected to one or more STRs [98] or to one or more plug flow reactors (PFR) [98, 99], through which the culture is circulated at a rate equivalent to a specified residence time. Typically, one compartment is controlled to offer "ideal" homogeneous conditions, whereas the others induce some stress (e.g. highly concentrated substrate, base, or acid). In a pulsing system, the stress inducer is injected into the unique bioreactor at specified intervals [100] or randomly [101]. This has revealed interesting phenomena with respect to the fast response of *E. coli*, for example [102]. These operation mechanisms produce zones similar to feeding and starvation zones in large-scale bioreactors and result in the periodic exposure of the culture to varying stresses. Scale-down techniques have been applied for the successful study of the impact of large-scale gradients for most industrially relevant organisms, with significant differences in process behavior compared with standard small-scale cultivations [66, 103–105].

The widespread use of parallel mini-bioreactor systems for strain screening has been adopted in many bioprocess development settings to help reduce product lifecycles and accelerate R&D [106]. The parallelization and automation of such cultivation platforms enable a fast screening of large libraries [107]. Several solutions have been developed recently to enable fed-batch-like conditions, such as the gradual supply of glucose to the culture through enzyme-based glucose release systems [108] and the application of micropumps for continuous feed supply [109–111]. These improvements bring the screening system closer to actual cultivation conditions such that scale-down experiments can run in high number in parallel. High-throughput experimental facilities that perform fed-batch experiments with frequent glucose perturbations have been reported [112] including parallel ambr15 TM cultivations of Chinese hamster ovary (CHO) cells for scale-up to a 15 000 l production-scale bioreactor [113] and the use of power per unit volume (P/V) as a scale-down criterion for *E. coli* [114]. These results show not only the importance of scale-down procedures for bioprocess development but also the need for model-based tools to properly design, operate, and extrapolate the results to large scales. Model-based operation of dynamic experiments in parallel systems has been addressed aiming at optimal parallel designs in an adaptive operating framework [115, 116]. Finally, Anane et al. [117] demonstrated that macro-kinetic models are a powerful tool to operate scale-down experiments in small-scale parallel systems for proinsulin expressed in *E. coli* as inclusion bodies.

8.6.4 Conclusions

The response of cell factories to different cultivation environments is arguably the biggest challenge in the bioprocess development. Efforts to tackle this issue purely with computer aided tools or solely using experimental data have encountered several challenges. It is hence essential to properly embed novel digital tools, advanced automated facilities, and expert's know-how to increase the robustness of scale-up in biomanufacturing [118]. It is by optimally designing and operating experiments for scale-up in high throughput, properly handling the generated data and learning for it, but most importantly finding tools to properly transfer the information throughout scales.

8.7 Digitalization as an Enabler for Continuous Manufacturing

With the advent of advanced sensor technology, high-throughput automated experimental systems, increased computational power, meticulous data management solutions, and a customized set of advanced data analytics techniques supported by a process engineering framework, the biopharma industry has a strong basis to master operational and technological bottlenecks. However, to excel along the entire value creation chain, a unified and smart platform capable of integrating all knowledge sources and establishing interaction among different unit operations is required. Consider the production of a biological with a continuous process, which is now possible both from technological [119, 120] and regulatory perspectives [121]. The continuous manufacturing of biologicals is attractive as it exhibits several advantages such as a significant lower footprint and more stable product quality profile [122]. Also, their development awards with considerable benefits, e.g. scale-up, can be achieved via numbering up, hence reducing development times significantly. With this possibility of continuous manufacturing, the number of manufacturing options to choose from during process development has increased. Depending on the anticipated drug product demand, the associated uncertainty in demand, the available existing facilities and platform processes, and the company's overall manufacturing strategy, continuous manufacturing might be a valuable choice.

At the beginning of the process development exercise, decisions need to be taken about which manufacturing route(s) should be investigated. While, in the past, these decisions were made from the experience, the increase in the complexity of modalities and the increase in manufacturing options have rendered these decisions more and more difficult. Provided a good notion of the demand and its uncertainty and an accurate overview of the available manufacturing capacities and utilization exists, economic modeling of the bioprocess helps evaluating different manufacturing routes and setting development targets for each unit operation. Technical risk analysis that takes prior knowledge and data and the experience of process experts into account can further be used to weigh the different routes. Consider, for instance, a case of a candidate drug for which a relatively low demand is expected, significant "batch" platform knowledge is available (lower technical risks as development needs and approach to tackle them are clear), while having significant "batch" capacity at disposal (little additional investment need while reaching a higher utilization of the existing assets) and while facing pressing timelines (reducing the amount of time available to lower the risk by experimentation on alternative, potentially more economic routes). In this case, the development of a continuous process might economically not make sense, even if generally the company strives to develop continuous processes. However, remove one factor of this example and it becomes much harder to judge from experience whether to produce continuously or not.

Both economic modeling and technical risk analysis require (i) process knowledge and (ii) data. It is evident that the greater the quality and quantity of both assets, the more accurate the manufacturing route analysis. While companies have been starting to invest into platforms to gather and access data, tools for systematic capturing of knowledge (other than documentation) are largely absent. To this end, the concept of the digital twin (DT) has recently found significant attention, i.e. using a digital representation of the process to systematically deposit knowledge, making it explorable on demand, e.g. for what-if scenario analysis. The business case for the adoption of digital twin technology in a biopharmaceutical manufacturing has been investigated by [123], concluding that "… in light of the technical feasibilities, financial opportunities and regulatory support, one can only conclude that those [companies] that systematically invest in Digital Twins now will gain an advantage over competition that will be difficult to overcome by late adopters."

The case for adoption of the DT concept in continuous manufacturing seems even stronger, since continuous manufacturing exhibits unique challenges specific to its operation (that can much more easily be addressed with DT) such as (i) increased need for online monitoring of the process state, (ii) increased need for understanding the process dynamics to be able to control them and bring them back to the operating set point, and (iii) increased need to understand interactions between process units such that one can counteract, should variations travel down the unit train [122].

While these challenges are not unique to continuous bioprocesses, the high complexity of the underlying biological system and product (biological) is strongly favoring the adoption of DT technologies that provide the means to embrace complexity yet being conceptually simple, practical, and easy to use.

In essence, the DT unites or hosts the before-detailed process modeling, monitoring, development, optimization and control approaches under one umbrella into a cohesive and consistent platform, i.e. all of the before-described approaches constitute one part of the digital twin. Yet, the platform is more than the sum of its parts, allowing for a seamless exchange of information and knowledge transfer vertically from small to large scale and horizontally from product to product.

The digital representation of the DT provides another advantage that could prove helpful in filing the process. To scoop the full operational potential of continuous processes, advanced process control algorithms will be sought to be applied. These algorithms require and will use the process dynamics to bring the process back on track, should the process deviations occur. To allow for such a "dynamic" operation, dynamic design spaces will have to be developed [122]. The DT can be directly interrogated to determine and represent the dynamic design space, complementing the representation with experimental evidence only where the DT's predictions are weak or where it is critical to assess the DT's prediction quality. Ultimately, one could even consider submitting the underlying mathematical model when filing the process, as it constitutes an interpretation-free manner to communicate the design space [6].

Overall, one can conclude that knowledge-centered digital approaches further intensify the value preposition of continuous manufacturing, rendering the technology to a cost-effective manufacturing option that is simple to develop and easy to use.

References

1 Walsh, G. (2018). Biopharmaceutical benchmarks 2018. *Nat. Biotechnol.* 36: 1136–1145.
2 David, E. et al. (2009). Pharmaceutical R&D: the road to positive returns. *Nat. Publ. Gr.* 8: 609–610.
3 BAK Basel Economics on Behalf of Interpharma. (2015). The importance of the pharmaceutical industry for Switzerland [Online]. https://www.bak-economics.com/fileadmin/documents/reports/BAKBASEL_Bedeutung_der_Pharmaindustrie_fuer_die_Schweizer_Volkswirtschaft.pdf (accessed 3 September 2018).

4 Sokolov, M. et al. (2018). Big data in biopharmaceutical process development: vice or virtue? *Chim. Oggi - Chem. Today* 36: 26–29.
5 How Novartis is embracing the digital and data revolution | MobiHealthNews [Online]. https://www.mobihealthnews.com/news/europe/how-novartis-embracing-digital-and-data-revolution (accessed 6 September 2020).
6 Narayanan, H. et al. (2020). Bioprocessing in the digital age: the role of process models. *Biotechnol. J.* 15: 1–10.
7 Papathanasiou, M.M. and Kontoravdi, C. Engineering challenges in therapeutic protein product and process design. *Curr. Opin. Chem. Eng.* 27: 81–88.
8 Craven, S. et al. (2013). Process model comparison and transferability across bioreactor scales and modes of operation for a mammalian cell bioprocess. *Biotechnol. Progr.* 29: 186–196.
9 Xing, Z. et al. (2010). Modeling kinetics of a large-scale fed-batch CHO cell culture by Markov chain Monte Carlo method. *Biotechnol. Progr.* 26: 208–219.
10 Kim, J.I. et al. (2008). Article: applied cellular physiology and metabolic engineering. *Biotechnol. Progr.* 24: 993–1006.
11 Poccia, M.E. et al. (2014). Modeling the microbial growth of two *Escherichia coli* strains in a multi-substrate environment. *Braz. J. Chem. Eng.* 31: 347–354.
12 Jones, K.D. and Kompala, D.S. (1999). Cybernetic model of the growth dynamics of *Saccharomyces cerevisiae* in batch and continuous cultures. *J. Biotechnol.* 71: 105–131.
13 Hutter, S. et al. (2018). Glycosylation flux analysis of immunoglobulin G in Chinese hamster ovary perfusion cell culture. *Processes* 6: 176.
14 Narayanan, H. et al. (2019). Decision tree – PLS (DT – PLS) algorithm for the development of process – specific local prediction models. *Biotechnol. Progr.* https://doi.org/10.1002/btpr.2818.
15 Melcher, M. et al. (2015). The potential of random forest and neural networks for biomass and recombinant protein modeling in *Escherichia coli* fed-batch fermentations. *Biotechnol. J.* 10: 1770–1782.
16 Li, Y. and Yuan, J. (2006). Prediction of key state variables using support vector machines in bioprocesses. *Chem. Eng. Technol.* 29: 313–319.
17 Masampally, V.S. et al. (2019). Cascade gaussian process regression framework for biomass prediction in a fed-batch reactor. *Proc. 2018 IEEE Symp. Ser. Comput. Intell. SSCI 2018* https://doi.org/10.1109/SSCI.2018.8628937.
18 Di Massimo, C. et al. (1992). Towards improved penicillin fermentation via artificial neural networks. *Comput. Chem. Eng.* 16: 283–291.
19 Franceschini, G. and Macchietto, S. (2008). Model-based design of experiments for parameter precision: state of the art. *Chem. Eng. Sci.* 63: 4846–4872.
20 von Stosch, M. et al. (2014). Hybrid modeling for quality by design and PAT-benefits and challenges of applications in biopharmaceutical industry. *Biotechnol. J.* 9: 719–726.
21 von Stosch, M. et al. (2016). Hybrid modeling as a QbD/PAT tool in process development: an industrial *E. coli* case study. *Bioprocess. Biosyst. Eng.* 39: 773–784.
22 Gnoth, S. et al. (2008). Product formation kinetics in genetically modified *E. coli* bacteria: inclusion body formation. *Bioprocess. Biosyst. Eng.* 31: 41–46.

23 Thompson, M.L. and Kramer, M.A. (1994). Modeling chemical processes using prior knowledge and neural networks. *AIChE J.* 40: 1328–1340.

24 Schubert, J. et al. (1994). Hybrid modeling of yeast production processes – combination of a-priori knowledge on different levels of sophistication. *Chem. Eng. Technol.* 17: 10–20.

25 Psichogios, D. (1992). A hybrid neural network-first principles approach to process modeling. *AIChE J.* 11: 337–346.

26 Narayanan, H. et al. (2019). A new generation of predictive models: the added value of hybrid models for manufacturing processes of therapeutic proteins. *Biotechnol. Bioeng.* 116: 2540–2549.

27 Narayanan, H. et al. (2020). Hybrid-EKF: hybrid model coupled with extended Kalman filter for real-time monitoring and control of mammalian cell culture. *Biotechnol. Bioeng.* https://doi.org/10.1002/bit.27437.

28 Yang, A. et al. (2011). Identification of semi-parametric hybrid process models. *Comput. Chem. Eng.* 35: 63–70.

29 Von Stosch, M. et al. (2011). A novel identification method for hybrid (N)PLS dynamical systems with application to bioprocesses. *Expert Syst. Appl.* 38: 10862–10874.

30 Bro, R. et al. (2008). Cross-validation of component models: a critical look at current methods. *Anal. Bioanal.Chem.* 390: 1241–1251.

31 Sokolov, M. et al. (2017). Enhanced process understanding and multivariate prediction of the relationship between cell culture process and monoclonal antibody quality. *Biotechnol. Progr.* https://doi.org/10.1002/btpr.2502.

32 Zürcher, P. et al. (2020). Cell culture process metabolomics together with multivariate data analysis tools opens new routes for bioprocess development and glycosylation prediction. *Biotechnol. Progr.* https://doi.org/10.1002/btpr.3012.

33 Sokolov, M. (2020). Decision making and risk management in biopharmaceutical engineering – opportunities in the age of covid-19 and digitalization. *Ind. Eng. Chem. Res.* https://doi.org/10.1021/acs.iecr.0c02994.

34 von Stosch, M. et al. (2016). Toward intensifying design of experiments in upstream bioprocess development: an industrial *Escherichia coli* feasibility study. *Biotechnol. Progr.* 32: 1343–1352.

35 von Stosch, M. and Willis, M.J. (2017). Intensified design of experiments for upstream bioreactors. *Eng. Life Sci.* 17: 1173–1184.

36 Bayer, B. et al. (2020). Hybrid modeling and intensified DoE: an approach to accelerate upstream process characterization. *Biotechnol. J.* 2000121: 1–10.

37 Krausch, N. et al. (2019). Monte Carlo simulations for the analysis of non-linear parameter confidence intervals in optimal experimental design. *Front. Bioeng. Biotechnol.* 7: 1–16.

38 Ejiofor, A.O. et al. (1994). A robust fed-batch feeding strategy for optimal parameter estimation for baker's yeast production. *Bioprocess. Eng.* 11: 135–144.

39 Manesso, E. et al. (2017). Multi-objective optimization of experiments using curvature and fisher information matrix. *Processes* 5: 1–15.

40 Kroll, P. et al. Model-based methods in the biopharmaceutical process lifecycle. *Pharm. Res.* 34: 2596–2613.

41 Bogaerts, P. and Hanus, R. (2006) *Macroscopic modelling of bioprocesses with a view to engineering applications.* In: Engineering and Manufacturing for Biotechnology, 77–109. Dordrecht: Springer.

42 Balsa-Canto, E., Alonso, A.A., and Banga, J.R. (2010). An iterative identification procedure for dynamic modeling of biochemical networks. *BMC Syst. Biol.* 4: 11. https://doi.org/10.1186/1752-0509-4-11.

43 Michalik, C. et al. (2010). Optimal experimental design for discriminating numerous model candidates: the AWDC criterion. *Ind. Eng. Chem. Res.* 49: 913–919.

44 Brendel, M. and Marquardt, W. (2008). Experimental design for the identification of hybrid reaction models from transient data. *Chem. Eng. J.* 141: 264–277.

45 Shahriari, B. et al. (2016). Taking the human out of the loop: a review of Bayesian optimization. *Proc. IEEE* 104: 148–175.

46 von Stosch, M. (2018). Hybrid models and experimental design. In: *Hybrid Modeling in Process Industries.* Boca Raton: CRC Press, Taylor & Francis Group.

47 Ferreira, A.R. et al. (2014). Fast development of Pichia pastoris GS115 Mut+ cultures employing batch-to-batch control and hybrid semi-parametric modeling. *Bioprocess. Biosyst. Eng.* 37: 629–639.

48 Teixeira, A.P. et al. (2006). Bioprocess iterative batch-to-batch optimization based on hybrid parametric/nonparametric models. *Biotechnol. Progr.* 22: 247–258.

49 Teixeira, A. et al. (2006). Dynamic optimisation of a recombinant BHK-21 culture based on elementary flux analysis and hybrid parametric/nonparametric modeling. *Microb. Cell Fact.* 5: S25.

50 Kahrs, O. and Marquardt, W. (2007). The validity domain of hybrid models and its application in process optimization. *Chem. Eng. Process. Process Intensif.* 46: 1054–1066.

51 Zhang, H. (2009). Software sensors and their applications in bioprocess. *Stud. Comput. Intell.* 218: 25–56.

52 Luttmann, R. et al. (2012). Soft sensors in bioprocessing: a status report and recommendations. *Biotechnol. J.* 7: 1040–1048.

53 Rathore, A.S. et al. (2010). Process analytical technology (PAT) for biopharmaceutical products. *Anal. Bioanal.Chem.* 398: 137–154.

54 Chéruy, A. (1997). Software sensors in bioprocess engineering. *J. Biotechnol.* 52: 193–199.

55 Baldwin, E.A. et al. (2011). Electronic noses and tongues: applications for the food and pharmaceutical industries. *Sensors* 11: 4744–4766.

56 Claßen, J. et al. (2017). Spectroscopic sensors for in-line bioprocess monitoring in research and pharmaceutical industrial application. *Anal. Bioanal.Chem.* 409: 651–666.

57 Whelan, J. et al. (2012). In situ Raman spectroscopy for simultaneous monitoring of multiple process parameters in mammalian cell culture bioreactors. *Biotechnol. Progr.* 28: 1355–1362.

58 Gautam, R., Vanga, S., Ariese, F. et al. (2015). Review of multidimensional data processing approaches for Raman and infrared spectroscopy. *EPJ Tech. Instrum.* 2: 8. https://doi.org/10.1140/epjti/s40485-015-0018-6.

59 Mohd Ali, J. et al. (2015). Review and classification of recent observers applied in chemical process systems. *Comput. Chem. Eng.* 76: 27–41.
60 Schneider, R. and Georgakis, C. (2013). How to NOT make the extended Kalman filter fail. *Ind. Eng. Chem. Res.* 52: 3354–3362.
61 Simon, D. (2006). *Optimal State Estimation: Kalman, H∞, and Nonlinear Approaches*. John Wiley & Sons.
62 Feidl, F., Garbellini, S., Luna, M.F. et al. (2019). Combining mechanistic modeling and raman spectroscopy for monitoring antibody chromatographic purification. *Processes* 7: 683. https://doi.org/10.3390/pr7100683.
63 Nicolaou, S.A. et al. (2010). A comparative view of metabolite and substrate stress and tolerance in microbial bioprocessing: from biofuels and chemicals, to biocatalysis and bioremediation. *Metab. Eng.* 12.
64 Ivarsson, M. et al. (2014). Evaluating the impact of cell culture process parameters on monoclonal antibody N-glycosylation. *J. Biotechnol.* 188: 88–96.
65 Tanzeglock, T. et al. (2009). Induction of mammalian cell death by simple shear and extensional flows. *Biotechnol. Bioeng.* 104: 360–370.
66 Simen, J.D. et al. (2017). Transcriptional response of Escherichia coli to ammonia and glucose fluctuations. *Microb. Biotechnol.* 10: 858–872.
67 Lin, H.Y. et al. (2001). Determination of the maximum specific uptake capacities for glucose and oxygen in glucose-limited fed-batch cultivations of Escherichia coli. *Biotechnol. Bioeng.* 73: 347–357.
68 Lemoine, A. et al. (2016). Performance loss of Corynebacterium glutamicum cultivations under scale-down conditions using complex media. *Eng. Life Sci.* 16: 620–632.
69 Brand, E. et al. (2018). Importance of the cultivation history for the response of *Escherichia coli* to oscillations in scale-down experiments. *Bioprocess. Biosyst. Eng.* 41: 1305–1313.
70 Shimizu, K. (2013). Regulation systems of bacteria such as *Escherichia coli* in response to nutrient limitation and environmental stresses. *Metabolites* 4: 1–35.
71 Neubauer, P. et al. (2013). Consistent development of bioprocesses from microliter cultures to the industrial scale. *Eng. Life Sci.* 13: 224–238.
72 Tsai, B.I. et al. (1969). The effect of micromixing on growth processes. *Biotechnol. Bioeng.* https://doi.org/10.1002/bit.260110206.
73 Fan, L.T. et al. (1971). Simultaneous effect of macromixing and micromixing on growth processes. *AIChE J.* https://doi.org/10.1002/aic.690170336.
74 Tsai, B.I. et al. (2007). The reversed two-environment model of micromixing and growth processes. *J. Appl. Chem. Biotech.* https://doi.org/10.1002/jctb.5020211008.
75 Bajpai, R.K. and Reuss, M. (1982). Coupling of mixing and microbial kinetics for evaluating the performance of bioreactors. *Can. J. Chem. Eng.* https://doi.org/10.1002/cjce.5450600308.
76 Vrábel, P. et al. (2001). CMA: integration of fluid dynamics and microbial kinetics in modelling of large-scale fermentations. *Chem. Eng. J.* https://doi.org/10.1016/S1385-8947(00)00271-0.

77 Nauha, E.K. et al. (2018). Compartmental modeling of large stirred tank bioreactors with high gas volume fractions. *Chem. Eng. J.* https://doi.org/10.1016/j.cej.2017.11.182.

78 Morchain, J. et al. (2017). A population balance model for bioreactors combining interdivision time distributions and micromixing concepts. *Biochem. Eng. J.* 126: 135–145.

79 Mantzaris, N.V. et al. (1999). Numerical solution of a mass structured cell population balance model in an environment of changing substrate concentration. *J. Biotechnol.*

80 Henson, M.A. (2003). Dynamic modeling of microbial cell populations. *Curr. Opin. Biotechnol.* 460–467.

81 Stamatakis, M. (2010). Cell population balance, ensemble and continuum modeling frameworks: conditional equivalence and hybrid approaches. *Chem. Eng. Sci.* https://doi.org/10.1016/j.ces.2009.09.054.

82 Quedeville, V. et al. (2018). A two-dimensional population balance model for cell growth including multiple uptake systems. *Chem. Eng. Res. Des.* https://doi.org/10.1016/j.cherd.2018.02.025.

83 Maluta, F. et al. (2020). Modeling the effects of substrate fluctuations on the maintenance rate in bioreactors with a probabilistic approach. *Biochem. Eng. J.* https://doi.org/10.1016/j.bej.2020.107536.

84 Haringa, C. et al. (2016). Euler–Lagrange computational fluid dynamics for (bio) reactor scale down: an analysis of organism lifelines. *Eng. Life Sci.* 16: 652–663.

85 Siebler, F. et al. (2019). The impact of CO gradients on *C. ljungdahlii* in a 125 m^3 bubble column: mass transfer, circulation time and lifeline analysis. *Chem. Eng. Sci.* https://doi.org/10.1016/j.ces.2019.06.018.

86 Haringa, C. et al. (2018). Computational fluid dynamics simulation of an industrial *P. chrysogenum* fermentation with a coupled 9-pool metabolic model: towards rational scale-down and design optimization. *Chem. Eng. Sci.* 175: 12–24.

87 Tang, W., Deshmukh, A.T., Haringa, C. et al. (2017). *A 9-pool metabolic structured kinetic model describing days to seconds dynamics of growth and product formation by Penicillium chrysogenum. Biotechnol. Bioeng.* 114(8): 1733–1743. PMID: 28322433. https://doi.org/10.1002/bit.26294.

88 Wang, G., Tang, W., Xia, J. et al. (2015) Integration of microbial kinetics and fluid dynamics toward model-driven scale-up of industrial bioprocesses. *Eng. Life Sci.* 15: 20–29. https://doi.org/10.1002/elsc.201400172.

89 Morchain, J. et al. (2014). A coupled population balance model and CFD approach for the simulation of mixing issues in lab-scale and industrial bioreactors. *AIChE J.* https://doi.org/10.1002/aic.14238.

90 Lapin, A. et al. (2004). Dynamic behavior of microbial populations in stirred bioreactors simulated with Euler–Lagrange methods: traveling along the lifelines of single cells. *Ind. Eng. Chem. Res.* 43: 4647–4656.

91 Canelas, A.B. et al. (2011). An in vivo data-driven framework for classification and quantification of enzyme kinetics and determination of apparent thermodynamic data. *Metab. Eng.* https://doi.org/10.1016/j.ymben.2011.02.005.

92 Vanrolleghem, P.A. et al. (1996). Validation of a metabolic network for *Saccharomyces cerevisiae* using mixed substrate studies. *Biotechnol. Progr.* https://doi.org/10.1021/bp960022i.

93 Schmalzriedt, S., Jenne, M., Mauch, K., and Reuss, M. (2003). Integration of physiology and fluid dynamics. *Adv. Biochem. Eng. Biotechnol.* 80: 19–68. PMID: 12747541. https://doi.org/10.1007/3-540-36782-9_2.

94 Hristov, H. et al. (2001). A 3-D analysis of gas-liquid mixing, mass transfer and bioreaction in a stirred bio-reactor. *Food Bioprod. Process. Trans. Inst. Chem. Eng. Part C* https://doi.org/10.1205/096030801753252306.

95 Lapin, A. et al. (2004). Dynamic behavior of microbial populations in stirred bioreactors simulated with Euler–Lagrange methods: traveling along the lifelines of single cells. *Ind. Eng. Chem. Res.* https://doi.org/10.1021/ie030786k.

96 Lapin, A. et al. (2006). Modeling the dynamics of *E. coli* populations in the three-dimensional turbulent field of a stirred-tank bioreactor-A structured-segregated approach. *Chem. Eng. Sci.* https://doi.org/10.1016/j.ces.2006.03.003.

97 Morchain, J. and Fonade, C. (2009). A structured model for the simulation of bioreactors under transient conditions. *AIChE J.* 55: 2973–2984.

98 Limberg, M.H. et al. (2016). Plug flow versus stirred tank reactor flow characteristics in two-compartment scale-down bioreactor: setup-specific influence on the metabolic phenotype and bioprocess performance of *Corynebacterium glutamicum*. *Eng. Life Sci.* https://doi.org/10.1002/elsc.201500142.

99 Glazyrina, J. et al. (2011). Two-compartment method for determination of the oxygen transfer rate with electrochemical sensors based on sulfite oxidation. *Biotechnol. J.* 6: 1003–1008.

100 Anane, E. et al. (2018). Modelling concentration gradients in fed-batch cultivations of *E. coli* – towards the flexible design of scale-down experiments. *J. Chem. Technol. Biotechnol.* https://doi.org/10.1002/jctb.5798.

101 Sunya, S. et al. (2013). Short-term dynamic behavior of *Escherichia coli* in response to successive glucose pulses on glucose-limited chemostat cultures. *J. Biotechnol.* 164: 531–542.

102 Anane, E. et al. (2017). Modelling overflow metabolism in *Escherichia coli* by acetate cycling. *Biochem. Eng. J.* 125: 23–30.

103 Li, J., Jaitzig, J., Lu, P. et al. (2015). Scale-up bioprocess development for production of the antibiotic valinomycin in *Escherichia coli* based on consistent fed-batch cultivations. *Microb. Cell Fact.* 14: 83. https://doi.org/10.1186/s12934-015-0272-y.

104 Delvigne, F., Boxus, M., Ingels, S. et al. (2009). Bioreactor mixing efficiency modulates the activity of a prpoS::GFP reporter gene in *E. coli*. *Microb. Cell Fact.* 8: 15. https://doi.org/10.1186/1475-2859-8-15.

105 Limberg, M.H. et al. (2017). pH fluctuations imperil the robustness of *C. glutamicum* to short term oxygen limitation. *J. Biotechnol.* 259: 248–260.

106 Tajsoleiman, T. et al. (2019). An industrial perspective on scale-down challenges using miniaturized bioreactors. *Trends Biotechnol.* 37(7): 697–706.

107 Back, A. et al. (2016). High-throughput fermentation screening for the yeast *Yarrowia lipolytica* with real-time monitoring of biomass and lipid production. *Microb. Cell Fact.* 15.

108 Krause, M. et al. (2016). The fed-batch principle for the molecular biology lab: controlled nutrient diets in ready-made media improve production of recombinant proteins in *Escherichia coli*. *Microb. Cell Fact.* 15: 1–13.

109 Gebhardt, G. et al. (2011). A new microfluidic concept for parallel operated milliliter-scale stirred tank bioreactors. *Biotechnol. Progr.* 27: 684–690.

110 Sawatzki, A. et al. (2018). Accelerated bioprocess development of endopolygalacturonase-production with *Saccharomyces Cerevisiae* using multivariate prediction in a 48 mini-bioreactor automated platform. *Bioengineering* 5: 101.

111 Philip, P. et al. (2018). Parallel substrate supply and pH stabilization for optimal screening of *E. coli* with the membrane-based fed-batch shake flask. *Microb. Cell Fact.* https://doi.org/10.1186/s12934-018-0917-8.

112 Vester, A. et al. (2009). Discrimination of riboflavin producing *Bacillus subtilis* strains based on their fed-batch process performances on a millilitre scale. *Appl. Microbiol. Biotechnol.* 84: 71–76.

113 Janakiraman, V. et al. (2015). Application of high-throughput mini-bioreactor system for systematic scale-down modeling, process characterization, and control strategy development. *Biotechnol. Progr.* https://doi.org/10.1002/btpr.2162.

114 Velez-Suberbie, M.L. et al. (2017). High throughput automated microbial bioreactor system used for clone selection and rapid scale-down process optimization. *Biotechnol. Progr.* 15: 1–11.

115 Bournazou, M.N.C. et al. (2017). Online optimal experimental re-design in robotic parallel fed-batch cultivation facilities. *Biotechnol. Bioeng.* 114: 610–619.

116 Barz, T. et al. (2018). Adaptive optimal operation of a parallel robotic liquid handling station. *IFAC-PapersOnLine* 51: 765–770.

117 Anane, E. et al. (2019). A model-based framework for parallel scale-down fed-batch cultivations in mini-bioreactors for accelerated phenotyping. *Biotechnol. Bioeng.* 116: 2906–2918.

118 Abt, V. et al. (2018). Model-based tools for optimal experiments in bioprocess engineering. *Curr. Opin. Chem. Eng.* https://doi.org/10.1016/j.coche.2018.11.007.

119 Rathore, A.S. et al. (2015). Continuous processing for production of biopharmaceuticals. *Prep. Biochem. Biotechnol.* 45: 836–849.

120 Konstantinov, K.B. and Cooney, C.L. (2015). White paper on continuous bioprocessing May 20–21, 2014 continuous manufacturing symposium. *J. Pharm. Sci.* 104: 813–820.

121 https://www.govinfo.gov/content/pkg/CFR-2014-title21-vol4/pdf/CFR-2014-title21-vol4.pdf (accessed 24 September 2020).

122 von Stosch, M. et al. (2020). Working within the design space: do our static process characterization methods suffice? *Pharmaceutics* 12: 1–15.

123 Portela, R.M.C., Varsakelis, C., Richelle, A., et al. (2021). When is an in silico representation a digital twin? A biopharmaceutical industry approach to the digital twin concept. *Adv. Biochem. Eng. Biotechnol.* 176: 35–55. PMID: 32797270. https://doi.org/10.1007/10_2020_138.

9

Digital Twins for Continuous Biologics Manufacturing

Axel Schmidt, Steffen Zobel-Roos, Heribert Helgers, Lara Lohmann, Florian Vetter, Christoph Jensch, Alex Juckers, and Jochen Strube

Clausthal University of Technology, Institute for Separation and Process Technology, Leibnizstr. 15, D-38678 Clausthal-Zellerfeld, Germany

9.1 Introduction

Actual discussion about biologics manufacturing points especially out that a major challenge is to cope in the process development and manufacturing technology with innovations needed for a more diverse molecular biologics pipeline [1]:

> The biopharmaceutical industry has been growing and evolving at a pace that's hard to match, especially in terms of manufacturing. Gone are the days when developers looked at monoclonal antibodies (mAbs), or antibodies that are made by identical immune cells cloned from a single cell, as a challenge – rather, mAbs have become an industry standard, with well-established research, development, and manufacturing processes that have proven to be robust and scalable. In the current biopharmaceutical landscape, developers are constantly encountering new and diverse biological molecules with complex structures. While these novel biologics are filling drug pipelines, they are also challenging existing manufacturing processes, both in upstream, downstream and analytical development.

As a consequence, analyzing methods available to solve those process development, engineering, and manufacturing operation challenges, the experts fall back on the quality-by-design (QbD) concept demanded by regulatory authorities [2] including process analytical technology (PAT) [3].

Thereby, it should be kept in mind that PAT is not reduced to in-line analytics but is a consistent approach to process control. In the QbD approach, a design space of operating parameters is defined to ensure specified quality attributes (QAs). This leads to multiparameter optimizations and a significant experimental effort. The modern approach for process development and quality assessment is shown in Figure 9.1.

Process Control, Intensification, and Digitalisation in Continuous Biomanufacturing, First Edition.
Edited by Ganapathy Subramanian.
© 2022 WILEY-VCH GmbH. Published 2022 by WILEY-VCH GmbH.

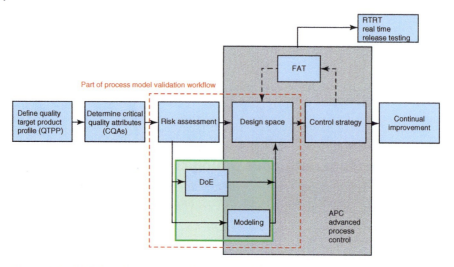

Figure 9.1 QbD-based process development workflow.

Also, process control itself will be seen for the future in autonomous manufacturing operation – as any mature technology finally ends up. Process control methods known and already successfully applied at other manufacturing branches are mainly based on advanced process control (APC) approaches. The general drawback of such sophisticated methods is in the efforts to gain a valid process model [4, 5].

However, especially here the last few years, the rapid success has been made to adopt well-known process modeling techniques from chemicals to biologics manufacturing. Main challenge has been analytical methods as engineers do need quantitative data for their work flow. Industrialization 4.0, Internet of things, artificial intelligence, and machine learning activities up to big data analysis have taken their share to solve the fundamental problems like component or at least group specific evaluation of spectroscopic data by partial least-squares (PLS) algorithms. Besides in-line analytics methods included in the PAT concepts, the key technology has been the generation of decisive validated digital twins based on process models [6–15].

The digital twin serves as interface between physical manufacturing operation toward the digital world. Thereby an integrated data availability during the whole life cycle could be mapped: from the product and process development over manufacturing and start-up to the predictive maintenance. New business models are implied for even small and medium entities supplying manufacturing skids and modules and big pharma end users. Digital twins market has double digit growth rates, and platforms are offered from all main automation players [16].

For process development (see Figure 9.2), a sophisticated PAT concept has to be developed parallel to upstream processing (USP) and DSP modeling to generate digital twins, later on supporting model validation [6–15], piloting, and production. Parallel to these three, the developed PAT method and the PLS system have to be further refined. Central key technology besides appropriate PAT is therefore the generation

9.1 Introduction

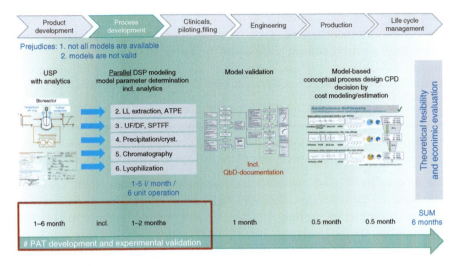

Figure 9.2 Process development strategy.

of digital twins for the whole process. It has been proven that for all unit operations, such distinct validated process models are available as digital twins [6–15].

In 2019, a joint industrial and academic working group on modeling has summarized their needs toward *in silico* chemistry, manufacturing, and controls (CMC) [17] based on modeling tools and methods from molecular to equipment design level. The final roadmap and somewhat SWOT analysis give space for innovation but especially on the molecular level; nevertheless classical engineering work steps are already available and ready to use.

Figure 9.3 depicts different methods that are applied for the different working steps described above. Missing gaps between risk assessment and process characterization are small-scale models, which are downscaled laboratory-scale experiments

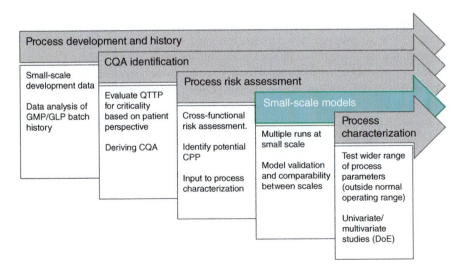

Figure 9.3 Process development including DoE.

for experimental model parameter determination in combination with process modeling. Process models need to be validated appropriately to predict process characteristics precisely enough to file data-driven decisions.

It states that QbD is more than design of experiments (DoE). In the following, the different QbD elements are demonstrated along the processing of mAbs:

(1) The risks of the preparation of plant materials are assessed.
(2) Based on this risk assessment, the impact factors on extraction are evaluated.
(3) The approach to process control is shown by deriving the design space for the final separation step and pointing out the linkage to prior unit operations.

The production of biopharmaceuticals in regulated industries is still predominantly based on the batch processes, although continuous processes offer decisive advantages regarding agility, flexibility, quality, cost, and societal benefits [18]. For those categories, US Food and Drug Administration (FDA) and Center for Drug Evaluation and Research (CDER) give specific examples of achievable improvements, e.g. continuous manufacturing (CM) enables expansion of production volumes without the current problems related to batch scale-up (agility). CM also tackles the challenge to rapidly increase production in the case of shortages or emergencies (flexibility). Currently, production supply chains are spread globally and therefore vulnerable in many ways. Furthermore, CM enables regional and intranational manufacturing (geography). Switching to CM also allows for the introduction of highly meaningful statistical process control (quality) and a decreased initial investment (cost). The environmental impact is generally lower. It triggers the demand for highly trained staff, and freed resources can be invested in new products (societal benefits) [18–20].

Continuous biomanufacturing has shown its benefits and potential also in many case studies – especially exemplified for the biologics workhorse mAbs – in the last decade [21–23]. The phase of piloting studies has successfully been overcome by many participants [21–23]; hence, the missing link for successful broad industrialization is the fear for complexity of operation to prevent any mistakes in continuous operation as the weakest part may limit the total success. However, it should be taken into account that all studies did point out the more robust product generation with regard to quality issues in relation to the batch lot operation. Nevertheless, autonomous operation would convince management more reliable for such innovative changes.

To promote the introduction of CM, FDA, European Medicinal Agency (EMA), International Council for Harmonization of Technical Requirements for Pharmaceuticals for Human Use (ICH), and some industrial working groups have started initiatives and published series on guidance, with the most prominent example being the QbD associated ICH Q8–Q11 (see Figure 9.4).

Key focus areas that still need to be explored further are Pharmaceutical Quality Systems (PQS), real-time release testing (RTRT), QbD, and PAT. There is a vast amount of published work for each category [24–31]; however precise examples and strategies that address the linkage of QbD principles and PAT in more detail are rare, especially regarding dynamic batch processes in contrast to semi-stationary

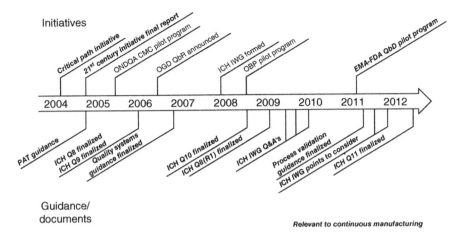

Figure 9.4 Quality-related guidance and initiatives.

continuous processes. To enable the transition from batch to continuous processing, the applicability of these guidances must be demonstrated.

Therefore, to close this gap with aid of PAT and digital twins toward autonomous operation, this chapter summarized the state-of-the-art successes.

9.2 Digital Twins in Continuous Biomanufacturing

In the following, it is pointed out that all unit operations are process models representing the digital twins available. It is quite possible to determine for any new efficient component system the model parameters in laboratory scale. Also, a consistent method with precise workflow has to be followed to generate distinct validated process models as digital twins that are needed to fulfill regulatory demands for reliable proven predictions.

The molecular complexity of biologics is much higher than that of botanicals, exemplified by about the factor 1000 in higher molecular weight. In addition, societal demand for biologics products are quite two magnitudes higher than botanicals. Therefore, revenue and, due to that, technical sophistication in manufacturing technology are much higher than in botanicals processing.

Nevertheless, drug candidates succeeding the tough development and approval process through conical failure and finally reaching the market are quite less. Critical in process development is the competitiveness with a given timeline of about six months from drug candidate to first test amounts. In that phase, the processing steps, i.e. the manufacturing process, are defined as the process describing the product quality in addition to analytical testing by regulation. Quality in process development needs to be therefore high, but due to the high number of still few hundred candidates, the development pipeline for one final blockbuster on the market needs to be realistically low. Such hunt for most efficient process development methods has resulted in platform definitions for classes of mAbs. Fortunately, many other

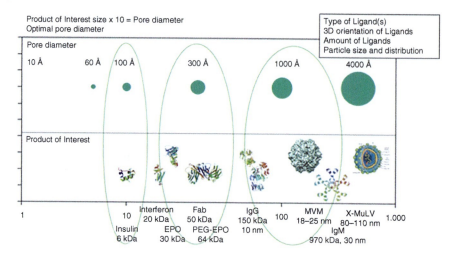

Figure 9.5 Product of interest and pore diameter with unhindered diffusion. Source: Based on Ditz [32].

molecular classes like fragments, peptide, proteins, virus-like particles, exosomes, and virus are in the meantime within the pipelines of pharmaceutical industry.

Such a high molecular complexity, caused by larger product variability of much larger three-dimensional molecules with appropriate glycosylation for bioavailability and efficacy, creates additional need for sophisticated separation mechanisms with high selectivity to handle small molecular variations and capacities for heterogeneous molecular classes. Figure 9.5 exemplifies the molecular complexity of those products by giving a structural picture and the related pore size of potential separation systems that need to be at least a factor of 10 larger to allow unhindered diffusion.

Due to such molecular variety, any platform approach to classes of mAbs is not feasible. The actual approach is the acceleration of process development by modeling in pharmaceutical industry following the success in the petrochemical and bulk and fine chemical industries.

It is recommended to start project work in parallel at manufacturing, piloting, and process design. The needed change in mindset is the highest in process design, therefore with highest efforts but higher long-term gains.

In the first month of the process development and design, USP fermentation starts at scales of a few liters. This feedstock is split for the different unit operations applicable to downstream processing such as LL extraction especially aqueous two-phase extraction (ATPE) and ultra/diafiltration (UF/DF) especially single-pass tangential flow filtration (SPTFF) for continuous operation, precipitation, or crystallization and all different chromatography units possible and final lyophilization preparing formulation. Each fermentation optimization runs over the process development time span of about six months and is fed into the model parameter determination and model validation concept runs of each unit in parallel. Analytics allow to take any relevant component group into account within the models. The separation sequence could be designed *in silico* a priori by various simulation studies that show the purification power of each step as quantified feed input of the next one. The benefits of appropriate USP and downstream process (DSP) integrations have been demonstrated before [23, 33–35]. Besides, USP modeling with Monod kinetics including reduced metabolomics [36–40] enables to integrate USP into total process design and lyophilization as the final steps toward formulation [41–43]. The process sequences are evaluated by cost estimation tools easily added in the process modeling [44, 45]. This results in a theoretical feasibility decision on best process in class. Afterward, this can be operated in mini or pilot scale for model validation toward QbD documentation and test amount supply. PAT concepts developed and applied in the process development are transferred to be of help in pilot and manufacturing scale as well. Calibration and maintenance are typical routines, which should be established already during the process development. Any sensor drift needs to be counteracted with aid of APC methods. In-line release approval is another challenge for regulatory affairs and quality assurance.

In industrial application of models, only quantitative validated models are of predictive use, because either hazard control or regulatory approval needs to be reliably data-driven documented – at least for safety and liability reasons. State of the art is an experimental feasibility study following the theoretical model-based feasibility study to validate the process modeling with the aid of miniplant data (Figure 9.6) [46–48]. Here, typical high-end miniplant under ATEX approval and TÜV conformity is shown. All unit operations are designed to operate simultaneously at a flow rate between 1 and 2 l/h [49].

Validating physicochemical models quantitatively is not that simple as the experimental error for model parameter determination and final miniplant validation operation has to be taken into account, as well as the model immanent error based on simplifying assumptions. Of course, the model intended needs to be implemented mathematically correct at first. Model accuracy and prediction needs to be not better than reality but needs to be within such variance to be valid for appropriate predictions. Variance and probability has to be given for each number to file valid decisions. A distinct workflow is given in Figure 9.7 and described in detail for nearly all unit operations in [6–11, 14, 15, 27, 50].

The last module missing is advance process control based on in-line measurement devices. PAT is the FDA- and EMA-demanded method including in-line

Figure 9.6 Miniplant unit operations. Source: Institute for Thermal Process Engineering, www.itv.tu-clausthal.de.

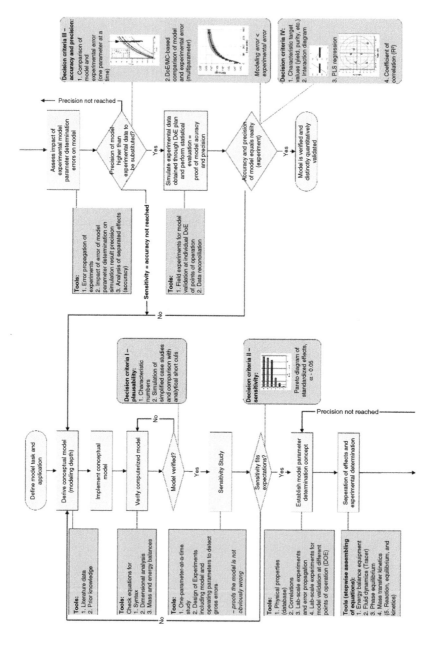

Figure 9.7 Model design, parameterization, and validation workflow.

measurement technologies. Besides density, UV, pH, and conductivity, nowadays turbidity, Raman, and FTIR are on hand for phytochemicals. Of course, classical GC and HPLC and high-end offline MS and NMR could be used for calibration and/or control of such methods.

9.2.1 USP Fed Batch and Perfusion

The modeling approach for fed-batch and perfusion cultivation processes is summarized in Figure 9.8. The main equations are represented by a Monod kinetic, considering the time-dependent alteration of substrate (e.g. glucose, glutamine, etc.), metabolite (e.g. lactate, ammonium, etc.), cell, and product concentration. The correlation between input (e.g. substrate concentration) and output (e.g. cell concentration) variables can be macroscopically determined by using empirical observations, such as yield coefficients, which are strongly dependent on the cell line [51].

In terms of (1) fluid dynamics (red-marked parameters), the determination of oxygen transfer rates according to the unsteady-state (dynamic) technique, mixing time with conductivity measurements, and residence time with tracer experiments leads to the characterization of the equipment.

To determine the (2) kinetic (green marked, e.g. maximum growth rate) and (3) equilibrium (blue marked, e.g. yield coefficients) parameters, cultivations and analysis of substrate, metabolite, cell, and product concentration need to be performed. The substrate saturation constant (or substrate affinity constant) is equal to the concentration that supports half-maximum growth rate.

Similar experiments can be used for (4) validation. Furthermore, online model-assisted cultivation increases the gain in process information by integrating process data (e.g. turbidity) into the macroscopic kinetic model to extract information on process variables (e.g. glucose and lactate concentration) [36].

Efforts are 2–3 weeks for cultivation and analysis, i.e. 2–3 cultivations along with a 1–2 l scale [36].

In Figure 9.9, the model development results are summarized. Parameter sensitivity and distribution are checked, as well as significance. Only the significant parameter are varied in the Monte-Carlo simulation studies to prove accuracy and precision (see also [11]).

9.2.2 Capture, LLE, Cell Separation, and Clarification

The workflow to determine model parameters needed for a physicochemical liquid–liquid extraction (LLE) process model is shown in Figure 9.10. Here, the focus is on the extraction columns and mixer settlers as those cover most of the possible process implementations for LLE. For the sake of completeness, it should be stated that innovative technologies such as membrane-supported LLE or side technology such as centrifugal extractors also exist, for which correct physicochemical modeling however follows the same rationale and only has to account for the difference in fundamental geometry and fluid dynamics. This again is the

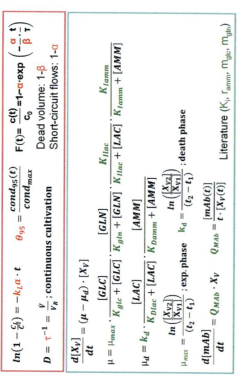

1. Equipment: Residence and mixing time, mass transfer for scale-up (τ, θ_{95}, $k_L a$)

Flow-through: τ, 1-β, 1-α

Batch: θ_{95}

$$\ln\left(1 - \frac{c_t}{c^*}\right) = -k_L a \cdot t \qquad \theta_{95} = \frac{cond_{95}(t)}{cond_{max}}$$

$$F(t) = \frac{c(t)}{c_0} = 1 - \alpha \cdot \exp\left(-\frac{\alpha}{\beta} \cdot \frac{t}{\tau}\right)$$

$$D = \tau^{-1} = \frac{\dot{V}}{V_R}; \text{continuous cultivation}$$

Dead volume: 1-β
Short-circuit flows: 1-α

$y = -0.00275 \cdot x$

$k_L a = 2.75 \cdot 10^{-3} \pm 5.24 \cdot 10^{-5} s^{-1}$

2. Kinetic: Cell and antibody conc. (μ_{max}, k_d, Q_{MAb})

$$\frac{d[X_V]}{dt} = (\mu - \mu_d) \cdot [X_V]$$

$$\mu = \mu_{max} \cdot \frac{[GLC]}{K_{glc} + [GLC]} \cdot \frac{[GLN]}{K_{gln} + [GLN]} \cdot \frac{K_{llac}}{K_{llac} + [LAC]} \cdot \frac{K_{Iamm}}{K_{Iamm} + [AMM]}$$

$$\mu_d = k_d \cdot \frac{[LAC]}{K_{Dlac} + [LAC]} \cdot \frac{[AMM]}{K_{Damm} + [AMM]}$$

$$\mu_{max} = \frac{\ln\left(\frac{X_{V2}}{X_{V1}}\right)}{(t_2 - t_1)}; \text{exp. phase} \qquad k_d = \frac{\ln\left(\frac{X_{V2}}{X_{V1}}\right)}{(t_2 - t_1)}; \text{death phase}$$

$$\frac{d[mAb]}{dt} = Q_{MAb} \cdot X_V \qquad Q_{MAb} = \frac{[mAb(t)]}{t \cdot [X_V(t)]}$$

Literature (K_i, r_{amm}, m_{glc}, m_{gln})

3. Yield coefficients: Substrate and metabolite conc. ($Y_{Xv/glc}$, $Y_{Xv/gln}$, $Y_{lac/glc}$, $Y_{amm/gln}$)

$$\frac{d[AMM]}{dt} = Y_{amm/gln} \cdot \left(\frac{\mu - \mu_d}{Y_{Xv/gln}}\right) \cdot X_V - r_{amm} \cdot X_V$$

$$\frac{d[GLN]}{dt} = -\left(\frac{\mu - \mu_d}{Y_{Xv/gln}} + m_{gln}\right) \cdot X_V \qquad \frac{d[GLC]}{dt} = -\left(\frac{\mu - \mu_d}{Y_{Xv/glc}} + m_{glc}\right) \cdot X_V$$

$$\frac{d[LAC]}{dt} = Y_{lac/glc} \cdot \left(\frac{\mu - \mu_d}{Y_{Xv/glc}} + m_{glc}\right) \cdot X_V \qquad Y_i^j = -\frac{\Delta i(t)}{\Delta j(t)}; \text{prefeed phase}$$

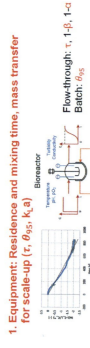

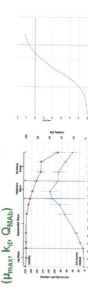

4. Validation: Model-assisted cultivation (PAT)

Integration of experimental data into model

$$\frac{dX_V}{dt} = (\mu - \mu_d) \cdot X_V$$

$$\frac{d[GLC]}{dt} = -\left(\frac{\mu - \mu_d}{Y_{Xv/glc}} + m_{glc}\right) \cdot X_V$$

Figure 9.8 Upstream processing (USP) – fermentation batch/fed-batch/perfusion – process modeling workflow. At first (red), equipment is characterized. Afterward, kinetic parameters (green) are determined, and the yield coefficients (blue) are measured. Finally, the model is validated. Sources: Based on Kornecki and Strube [11] and Kornecki et al. [12].

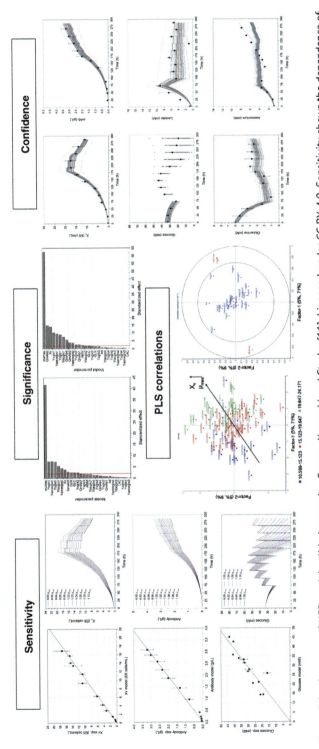

Figure 9.9 Summary of USP model validation results. Source: Kornecki and Strube [11]. Licensed under CC BY-4.0. Sensitivity shows the dependence of simulations from variation of single parameters. Significance is found through DoE-based simulations and statistical evaluation and is illustrated as Pareto diagrams. Accuracy and precision are gained by comparing simulations and experiments and their corresponding envelope curves (confidence). PLS correlation plots indicate the dependencies between input parameters and quality attributes.

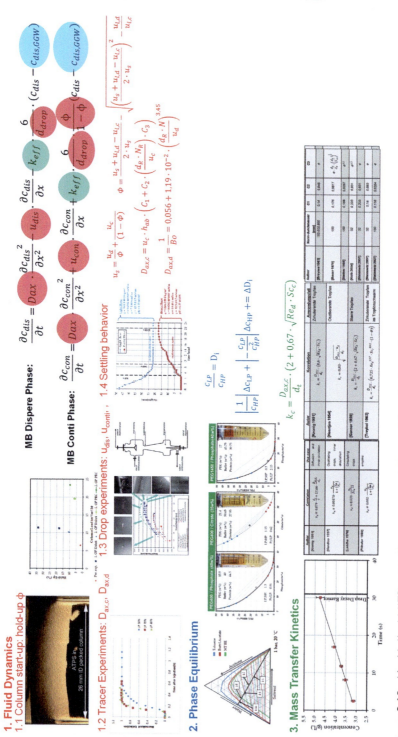

Figure 9.10 Liquid–liquid (ATP) extraction process modeling workflow. Again, model parameter determination starts with fluid dynamics (red). For LLE, this is done by tracer, droplet, and/or settling experiments or by observing the column start-up, depending on the liquid–liquid extraction device to be simulated. The second step (blue) evolves the phase equilibrium, mostly ternary diagrams. In the last step (green), mass transfer kinetics are measured. Source: Schmidt and Strube [8]. Licensed under CC BY-4.0.

striking advantage of this model type, if compared with the other (cost/shortcut, statistical/observer models). As for any unit operations, the sequence to determine these parameters are ordered according to their importance/impact on the process result.

9.2.2.1 Fluid Dynamics (Red)

This is the most important group of effects, which is predominantly responsible not only for the differences in the performance of different types of extraction equipment but also for the differences within the same equipment type (e.g. stirred vs. pulsed vs. static extraction columns).

Most important phenomena are the axial dispersion behavior of the system, characterized by the axial dispersion coefficient, and the holdup, characterized by drop rise velocity, mean droplet diameter, and throughput ($m^3/m^2/h$). Axial dispersion has to be determined for the specific equipment geometry, but only once as it is dependent on the geometry, but not the different types of systems that can be processed/modeled. The determination of axial dispersion behavior needs around 2–5 l of total system volume depending on the investigated scale but does not require the usage of actual feed material. This procedure can be finished within one day by an experienced operator. To keep the resource and time benefit of physicochemical model-based process design, the equipment size should not exceed DN26/32 for extraction columns and DN50 for mixer settlers, which is typical miniplant scale.

Drop rise velocity depending on the mean Sauter diameter is to be determined in a droplet measurement cell; few ml up to 50 ml is sufficient to determine these parameters in triplicate for a range of 3–5 points of drop size within 1–2 days.

9.2.2.2 Phase Equilibrium (Blue)

Determination of binodal, tie lines, and distribution coefficients of target and main side components is done by shaking-flask experiments. This standard procedure can be downscaled to system volumes of 5–10 ml each. Further scale-down is only recommended if data for interfacial tension, viscosities, and densities for both phases are known or accessible by reliable database or correlation. Time and feed material consumption can be drastically reduced by narrowing the relevant system combinations by rationale such as kosmotropic/chaotropic properties of the phase-forming salts of choice and the hydrophobic free volume excluding the effects of the polymers of choice, which are mostly dependent on the molecular weight and relevant pH range for the target component.

The total number of investigated system points should be narrowed by DoE. This leads to 3 systems with 5 distributed points of investigation. Thus, 75 ($3 \times 5 \times 5$ ml) up to 150 ($3 \times 5 \times 10$ ml) ml of system volume are sufficient for the determination and can be executed within two days.

9.2.2.3 Kinetics (Green)

In LLE, the most important kinetic parameter is the mass transfer coefficient. The lower this parameter, the more time for component separation required, which if not implemented correctly in any model type can lead to an incomplete separation (bad purity/yield) due to the underestimated kinetic limitations of the system. This parameter can easily be determined parallel to the drop measurements in the drop measurement cell during the 1- to 2-day period (see Section 9.2.2.1) and thus only increases the analytical efforts.

The total effort for a complete model parameter determination as described above is around 3 days up to 1 week and requires only 200–300 ml feed material [34, 36, 52–54].

The results of the modeling and simulation workflow for LLE are summarized in Figure 9.11. DoE simulations and sensitivity studies covering the whole possible space of operation are interpreted statistically to obtain the significant parameters. Monte-Carlo simulations around the possible point of operation with uncertainties given by model parameter determination error lead to envelope curves and allow for a comparison of accuracy and precision between simulations and experiments (see also [8]).

9.2.3 UF/DF, SPTFF for Concentration, and Buffer Exchange

To create a model for UF/DF module- and solution-based information is required, which is shown in Figure 9.12.

The red-marked experiments are needed to characterize (1) the membrane-module fluid dynamics comprising effective membrane area, hydrodynamic diameter, and membrane resistance. They are independent from the filtration system and are applicable on other systems as well. The sequential (2) step is the measurement of solution properties (blue). Due to the changing density, osmotic pressure, and viscosity with increasing protein concentration, filtration behavior varies over time, and the values for different protein concentrations have to be quantified as (3) mass transfer kinetics. After these steps, the filtration experiments are performed at different flows, pressures, and starting concentrations to (4) validate the model.

Efforts are 3 days up to 1 week, requiring 1–100 g feed material [14, 15, 55–57].

The continuous SPTFF version is analogous, but due to the complexity of additional setup variations [14, 15].

For SPTFF (Figure 9.13), the approach is comparable with the batch filtration. The used membrane-module fluid dynamics have to be (1) characterized (red), and (2) the solution properties (blue) must be determined likewise to batch filtration. In addition to the batch approach, the influence of stacked membranes has to be investigated. Furthermore, the SPTFF is dependent on the length, while batch filtration depends on time. This makes (3) the pressure drop a critical factor, which needs to be investigated.

For the research on an SPTFF, a separation of different filtration stages (parallel membrane stacks) is favorable. This provides the possibility to measure the development of process variables over the different stages and enhances process understanding. Validation has to be performed, as well as batch filtration for different flows, pressures, and concentrations and for different setups of filtration stages.

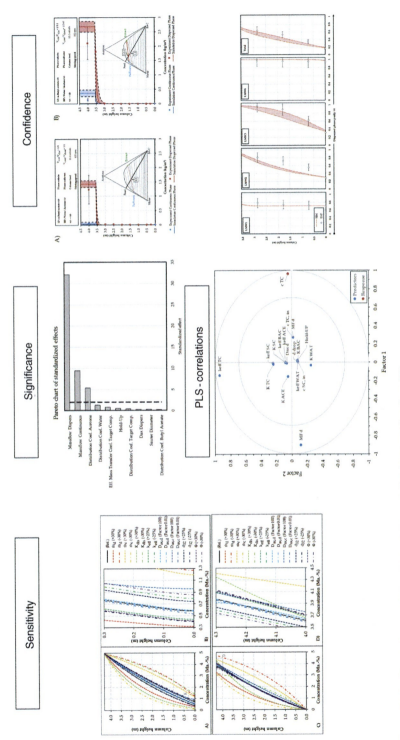

Figure 9.11 Simulation workflow results for liquid–liquid extraction from [8].

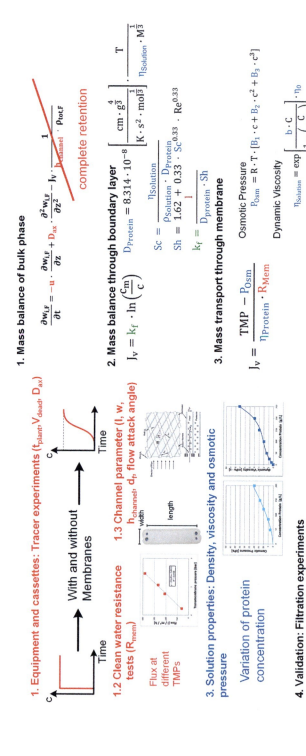

Figure 9.12 Ultrafiltration/diafiltration-batch process modeling workflow, recipe, and SOP. The first step is meant to measure fluid dynamic parameters (red) with tracer experiments. The second step is to evaluate the mass transfer through the boundary layer and through the membrane. Here, solution parameter are needed (blue). Source: Huter and Strube [15]. Licensed under CC BY-4.0.

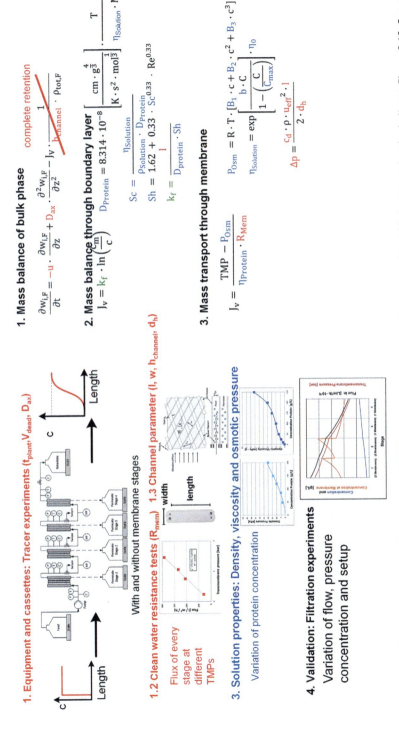

Figure 9.13 Single-pass tangential flow filtration (SPTFF) process modeling workflow, recipe, and SOP. The procedure is similar to Figure 9.12. See also [14]. Source: Huter and Strube [15]. Licensed under CC BY-4.0.

Efforts range from 3 days to 2 weeks to properly measure tracer velocities and reproducibility, channel geometries, concentrations, feed pressures, and setup variation [14, 15].

9.2.4 Precipitation/Crystallization

This modeling approach for precipitation, which is shown in Figure 9.14, is based on a former approach for crystallization [58], but with additional agglomeration and breakage kinetics. Due to the fact that precipitation also is the result of a shift in solubility, (1) the apparatus fluid dynamics is likewise characterized by residence and mixing time and energy balance and solubility of the target component in the system utilized (red).

Furthermore, (2) mass balance and kinetics of precipitation such as growth, nucleation, agglomeration, and disruption have to be implemented in the model for description of the actual mechanism during precipitation (green). Subsequently, (3) missing parameters and coefficients are determined by the evaluation of experimental data (blue).

Finally, (4) model validation is carried out by using PAT to identify measurable process parameters that can be monitored online during operation. These parameters also provide the basis for scale-up to the desired benchmark (black). Estimated time for precipitation runs and analysis are between 1 and 2 weeks.

9.2.5 Chromatography and Membrane Adsorption

The biologic pharmaceutical boom is not only apparent regarding the ever-growing market size but steadily increases molecular variety [12, 59–61]. Over the last decades, biologic variety started in the 1970 with peptides, and shortly followed by mAbs, agent variety increased with the upcoming vaccine technologies, like virus such as particles for human papilloma virus or mRNA as a readily available vaccine supporting global effort against COVID-19.

To date, downstream processing in biopharmaceutical processes is mostly chromatography driven [12, 59, 62, 63]. For total process design and optimization, a digital twin of the process is needed to guarantee QbD as demanded by regulatory authorities [6]. Thus, the mathematic models and experimental methods to successfully model the chromatographic process are needed.

In the literature, there are many existing proposals and results for different systems available to be utilized [64–68]. Also, many different methodologies to measure the needed parameters have been proposed by different working groups [68–70].

9.2.5.1 General Rate Model Chromatography

The simulation of chromatographic processes is usually done using either a general rate model or a lumped pore diffusion model [7, 68, 70]. Reported deviations between the models relatively small and for most applications lumped pore diffusion models are used, owing to their faster calculation times [7, 70]. General rate however models the pore diffusion, allowing a more accurate modeling of the real transport processes.

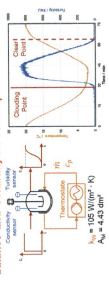
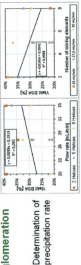
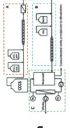

Figure 9.14 Precipitation process modeling workflow, recipe, and SOP. The procedure starts with equipment and fluid dynamic characterization (red boxes and parameters) followed by precipitation kinetics (green) and driving force (blue) experiments. Source: Lohmann and Strube [10]. Licensed under CC BY-4.0.

The mass balance of the stationary phase for general rate (Eq. (9.1)) and the lumped pore diffusion (Eq. (9.2)) model is [68]:

$$\varepsilon_{p,i} \cdot \frac{\delta c_{p,i}}{\delta t} + (1 - \varepsilon_{p,i}) * \frac{\delta q_i}{\delta t} = \frac{1}{r^2} \frac{\delta}{\delta r} \left[r^2 \left(\varepsilon_{p,i} \cdot D_{p,i} \cdot \frac{\delta c_{p,i}}{\delta r} + (1 - \varepsilon_{p,i}) \cdot D_{S,i} \frac{\delta q_i^*}{\delta r} \right) \right] \quad (9.1)$$

$$\varepsilon_{p,i} \cdot \frac{\delta c_{p,i}}{\delta t} + (1 - \varepsilon_{p,i}) * \frac{\delta q_i}{\delta t} = \frac{6}{d_p} \cdot \frac{(1-\varepsilon_S)}{\varepsilon_S} \cdot k_{\text{eff},i} \cdot (c_i - c_{p,i}) \quad (9.2)$$

where $\varepsilon_{p,i}$ is the inner porosity for the component, $c_{p,i}$ is the concentration of the component in the pores, t is the time, q_i is the loading, d_p is the mean diameter of the resin particle, ε_S is the voidage, $k_{\text{eff},i}$ is the effective mass transport coefficient, and c_i is the concentration in the continuous phase.

The mass transfer coefficient $k_{\text{eff},i}$ is given by Eq. (9.3). Here, $k_{f,i}$ is the film mass transfer coefficient, r_p is the particle radius, and $D_{p,i}$ is the pore diffusion coefficient. The pore diffusion coefficient and the film mass transfer coefficient correlations of Carta and Wilson are available and used widely [71, 72]:

$$k_{\text{eff},i} = \frac{1}{1/k_{f,i} + r_p/D_{p,i}} \quad (9.3)$$

As mentioned above, the inner porosity for the components is needed. This can be calculated with the voidage and the total porosity for the component. Voidage and total porosity can be obtained from tracer experiments or inverse size-exclusion chromatography (SEC) described in the literature [70, 73].

The relation of inner porosity, voidage, and total porosity is given in Eq. (9.4):

$$\varepsilon_{t,i} = \varepsilon_{S,i} + (1 - \varepsilon_{S,i}) \cdot \varepsilon_{p,i} \quad (9.4)$$

9.2.5.2 SEC

If the load $\frac{\delta q_i}{\delta t} = 0$, an SEC can be modeled, since here no adsorption of the target component to the resin occurs. To describe the other forms of chromatography, adsorption models are needed. Classically Langmuir, competitive Langmuir, or in some cases Freundlich models were used and offer good descriptions for ion-exchange (IEX) chromatography, hydrophobic interaction chromatography (HIC), and affinity chromatography (AC) [67, 68, 70, 74–76].

9.2.5.3 Adsorption Mechanism

The aforementioned adsorption models fit the adsorption of small- to medium-sized molecules very well and give generally good results for biologics inside the observed area. However, the Langmuir model makes assumptions that do not meet the expected behavior for larger molecules and proteins, e.g. that every protein only binds to one binding site. In the steric mass action (SMA) model, to our knowledge firstly described by Brooks and Cramer [65], more effects of protein adsorption are taken into account.

9.2.5.4 IEX-SMA

IEX chromatography is the most often applied form of chromatography due to their combination of high selectivity, strong binding mechanism, and high binding capacities. Besides its uses in water treatment (e.g. heavy metal adsorption), IEX is a main workhorse in most platform processes for the production of IgG, which is to date the most commonly used biologic pharmaceutical. For other biologics, this is a promising candidate for a cheap and readily available alternative to affinity chromatography.

As mentioned above, SMA models enable of a more complex adsorption behavior. The main reason for their development was to model the steric hindrance in IEX due to the three-dimensional structure of proteins. The other goal of SMA was to model the adsorption of one protein, which normally carries multiple charges to multiple ligands. The SMA adsorption mechanism is shown in Eq. (9.9) [65]. Here, P is the protein, S is the salt, L is the ligand, and v is the protein binding load ratio:

$$P + vSL \Leftrightarrow PL_v + vS \tag{9.5}$$

From this mechanism, the adsorption equilibrium can be deducted as the following system [66]:

$$\frac{q_{p,i}}{c_{p,i}} = A_i * \left(1 - \sum_{j=1}^{m} \frac{q_{p,j}}{q_{p,j}^{max}}\right)^{v_i} \tag{9.6}$$

$$A_i = \widetilde{K}_{eq,i} \left(\frac{\Lambda}{z_S c_S}\right)^{v_i} \tag{9.7}$$

In this model, A_i represents the initial slope of the isotherm if $q_{p,i} \rightarrow 0$. A composes of the equilibrium constant $\widetilde{K}_{eq,i}$ and the relation between the ligand density Λ, the salt ion charge z_S, and the salt concentration c_S.

9.2.5.5 HIC-SMA

HIC is most often used for polishing of biologics. Compared with reversed-phase (RP) chromatography, this is less hydrophobic and thus can be eluted, reducing salt concentration in the eluent. The absence of organic solvents leads to a gentler process and reduces solvent regeneration expenses. HIC has the same competitive adsorption system as in IEX occurs; see Eq. (9.6). However, the initial slope of the isotherm changes to [66]

$$A_i = \widetilde{K}_{eq,i} \left(\frac{\Lambda}{c}\right)^{v_i} \tag{9.8}$$

where c represents the molar concentration of the protein inside the pores.

9.2.5.6 Modified Mixed-Mode SMA

Under mixed-mode chromatography (MMC), generally a chromatography employing both hydrophobic interaction and IEX is understood [77–79]. Besides these two modes, there are several other combinations of chromatographic methods labeled as MMC. One of these is hydroxyapatite chromatography, using both cation-exchange and anion-exchange interaction as an adsorption mechanism [78]. Another example are hydrophobic resins, which are eluted using a pH shift inducing a charge in the resin, also called hydrophobic charge induction chromatography [78]. An interesting point to make about MMC is that in the past, IEX resins often showed hydrophobic interaction as well. However, this effect was not desired and had its origin in the less developed production technology [78].

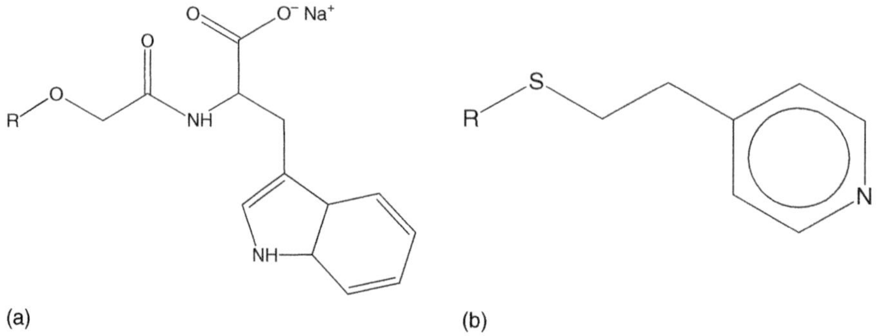

Figure 9.15 Examples of different mixed mode resin ligands. (a) is a tryptophan group. (b) is 4-mercapto-ethyl pyridine.

To date several different molecules are used to enable a hydrophobic exchange effect besides an ionic group. Examples are given in Figure 9.15. In (a), the ligand used in MMC is tryptophan operating as a weak cation exchanger and for hydrophobic interaction. In (b), 4-mercapto-ethyl pyridine is used as the ligand. This ligand under neutral pH shows hydrophobic effects and can be positively charged at lower pH, eluting the also positively charged proteins. It is apparent that different suppliers employ different strategies when it comes to designing the resin's ligand. Besides, the groups for IEX mechanism cation or anion exchange and strong or weak ionic groups are used. For hydrophobic interaction, predominately benzyl or other aromatic ring structures are used.

The promised effect of these multifunctional groups is that selectivity is greatly increased since an adsorbent from solution needs to have a combination of hydrophobic and ionic properties to allow adsorption. This highly selective process is promising for the capture of proteins being employed in the industrial manufacturing of rhVEGF [80] and was investigated as a potential alternative to protein A capture of antibodies [81, 82]. Several other examples can be found in the literature [83]. As gradients pH and salt gradients can be used [82, 84], the protein load is adjusted using the pH, and the salt gradient either elutes the protein from the resin according to the IEX mechanism or allows adsorption to the resin at high salt concentrations according to the hydrophobic interaction mechanism [85].

Modeling MMC is challenging since salt and pH gradients are employed in practice. The difficulties arise from the fact that the protein adsorption cannot be described continuously with a single-component or competitive Langmuir, as for different pH values different coefficients have to be measured [79, 86, 87]. As an alternative to Langmuir adsorption, to describe adsorption in MMC, SMA models have been successfully employed [88, 89]. SMA models for MMC are mainly based on Mollerup's thermodynamic framework for SMA model of IEX chromatography and hydrophobic exchange chromatography [66]. The adsorption for MMC is described as [85]

$$P + vSL + nL \Leftrightarrow PL_n + vS \tag{9.9}$$

where P is the protein, S is the salt, L is the resin ligand, n and v are the steric hindrance coefficients, and v is the binding charge ratio of the protein and the salt $v = z_P/z_S$. The adsorption equilibrium is generally given as [85]

$$\frac{q_{p,i}}{c_{p,i}} = A_i * \left(1 - \sum_{j=1}^{m} \frac{q_{p,j}}{q_{p,\text{IEX},j}^{\max}}\right)^{v_i} * \left(1 - \sum_{j=1}^{m} \frac{q_{p,j}}{q_{p,\text{HIC},j}^{\max}}\right)^{n_i} \qquad (9.10)$$

where q is the protein load, c_p is the protein concentration, and $q_{\max}$ is the maximum loading capacity. A_i is the initial slope of the isotherm for component i and is given by [85]

$$A_i = \widetilde{K}_{eq} \left(\frac{\Lambda_{\text{IEX}}}{z_S c_S}\right)^v \left(\frac{\Lambda_{\text{HIC}}}{c}\right)^n \left(1 - \frac{q_p}{q_{p,\text{IEX}}^{\max}}\right)^v \left(1 - \frac{q_p}{q_{p,\text{HIC}}^{\max}}\right)^n \tilde{\gamma}_P \qquad (9.11)$$

Λ_{IEX} and Λ_{HIC} are the ligand densities of the resin. For most mixed-mode resins, $\Lambda_{\text{IEX}} = \Lambda_{\text{HIC}} = \Lambda$, and therefore $q_{p,\text{IEX}}^{\max} = q_{p,\text{HIC}}^{\max} = q_{p,\text{MM}}^{\max}$. Using this model, adsorption can be modeled including both HIC and IEX effects. The pH value can be implemented via the binding charge ratio v. Saleh et al. [88] model v_i and $K_{eq,i}$ as a function of pH using the correlations of Hunt et al. [90] shown in Eqs. (9.12) and (9.13):

$$K_{eq,i}(\text{pH}) = K_{eq0,i} e^{K_{eq,1,i}*\text{pH} + K_{eq,2,i}*\text{pH}^2} \qquad (9.12)$$

$$v_i(\text{pH}) = v_{0,i} + \text{pH} * v_{1,i} \qquad (9.13)$$

Another correlation found in the literature by Mollerup is given in Eq. (9.14) [66], where a_1 and a_2 are component specific coefficients:

$$v_i = a_{1,i} \cdot \ln(\text{pH}) + a_{2,i} \qquad (9.14)$$

If intended to leave a correlation-based model and move toward a more physicochemical descriptions, a modification is possible. Good correlations for q can be achieved when v is calculated according to Eq. (9.15), with f being the form factor describing the ratio of the protein surface load and the protein binding load. z_P is the protein load calculated with a modified Henderson–Hasselbalch [91] shown in Eq. (9.16), with c_{AA} being the cutoff factor of the amino acid, describing their distribution over the protein, and N_{AA} being the number of the amino acid. AA stands for amino acid:

$$v_i = f * z_P/z_S \qquad (9.15)$$

$$z_P = \sum_{AA} -\frac{c_{AA} * N_{AA}}{1 + 10^{\text{pKa}_{AA} - \text{pH}}} \qquad (9.16)$$

9.2.5.7 Modified HIC-SMA Process Model Exemplification by mab Purification

The cutoff factor of z_P was calculated from with the least-squared error method from the literature data [92] according to Eq. (9.16). The results for this are shown in Figure 9.16. The protein load can be satisfactory described with this model. In this calculation, R^2 is 0.98 and therefore a strong correlation for a fit to experimental data.

Using adsorption data given in the literature for IgG, SMA adsorption model parameters were estimated using a Nelder–Mead estimation routine [92, 93]. The

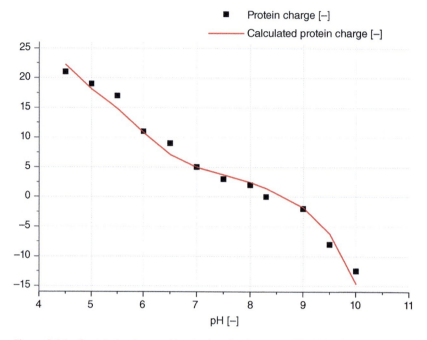

Figure 9.16 Protein load over pH, calculated using a modified Henderson–Hasselbalch equation.

Table 9.1 Estimated SMA parameters for chromatography.

Estimated variable	Value
$\widetilde{K}_{eq}$	250.04
f	0.0866
q_p^{max}	37.862
K_p	0.0674
K_S	−2.3918
n	0.7507

results of the estimation are given in Table 9.1. The optimization function was maximum logarithmic likelihood, since better convergence was achieved than with least error squares. Optimization tolerance should be reasonably low for this estimation. In our SMA system, best fit was achieved using 10^{-6}.

The overall variation explained was 94.7%. This corresponds to a R^2 of 0.90. This can be considered satisfactory, based on the data used to correlate z_p and the adsorption isotherm. The major error source is that the z_p calculation and the isotherm are based on different antibodies. Another reason for deviation are errors resulting from the experiments.

Figure 9.17 Simulated chromatograms employing different gradients. In (a) a salt gradient is applied eluting the product in a narrow peak, (b) shows the influence of a pH gradient, which lead to peak broadening and also a earlier elution. In (c) Salt and pH gradient are combined to allow for an early elution in a narrow peak.

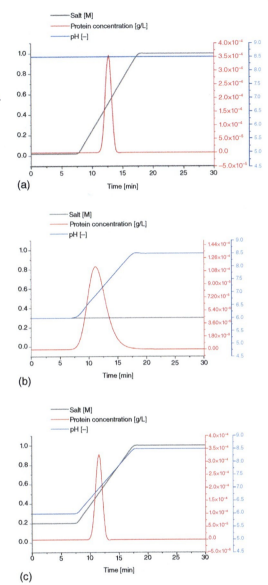

However, the goal of this example is to show the implementation of a pH and salt gradient. In Figure 9.17, solutions of the model for different gradients are given. It is apparent that salt gradients and pH gradients can be represented using the model given above (Figure 9.17).

9.2.5.8 Model Parameter Determination

The parameters needed for a successful modeling of a chromatographic process can be separated into three different groups that are determined experimentally stepwise. These are illustrated in Figure 9.18 such that the model is assembled step by step by its equation parts with regard to their function of fluid dynamics, equilibrium, and mass transfer kinetics description [94].

9.2.5.8.1 Fluid Dynamics

Parameters that need to be determined regarding the fluid dynamics are the plant dead volume V_{dead}, the axial dispersion coefficient D_{ax}, the total porosity ε_T, and the voidage ε_S. To determine the plant dead volume and the total porosity, two tracer experiments are needed, one with the column and one without the column. As a tracer substance, a small, nonbinding, but UV-active molecule should be chosen. For IEX, acetone is a typical example.

To determine the voidage and the dependency of porosity over molecular weight, multiple injections of different molecule sizes on the column can be used. In IEX, pullulan or dextran standards are commonly used. From these results, the total porosity ε_T can be determined for $M \rightarrow 0$, and the voidage ε_S for $M \rightarrow \infty$. The particle porosity can then be calculated using Eq. (9.4) (Figure 9.19).

The axial dispersion coefficient D_{ax} can be determined from the tracer experiment. However, D_{ax} is highly dependent on the molecular diameter. To measure the axial dispersion coefficient of the target component, it should be injected on the column under nonbinding conditions. The axial dispersion coefficient can be determined according to the Levenspiel model using the following equations [95]:

$$\frac{\sigma^2}{\bar{t}^2} = 2\frac{D_{ax}}{v\,l} - 2\left(\frac{D_{ax}}{v\,l}\right)^2 \left[1 - e^{-\frac{v\,l}{D_{ax}}}\right] \tag{9.17}$$

$$\frac{\sigma^2}{\bar{t}^2} = 2\frac{D_{ax}}{v\,l} + 8\left(\frac{D_{ax}}{v\,l}\right)^2 \tag{9.18}$$

9.2.5.9 Phase Equilibrium Isotherms

With the determined fluid dynamic parameters, the process model equation setup can be utilized to calculate phase equilibrium experiments by neglecting mass transfer resistance at sufficient long residence time.

For isotherms, many different determination concepts exist. They can be separated in static and dynamic binding methods [96].

To determine a point on the isotherm by a static shaking experimental setup, a known amount absorbent and feed are mixed. This mixture is shaken to ensure the equilibrium between feed medium and pore interior. After that, a sample from the excess fluid is taken and analyzed. This sample holds the equilibrium concentration. The loading can be determined by Eq. (9.19) [70]:

$$q_i = \frac{(c_{Feed,i} - c_{eq,i}) \cdot V_{Feed}}{V_{Ads}} \tag{9.19}$$

With this procedure, around 5–10 experiments are needed to achieve an acceptable error in a modeled isocratic chromatography.

If a gradient is to be used, this needs to be repeated 5 times at minimum with different modifier concentrations, e.g. on IEX or HIC media. The experimental plan of concentration and fluid amount in relation to adsorbent mass has to be appropriate for analytical accuracy [97–100].

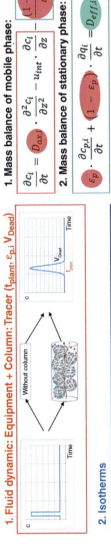
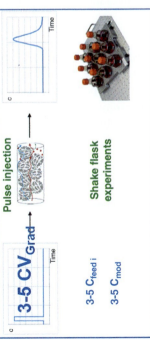

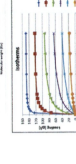

Figure 9.18 Overview of the model parameter determination. Source: Zobel-Roos [70].

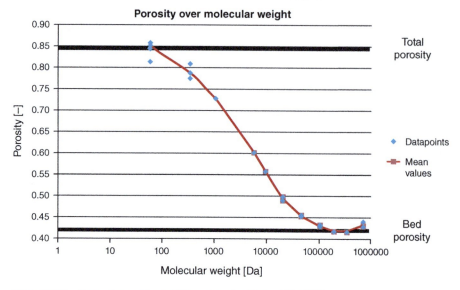

Figure 9.19 Dependency of porosity over molecular weight [70].

A vast variety of dynamic methods is explained in the literature, where the most commonly used are elution at the characteristic point, breakthrough curves, frontal analysis, and peak fitting [69]. The advantage of these methods are shorter acquisition times. On the other hand, especially for breakthrough curve measurements or frontal analysis, a relatively large amount of feed is required [70, 101].

9.2.5.10 Mass Transfer Kinetics

After these two first steps, the process model consists of appropriate fluid dynamic and phase equilibrium description. Finally, describing as well the mass transport from the continuous fluid phase into the pores and back, the mass transfer kinetics are needed to model chromatography – especially for large diffusion-limited biomolecules. The parameters to describe this transport process are the film transfer coefficient k_f, the molecular diffusion coefficient D_m, and the pore diffusion coefficient D_p.

The kinetic parameters can be easily correlated by different relations. The mass transfer coefficient k_f can be determined for biomolecules using the correlation shown in Eq. (9.20) in the interval of $0{,}0016 < \mathrm{Re} < 55$ [72]:

$$k_{f,i} = \frac{1{,}09}{\varepsilon_s d_p} D_{m,i} \left(\frac{u_{int} d_p}{D_{m,i}} \right)^{\frac{1}{3}} \qquad (9.20)$$

The molecular diffusion coefficient D_m required for this equation can be determined with the following correlation found by Young et al. [102]:

$$D_{m,i} = 8{,}34 \cdot 10^{-10} \frac{T}{\eta \sqrt[3]{M_i}} \qquad (9.21)$$

The pore diffusion coefficient D_P can also be determined using a correlation like suggested by Carta and Rodrigues [71]:

$$D_{p,i} = \frac{\varepsilon_s D_{m,i}}{\tau_i} \psi_{p,i} \qquad (9.22)$$

This correlation calls for the tortuosity factor τ_i that needs to be determined with a static binding experiment. Also, the diffusive hindrance factor $\psi_{p,i}$ can be determined with other correlations [71].

9.2.6 Lyophilization

Parameters such as the overall heat transfer coefficient, product resistance, and stationary product temperature must be determined in one previous characterizing experiment. The procedure to obtain these values is described in this chapter. Other necessary input values and their determination method are shown in Figure 9.20.

9.2.6.1 Thermal Conductivity of the Vial

The vial heat transfer coefficient k_{vial} is determined by an experiment in analogy to literature that is stopped right before the end of primary drying [103]. The mass of sublimated water is known by weighting, which enables calculation of total heat necessary to sublime. This value ΔQ in combination with process duration Δt gives the heat flow. Before weighing, the vials are thawed at room temperature. The temperature values for the measured vials were taken from the closest vial that was probed with a WTMplus. Now, k_{vial} can be calculated by heat conduction:

$$k_{vial} = \frac{\Delta Q / \Delta t}{A_{vial} \cdot (T_{S,PD} - T_{product,av})} \qquad (9.23)$$

where A_{vial} is the cross-sectional area of the vial, $T_{shelf,PD}$ is the shelf temperature during primary drying, and $T_{product,av}$ is the average product temperature during primary drying. The latter is measured by wireless temperature measurement sensors. To obtain the thermal conductivity of the vial λ_{vial}, k_{vial} has to be multiplied by the thickness of the vial bottom. The thermal conductivity of the vial has three main contributions [104]:

$$k_{vial} = k_{cond} + k_r + k_{gc} \qquad (9.24)$$

k_{cond} quantifies the contribution from direct conduction of the shelf to the vial, whereas k_r and k_{gc} describe the contributions of heat transfer by radiation and gas conduction, respectively [104]. Values of k_{vial} depend on chamber pressure, vial location, container type, and possibly the used freeze dryer equipment [105].

9.2.6.2 Product Resistance

Product resistance K is determined by the same experiment as the vial heat transfer coefficient. The calculated mass flow $\dot{m}_{PD}$ is inserted in Eq. (9.6), which is extended by the average vapor density $\rho_{w,g,av}$ and the cross-sectional area of the vial A_{vial}:

$$u_{W,av} \cdot \rho_{W,g,av} \cdot A_{vial} = \dot{m}_{PD,av} = \frac{p_{C,PD} - p_{subl,av}}{\eta_{W,av} \cdot K} \cdot \rho_{W,g,av} \cdot A_{vial} \qquad (9.25)$$

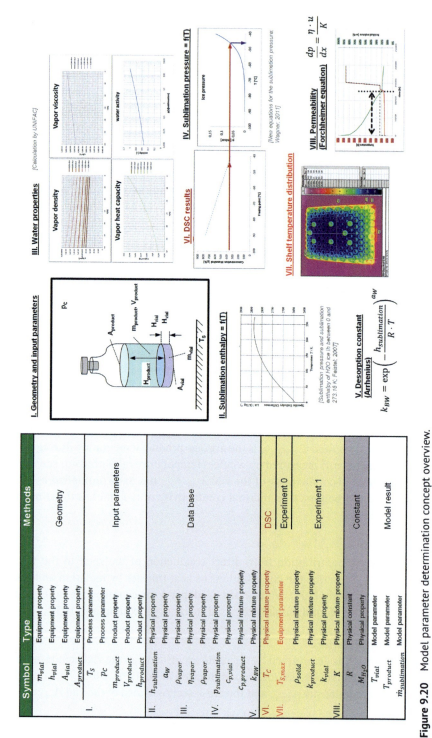

Figure 9.20 Model parameter determination concept overview.

where $u_{W,av}$ is the average gas velocity during primary drying, $p_{C,PD}$ is the primary drying chamber pressure, $p_{subl,av}$ is the sublimation pressure as function of the average product temperature during primary drying, and $\eta_{W,av}$ is the average dynamic viscosity.

The workflow for the determination of the parameters is shown in Figure 9.21.

9.2.6.3 Product Temperature

The maximal product temperature is a function of vial position in the freeze dryer. This effect is well known and described in the literature, mainly attributed to the radiation effects. This work uses the highest measured product temperature as position-dependent shelf temperature.

9.2.6.4 Water Properties

The water properties such as density and dynamic viscosity are calculated as a function of pressure and temperature by Universal Quasichemical Functional Group Activity Coefficients (UNIFAC). T_g is determined by differential scanning calorimetry (DSC), and $T_{collapse}$ has been determined by freeze-dry microscopy. The used formulation has a T_g of $-34\,°C$ and a $T_{collapse}$ of $-29\,°C$ [106].

9.3 Process Integration and Demonstration

Process integration is essential to provide a comprehensive solution, not leaving weak spots, which may be causing approval troubles. Appropriate process integration has to be achieved within process development before generating first clinical trial amounts. Afterward, the process is fixed, and any changes would cause additional clinical trials – which is too costly in most cases. Different academic working groups have recently proven the concepts to be valid in cooperation with industrial partners [107–110].

"Invention" with the largest impact at the moment is the transfer from batch to continuous operation. In addition, some studies combine continuous operation with a modification of the unit operations integrating ATPE and/or precipitation and/or membrane adsorbers [111–115].

Figure 9.22 summarizes the process schemes and performance for purity, yield, and concentration per unit operation at each step along with a cost analysis. Potential for improvement is in the magnitude of factor 10 in the cost of goods. Although all those processes have so far only been operated successfully in lab scale, it should be kept in mind that laboratory scale is the manufacturing scale in those cases of specialized product for patient groups in a lower kg per year scale. Technology readiness level is in the magnitude of 5–6, so prototyping with regulatory approval is the only step missing but seems to present no major technical obstacle.

To conclude the industrial status, all companies work on some approaches [112]. Funded projects deal with next-generation downstream processing [116], focusing on flow-through continuous-type operation as well [116–119].

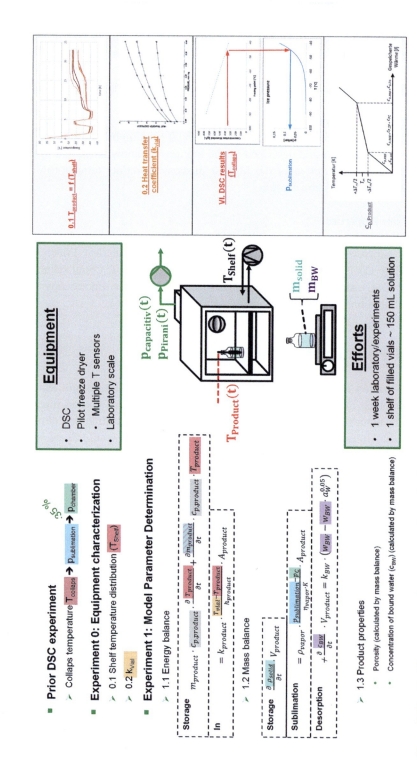

Figure 9.21 Workflow summary for parameter determination.

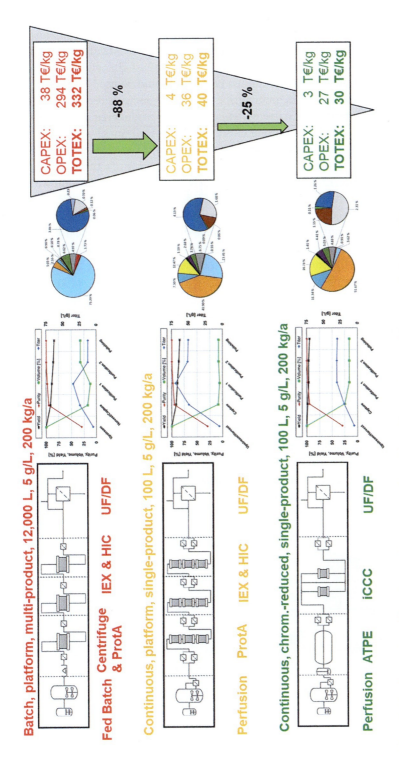

Figure 9.22 Comparison of batch to continuous operation with platform and chromatography-reduced processes.

Whether to make those equipment devices single use or disposable or stainless steel is an open question. Engineering companies do see an increasing demand [120], and work groups [121] have generated efficient approaches, standardizations, and decision-making tools. For process development with fast changing products, the cleaning efforts vanish, and well-trained personnel could be used for higher-value creative work. For custom manufacturing organizations with flexible multiproduct multipurpose facilities to cope with fast changing product requests, it seems to be a valid option to reduce cleaning effort and plant dead time by being fast with single-use equipment and passing the related costs on to customers, who are predominantly under project time pressure. However, for manufacturing in continuous operation mode, one major benefit is nearly optimal plant utilization with less cleaning efforts. Here, stainless steel should still be efficient and economic, although the manufacturing equipment has standard laboratory – or at least pilot-scale size. Additionally, some buffer handling and storage utilities may be single use.

A group of authors actually conclude that container-based modular manufacturing plants are for small and medium-sized enterprises (SME) to be used for fully developed manufacturing at any scale, e.g. tens to hundreds of kg per year (Figure 9.23).

Figure 9.23 exemplifies a 3D model of engineered container-based plants mounted flexibly for any process sequence with modular skids in standard configuration with regard to size and process control system (PCS) connection. Buffer handling is separated, as well as any option for purified water and for injection water preparation or recycling needed in harsh environments. Normal laboratory/pilot-scale skids could be used that are operated at flow rates of a few hundred ml/min. USP and DSP are integrated, as USP is housed and clean air conditioned separately. Clean room facility is implemented under S1 genetic safety level; nevertheless, all units are interconnected in continuous operation in a closed piping, i.e. without any handling of open product. Options for PAT in-line analytics are implemented for PCS, as well as additional options to draw samples for QA at each unit, all of which are validated under cGMP.

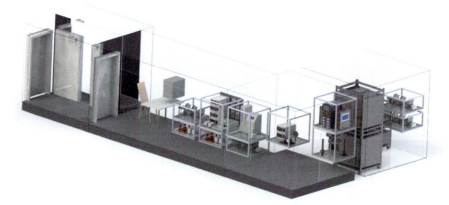

Figure 9.23 Container-based modular plant design for biologics in few hundred kg/a scale under continuous operation.

Figure 9.24 Cleaning cost for different process alternatives.

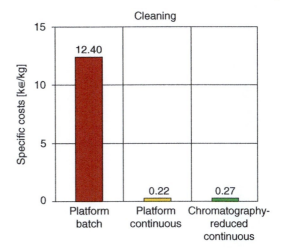

PCS of the container infrastructure is separated from the manufacturing part in a different room, as well as access for the personnel. However, PCS of the process is located directly at the units for any visual inspections, normally with an industrial standard Siemens S7.

Keep in mind that cost distribution for CM reduces cleaning and dead-time costs drastically – which are dominant cost drivers in batch operation and efforts of highly skilled and rarely found manpower; see Figure 9.24. Therefore, any benefit from disposable/single-use technologies disappears, almost by 98% total cost reduction of batch process cleaning costs.

Equipment skids in classical stainless steel are state of the art and the best choice based on cost calculations. Nevertheless, single-use devices like membranes or chromatography columns could be utilized within these skids and disposable buffer handling options, provided that robustness allows the continuous operation, which of course needs to be proven.

First "manufacturing devices" are established directly at hospitals for therapies that need direct and fast access to patient, i.e. for stem cell treatment or highly toxic anticancer therapies like antibody drug conjugates to minimize contamination potential, or treatment with personalized medicines in mg amounts [122, 123].

An experimental feasibility study on continuous bioprocessing in pilot scale of 1 l/day cell supernatant, that is, about 150 g/year product (monoclonal antibody) based on Chinese hamster ovary (CHO) cells for model validation, is performed for about six weeks including preparation, start-up, batch, and continuous steady-state operation for at least two weeks' stable operation and final analysis of purity and yield (see Figure 9.25) [12]. A mean product concentration of around 0.4 g/l at cell densities of 25×10^6 cells/ml was achieved. After perfusion cultivation with alternating tangential flow (ATF) filtration, an ATPE followed by UF/DF toward a final integrated countercurrent chromatography (iCCC) purification with an IEX and an HIC column prior to lyophilization was successfully done. In accordance to prior studies, continuous operation is stable and feasible. Efforts of broadly qualified operation

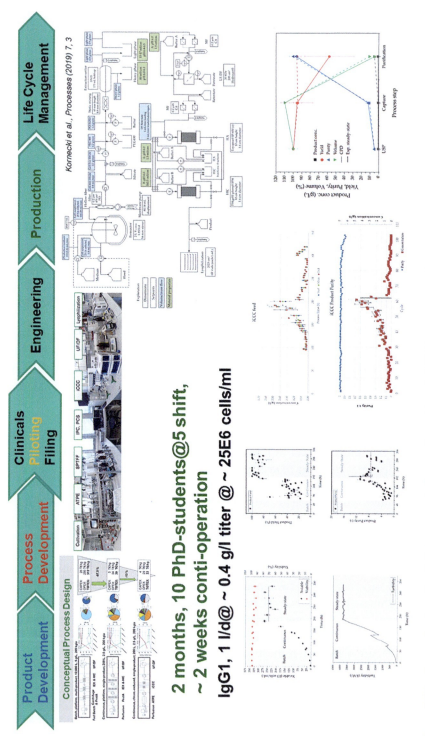

Figure 9.25 Summarized results of pilot case study. Source: Kornecki et al. [12]. Licensed under CC BY-4.0.

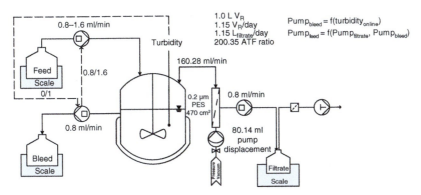

Figure 9.26 Schematic overview of the turbidostat perfusion cultivation including control strategy for viable cell concentration via turbidity, feed, and bleed pump.

personal and the need for an appropriate measurement and process control strategy are shown evidently.

9.3.1 USP Fed Batch and Perfusion

A schematic overview of the perfusion cultivation can be seen in Figure 9.26. At high viabilities (>95%), turbidity can be used for the online monitoring of viable cell concentration. Therefore, an online control was implemented by a turbidity threshold. One operating point (4800 FAU) was chosen for the control of viable cell concentration, which is 25×10^6 cells/ml. As soon as turbidity surpassed the respective threshold, a bleed pump was automatically triggered. The sum of permeate and bleed volume flow resulted in a higher feed volume flow to maintain the culture volume. An ATF ratio (ratio between permeate volume flow and ATF flow rate) of 200.35 was reached at an ATF flow rate of 160.28 ml/min by a pump displacement of 80.14 ml at one pump cycle per minute (one cycle is defined as vacuum exhaust and pressurized displacement of the entire pump volume). The viable cell concentration (1×10^5 cells/ml) and turbidity formazin attenuation units (FAU) are depicted in Figure 9.27.

Moreover, the correlation between viable cell concentration (1×10^5 cells/ml) and turbidity (FAU) is shown in Figure 9.28. Viable cell concentration can be expressed in the form of a simple linear regression depending on the online turbidity. A coefficient of determination of 0.969 can be achieved. The offline determined viable cell concentration can be described by online turbidity data. The high error above 3500 FAU is due to the measurement error of the probe ($\pm 3\%$). This can be circumvented by implementing a more precise measurement probe. Nevertheless, the quasi-stationary state of this turbidostat is clearly visible and increases process understanding, control, and data quantity for an advanced mammalian cell culture.

Process-related product concentration (a) and purity (b) can be seen in Figure 9.29. The perfusion cultivation achieved an antibody concentration of 0.34 ± 0.05 g/l (end of batch phase), 0.70 ± 0.06 g/l (start of steady state), and 0.50 ± 0.03 g/l (as of 240 hours cultivation). The lower product concentration may be due to the shorter process time in comparison with conventional perfusion cultivations (14–60 days)

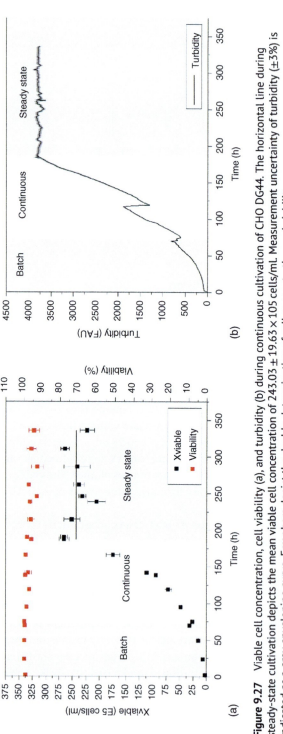

Figure 9.27 Viable cell concentration, cell viability (a), and turbidity (b) during continuous cultivation of CHO DG44. The horizontal line during steady-state cultivation depicts the mean viable cell concentration of $243.03 \pm 19.63 \times 10^5$ cells/mL. Measurement uncertainty of turbidity ($\pm 3\%$) is indicated as a gray enveloping curve. Error bars depict the double determination of cell concentration and viability.

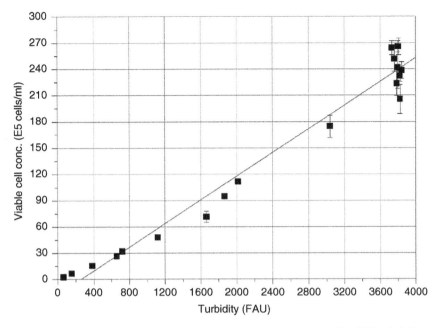

Figure 9.28 Linear correlation between viable cell concentration (1×10^5 cells/ml) and turbidity (FAU) with a slope of 0.068 ± 0.003 and an offset of -17.40 ± 8.01. Error bars depict the double determination of cell concentration.

(e.g. lower investment expenditures, long-term continuous cultivation), enhancing the perfusion systems.

However, even at lower product concentration, the volumetric productivity reaches 700.34 ± 56.85 mg/l/d (start of stationary phase) and 503.20 ± 28.45 mg/l/d (as of 240 hours cultivation). This high productivity among other process characteristics.

Figure 9.30 depicts offline conductivity and viability measurements (a) and glucose and lactate concentrations (b) during respective phase of the cultivation. As can be seen, the conductivity is except for one outlier (120 hours) in correlation with the cell viability. The outlier can be explained by evaluating the turbidity data in Figure 9.6. At approximately 120 hours, a significant decrease in turbidity can be observed. This is based on a prior concentration (starting at approximately 100 hours) of the cell concentration due to an error of the feed pump. This led to a decreased medium volume flow and subsequently to a decreased cell culture volume. Due to the cell retention system, the cell concentration increased. After 20 hours the error of the feed pump was eliminated, and the culture volume was increased manually to the initial volume of 1 l. However, the conductivity measurements depict no significant increase based on decreasing viability and release of cell lysate.

The glucose and lactate concentrations in Figure 9.30 indicate typical progression during the batch phase. The glucose and lactate concentrations decrease and level off, respectively, as soon as the continuous operation mode was initiated. This is due

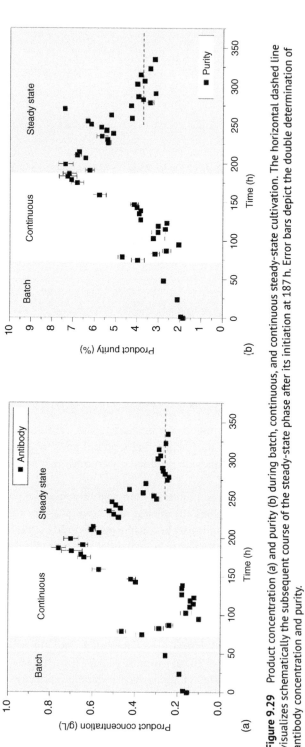

Figure 9.29 Product concentration (a) and purity (b) during batch, continuous, and continuous steady-state cultivation. The horizontal dashed line visualizes schematically the subsequent course of the steady-state phase after its initiation at 187 h. Error bars depict the double determination of antibody concentration and purity.

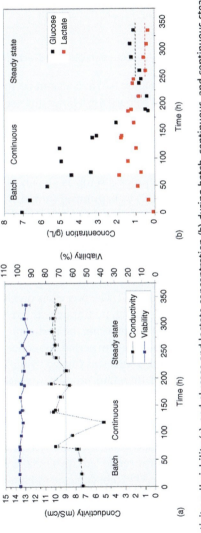

Figure 9.30 Conductivity, cell viability (a), and glucose and lactate concentration (b) during batch, continuous, and continuous steady-state cultivation. The horizontal solid line depicts the mean conductivity of 9.0 ± 1.4 mS/cm. The horizontal dashed line visualizes schematically the subsequent course of the steady-state phase after its initiation at 187 h. Error bars depict the double determination of conductivity, viability, glucose and lactate concentration.

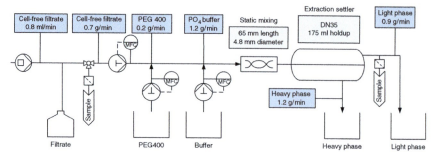

Figure 9.31 Schematic overview of the aqueous two-phase extraction (ATPE) process.

to an increasing cell concentration (Figure 9.6) at constant perfusion rates. As soon as the glucose concentration reached below 0.5 g/l, bolus glucose concentration was fed to achieve a respective concentration of 1.0 g/l.

9.3.2 Capture, LLE, Cell Separation, and Clarification

A schematic overview of the ATPE process setup is shown in Figure 9.31. Polymer, buffer, and cell culture fluid (CCF) were pumped from their respective storage vessels into the static mixing line. After thorough mixing, the two-phase system separates into a light phase (polymer rich) that contains the product and a heavy phase (salt rich) that contains impurities, such as DNA and lower-molecular-weight components (LMWC).

The product concentration (a), volume flow (b) of the light phase, and conductivity and pH value (c) is shown in Figure 9.32. The process is divided into three major stages. During batch phase of the bioreactor, no filtrate was harvested; hence ATPE was not operated. After 70 hours cultivation time, the perfusion begins, and the filtrate was fed into the ATPE. During this start-up period, the concentration was as low as the concentration in the bioreactor, which was caused by the low cell concentration. As the concentration of cells and, therefore, product increases in the bioreactor, the same trend was observable during ATPE. As the perfusion reaches steady state after a peak in product concentration at 192 hours, the concentration remained at a constant value of approximately 0.5 g/l. The conductivity and pH value are depicted in Figure 9.11 (right). In steady-state operation, the conductivity remained at a value of 16.89 ± 0.74 mS/cm, and the pH value remained constant at a value of 6.45 ± 0.04.

Yield (a) and purity (b) achieved in ATPE are depicted in Figure 9.33. The yield in the start-up phase of continuous operation (70–192 hours process time) is low and shows larger deviations, with averaging values of $32.46\% \pm 16.48\%$. This might be due to the low concentration in this process stage in the bioreactor and the factor that losses in concentration during ATPE have a stronger effect on yield, when the overall concentration is already very low.

However, in steady-state operation (192–336 hours process time), when the concentration in the bioreactor is constantly high at approximately 0.5 g/l, the yield in ATPE levels out at $87.22\% \pm 12.85\%$. Since the concentration in this phase was higher, the loss during the start-up period can be easily compensated.

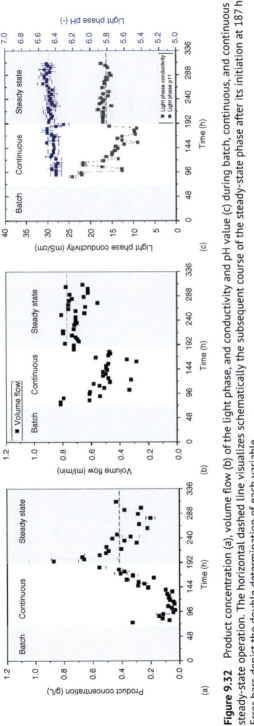

Figure 9.32 Product concentration (a), volume flow (b) of the light phase, and conductivity and pH value (c) during batch, continuous, and continuous steady-state operation. The horizontal dashed line visualizes schematically the subsequent course of the steady-state phase after its initiation at 187 h. Error bars depict the double determination of each variable.

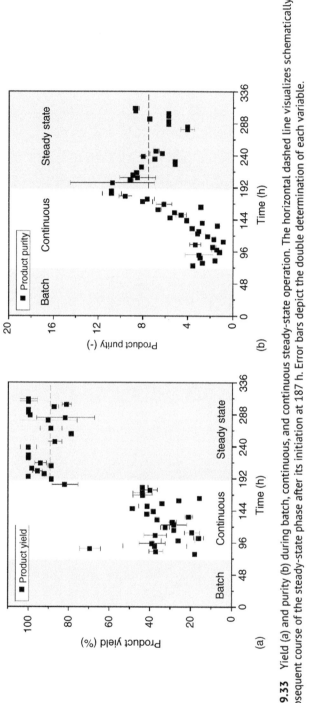

Figure 9.33 Yield (a) and purity (b) during batch, continuous, and continuous steady-state operation. The horizontal dashed line visualizes schematically the subsequent course of the steady-state phase after its initiation at 187 h. Error bars depict the double determination of each variable.

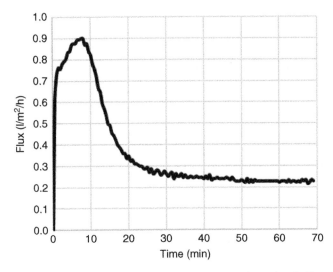

Figure 9.34 Exemplary flux development over time of a diafiltration step.

9.3.3 UF/DF, SPTFF for Concentration, and Buffer Exchange

The diafiltration step is necessary for changing the buffer of the target protein to low salt concentrations and additionally the removal of polyethylene glycol (PEG). A typical behavior of flux vs. time of a diafiltration step in this study is shown in Figure 9.34.

Obviously, the flux increases due to the removal of PEG, which increases the viscosity and lowers the filtration performance. After 10 minutes, the flux reaches its maximum. From this point, the filtration performance starts to decrease strongly. This is a hint for fouling owing to the side components blocking the pores of the membrane and possible precipitation of the target protein. The observed concentrations and purities of all diafiltration steps are depicted in Figure 9.35.

While the general concentration development follows the tendencies of previous process steps, the concentration in general was lower, compared with the results from ATPE. Furthermore, the yield of each diafiltration step shows no clear course. The average value was 52% ± 32%. This great spectrum of yield was probably influenced by handling error possibilities resulting from four to five different shifts and various operators, which could be solved by appropriate training in the future.

Besides the exchange of buffer, the diafiltration step also excludes side components by size. With the used membrane cassettes, all molecules smaller than the cutoff of the membrane (30 kDa) pass into the permeate and are separated.

The usage of filtration methods to exchange the surrounding buffer reaches an average purity of 45% ± 18%. It also increases the purity up to 40% ± 20% by removing smaller molecules (e.g. DNA and proteins). Therefore, the role of the diafiltration not only is necessary to change the buffer for subsequent unit operations but also has a possibility to increase the general purity of the target molecule.

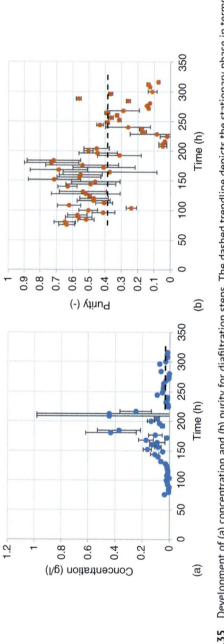

Figure 9.35 Development of (a) concentration and (b) purity for diafiltration steps. The dashed trendline depicts the stationary phase in terms of product concentration in (a) and the mean purity after diafiltration in (b). Error bars depict the double determination of each variable.

9.3 Process Integration and Demonstration | 311

The performed diafiltration experiments show a broad distribution of resulting process variables such as yield. A continuous filtration/diafiltration method, however, would reduce the problem of handling errors. For this, SPTFF systems could lead to more consistent results and developments [124].

9.3.4 Precipitation/Crystallization

In Figure 9.36, the achieved recovery of mAb after precipitation is shown, as well as the deviation between various batches. Furthermore, a distinction between precipitation with and without a subsequent washing step after loading became obvious. It seemed that the washing step with PEG 4000 solution is disturbed following dissolution and is not target-aimed in this context. Moreover, it was tested whether or not a precipitation step, as a capturing step, directly after cultivation is feasible. A complete precipitation of mAb from CCF could be accomplished because no mAb was found, neither in waste nor in supernatant, but in this instance it was not feasible to recover the mAb from precipitate. This can be seen easily in the green column in Figure 9.15, which displays 0% recovery of mAb. Furthermore, the concentration profile of mAb in the supernatant and its development throughout the precipitation process were analyzed with protein A chromatography. It was determined that mAb precipitation occurs very rapidly within one or two minutes. On the basis of results from preliminary experiments, the precipitation time needed was set to 120 minutes, due to the turbidity signal. In this case, a turbidity probe is not useful to display precipitation progress, because it is not selective enough to detect the precipitation of the target protein only.

To determine the host cell protein profile before and after precipitation, an SEC was used. Figure 9.37 shows the side component spectrum of the light phase

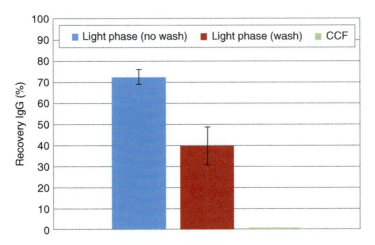

Figure 9.36 Recovery of mAb after dissolution with varying base materials. Dissolution without a previous washing step resulted in the approximately 72.59% (±3.57%) mAb (blue) recovery from precipitate and with subsequent washing resulted in the 39.68% (±8.92%) recovery (red). Dissolution of precipitated CCF (cell culture fluid) was not possible (green). Error bars depict the double determination of product recovery.

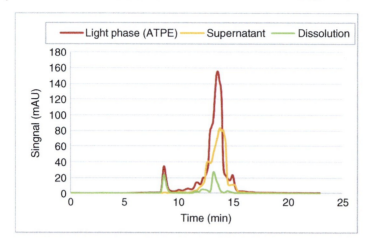

Figure 9.37 Recovery of mAb after precipitation. Precipitation was conducted, with light phase (LP) from aqueous two-phase extraction (ATPE) used as the base material and the absence of a washing step.

(base material) supernatant after precipitation and the IEX buffer with redissolved mAb. The experiment displayed in Figure 9.16 did not include a subsequent washing step after loading. It can be seen that the recovery of mAb was achieved and a further reduction of host cell proteins (HCPs) was accomplished. Most of the HCPs remained in the supernatant and were separated from the product by filtration.

In contrast, the side component spectrum of the second precipitation attempt, including a washing step, is depicted in Figure 9.38. Both figures indicate that HCP spectra of supernatant in the precipitation procedure with and without washing step are highly comparable. A huge difference can be seen in the purity of product obtained from precipitation. Product purity was improved, but, simultaneously, the recovery yield of mAb was reduced (39.68% ± 8.92%). This leads to the assumption that, on the one hand, the washing step had a positive effect on the purification of mAb because more HCPs were separated from the base material but, on the other hand, it had a negative influence on dissolution.

Overall, it can be said that the reduced recovery of target protein might be affiliated to the suboptimal properties of the utilized IEX buffer for dissolution. The buffer can be optimized by a pH shift to 5.5 or even 5 to enhance solubility of mAb, owing to its isoelectric point (IP) of around pH 8.4. Further, an increase of ionic strength of up to 50 or 100 mM will lead to a salting-in effect, which might increase solubility and stability of protein.

Additionally, it was tested whether complete mAb dissolution under optimal conditions given by a phosphate-buffered saline (PBS) buffer (pH 7.4) is feasible. Hence, the two different experiments described above were carried out again, but dissolution was conducted with PBS. Figure 9.39 shows that dissolution after precipitation with (red) and without (green) a washing step is within the deviations, which are a lot smaller than dissolution experiments conducted with IEX buffer. This shows in general that dissolution after precipitation is feasible and the two

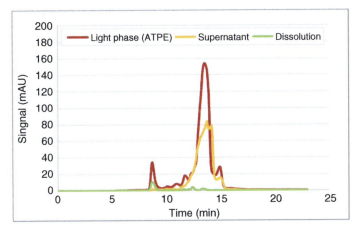

Figure 9.38 Recovery of mAb after precipitation. Light phase (LP) from aqueous two-phase extraction (ATPE) was used as base material, including a washing step after loading the hollow fiber.

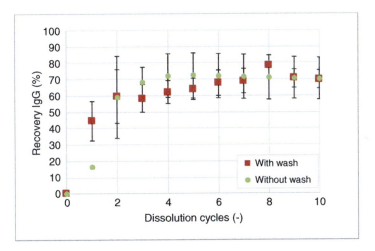

Figure 9.39 Feasibility study of dissolution of mAb after precipitation from light phase (LP) using phosphate-buffered saline (PBS) buffer (pH 7.4) and the experimental setup shown above. Error bars depict the double determination of product recovery. IgG, immunoglobulin G.

working procedures lead to similar results (washing, 70.22% ± 5.75%; no washing, 70.95% ± 13.10%). It is shown that dissolution kinetics with a previous washing step are slightly faster in the beginning than no washing. In the end, both working procedures reach an analogous recovery of mAb. The missing percentages could be recovered in a second washing step with H_2O. To do so only one washing cycle was necessary. Respective to these results, precipitation with a subsequent washing step is to be favored, owing to similar yields in dissolution and higher depletion of HCPs.

Furthermore, it can be said that complete dissolution after precipitation is feasible, but it is highly dependent on buffer conditions.

9.3.5 Chromatography and Membrane Adsorption

At first, some preliminary tests were carried out. For this, buffers with different pH values and different gradient slopes were tested. As shown in Figure 9.40, buffers with pH values of 5.0, 6.0, and 7.0 were investigated, where the one with pH 6.0 showed the best separation. Therefore, this pH value was selected.

Subsequently, various gradient slopes were tested with the designated pH value of 6.0 by varying the gradient lengths of 3, 5, and 7 CV. In Figure 9.41 it can be seen that even with the 3 CV gradient, a decent separation was achieved. It can also be detected that there is a potential reduction of the operating time.

For the iCCC method, the wash step on both columns was eliminated. In the case of the IEX, the gradient separation was carried out up to one third of the gradient length. Thereafter, a step gradient to regenerative conditions was carried out. However, in case of the HIC method, the last one third of the gradient were cut, so the regeneration could be started sooner. In addition, it was tested to cut the first one third of the gradient, since the separation takes place in the middle third of gradient elution. In preliminary iCCC test runs, it was observed that a concentration occurs, which results in a widening of the product peak. Due to this effect, it is feasible to start elution sooner, and, therefore, the peak was influenced by a step in the first third of the elution. Because of this, the linear elution gradient begins at 1 M ammonium sulfate (see Figure 9.42).

In Figure 9.43, the concentration of the iCCC feed is shown. For this, different DF batches (output of diafiltration) were pooled. The figure depicts three data points. The first dataset ("Start") is the concentration directly after the addition of a new feed. The second measurement ("End") was carried out before new feed was added. With this measurement it can be investigated if the product concentration will drop within the storage time. The third data point ("Mean") is the mean value of both measured concentrations. With this, the mean feed concentration for the time step could be achieved and was used to calculate the yield of each cycle.

Figure 9.44 shows the concentration of the product fraction of each iCCC cycle. Each fraction was analyzed by the SEC to determine the final purity. The analysis is shown in Figure 9.23. It can also be seen that the purity of all cases exceeds 97%. Due to a slight shift in the retention times of the HIC, a decrease of purity can be noticed. After adapting the cut points, a purity of over 99% was achieved. For all iCCC cycles, a yield of 73% ± 38% was reached.

9.3.6 Lyophilization

The IgG standard solutions used in the preliminary tests were reconstituted and analyzed by protein A chromatography. The results are shown in Figure 9.45.

With average yields of 96% ± 0.4%, the freeze-drying cycle is considered sufficient for the drying of IgG. The only difference in the study shown here is the composition of the solution. During the tests, the IgG standard solution consisted only of the protein solution with stabilizing substances (see Section 9.2.6) and purified water with 25 g/l sucrose.

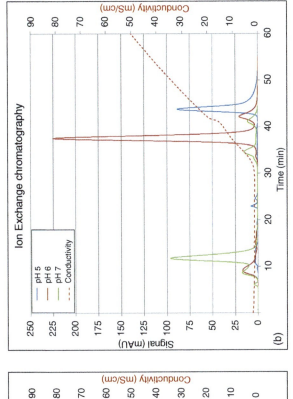

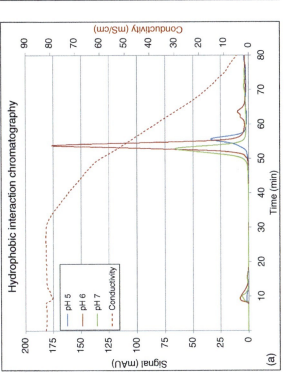

Figure 9.40 Influence of pH values for the hydrophobic interaction chromatography (a) and ion-exchange chromatography (b).

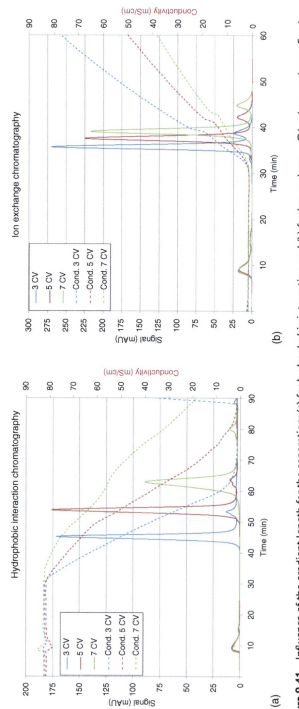

Figure 9.41 Influence of the gradient length for the separation: (a) for hydrophobic interaction and (b) for ion exchange. CV, column volume; Cond., conductivity.

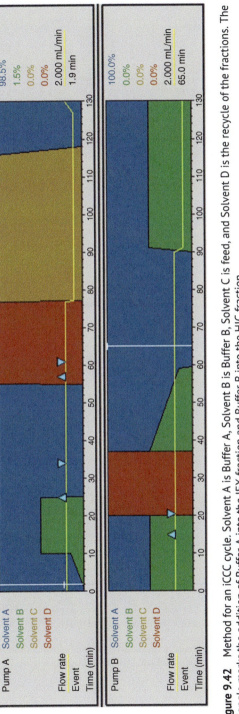

Figure 9.42 Method for an iCCC cycle. Solvent A is Buffer A, Solvent B is Buffer B, Solvent C is feed, and Solvent D is the recycle of the fractions. The triangle marks the addition of Buffer A into the IEX fraction and Buffer B into the HIC fraction.

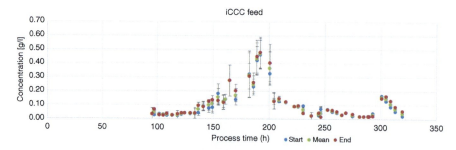

Figure 9.43 Feed concentration of the feed for the integrated counter-current chromatography (iCCC). Error bars depict the double determination of the product concentration.

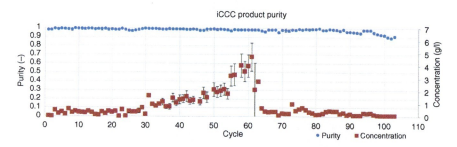

Figure 9.44 Product concentration and purity of the iCCC cycles. Error bars depict the double determination of the purity.

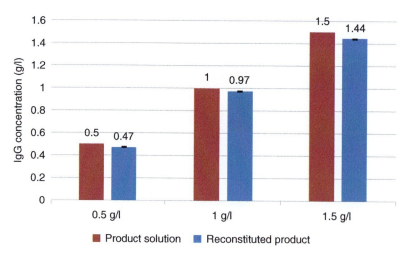

Figure 9.45 Preliminary study results for lyophilization. Error bars depict the double determination of the product concentration. IgG, immunoglobulin G.

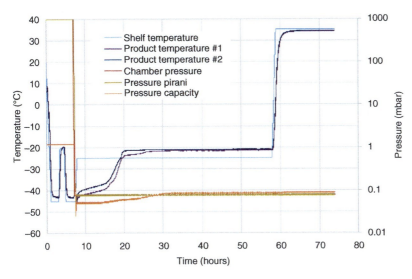

Figure 9.46 Temperature and pressure course during lyophilization.

Temperature and pressure profiles of the experiments are shown in Figure 9.46. The product temperature reached values higher than the shelf temperature. This can be explained by heat radiation from the surroundings. Additionally, because of that, this effect cannot be observed during secondary drying. From the convergence of the pressure measurements and the constant product temperature from 15 hours, it can be concluded that drying has already been completed at this point. The dried cake showed no optical errors.

9.3.7 Comparison Between Conceptual Process Design and Experimental Data

Based on the conceptual process design (CPD) in Figure 9.22, a comparison of yield (%), purity (%), volume (%), and product concentration (g/l) between the experimental data of this case study and the CPD data can be seen in Figure 9.47. Yield, purity, volume, and product concentration of the CPD were scaled to the laboratory experiments conducted in this case study (i.e. 1 l cell culture, mean product concentration of 0.5 g/l in cultivation supernatant, 150 $g_{Product}$/a). Errors of ±5% yield, ±5% purity, ±10% product concentration, and ±10% volume were implemented, owing to the typical uncertainties in CPD.

As can be seen in Figure 9.47, the product concentration and the product purity increase with each process step (i.e. capture and purification). In addition, the volume decreases to 11.0% ± 0.6% (CPD) and 4.8% ± 0.1% (experimental steady state). Product purity reaches approximately 99.9% ± 5.0% (CPD) and 96.8% ± 2.3% (experimental steady state). Moreover, the product concentration increases from 0.50 ± 0.05 to 5.00 ± 0.50 g/l (CPD) and 0.44 ± 0.15 to 0.78 ± 1.20 g/l (experimental steady state). This discrepancy between product concentration based on the CPD in Figure 9.1 and experimental data is mainly due to the modified product concentration during

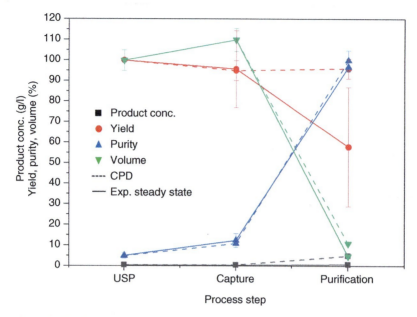

Figure 9.47 Comparison between conceptual process design (CPD) (solid line, —) and steady-state experimental data (Exp.) (dashed line, ---) in terms of product concentration (black, g/l), yield (red, %), product purity (blue, %), and volume (green, %). Purity is well met; only yield and product concentration differ due to the described handling and modified product concentrations during cell cultivation. Error bars depict the double determination of each variable based on experimental data and uncertainties of the CPD.

experimental cultivation (Figure 9.8), despite the continuous turbidostat cultivation procedure. However, even at modified product concentrations and increasing volumes during capture, the iCCC is able to significantly decrease the volume and achieve a high purity during continuous and steady-state operations [113, 125].

9.4 PAT in Continuous Biomanufacturing

In general, the PAT is not limited to in-line analytics but is a consistent technology approach that is integrated into the QbD philosophy demanded by regulatory authorities. It includes process control to gain RTRT as a benefit in QA effort reduction as improved product quality. RTRT has to correlate to critical product QAs like bio-efficacy by titer, purity, and bioactivity. State-of-the-art QAs are offline analytical methods such as protein A and SEC, enzyme-linked immunosorbent assay (ELISA), infrared spectroscopy, and glycosylation analytics via HPLC or HPLC/mass spectrometry [126–128]. The feasibility of RTRT by online PAT tools needs to be proven.

For process development, a sophisticated PAT concept has to be developed in parallel to USP and DSP modeling, followed by supporting model validation [6, 8–10, 15, 27], piloting, and production, as shown in Figure 9.2. In parallel to the model validation, piloting, and production, the developed PAT method and the partial least-squares regression (PLSR) system have to be further refined. In addition to appropriate PAT, digital twins for the whole process are a central key technology for achieving RTRT. It has been proven that for all unit operations, such distinct validated process models are available as digital twins [7–14, 50].

9.4.1 State-of-the-Art PAT

An overview of different spectroscopic methods is given in Table 9.2. Raman spectroscopy is a promising candidate to enable measurements in impure samples, e.g. by using an in situ probe in USP [130, 139, 140]. In Raman spectroscopy, due to the inelastic impacts of photons with the analytes, the photons are scattered [140]. The scattering pattern is molecule specific, and therefore the identification and quantification of different components are possible. Raman spectroscopy was successfully employed in the USP for the quantification of substrates (e.g. glucose), metabolites (e.g. lactate), and mAb [130, 141]. In contrast to alternative processes such as ATPE and precipitation in which barely any spectroscopic methods have been published for explicit use as PAT, various concepts have already been demonstrated for established processes such as chromatography, UF/DF, and lyophilization [142, 143].

Fourier-transform infrared (FTIR) spectroscopy measures the absorption of photons, typically between 4000 and 400 cm^{-1}. Biomolecules contain many amine, carbonyl, and hydroxy groups, which absorb photons in this low-energy range by inducing stretching, scissoring, and bending of molecule bonds [144]. In contrast to Raman spectroscopy, in IR spectroscopy, the absorption from water is considerably strong and may interfere with concentration measurement of biomolecules in aqueous solution [145]. Nonetheless, IR measurement was successfully applied in the USP for the prediction of mAb concentration [133]. While applicable in early process stages, FTIR is also a viable technique for quality evaluation [126, 146]. It was successfully applied in protein detection in ATPE [147], monitoring of mAb purification in chromatography [131], and in-line concentration measurement in ultrafiltration [143]. Raman and NIR can be used as PAT technology in lyophilization processes [29, 148–153]. Raman spectra can indicate different critical product and process characteristics, e.g. water to ice conversion, product crystallization, annealing steps, solid-state characteristics of intermediate and end

Table 9.2 Overview of measurement parameters for Raman, FTIR DAD, and fluorescence from literature and manufacturers.

Detector	Measurement range	In situ probes	Flow cell	LOD	Acquisition time (s)	Averaged scans
Raman	4000–400 cm^{-1}	Yes	380 µl	>50 mg/l [129]	10 [130]	75 [130]
FTIR	4000–400 cm^{-1}	Yes	—	>700 mg/l [131]	4 [132]	16–64 [132, 133]
DAD	190–520 nm	No	8 µl	>10 mg/l [134]	0.1 [135]	—
Fluorescence	280–900 nm	No	16 µl	>40 fg/l [136]	3 [137]	10 [138]

products, and kinetics of polymorphic transitions [29, 153]. NIR can also indicate critical product and process characteristics [153] such as the secondary structure of lyophilized proteins and the residual moisture [29]. NIR and Raman have the ability to determine the endpoint of primary drying. Water and ice produce weak Raman signals but have high absorption in NIR spectra. Both measurements determine the endpoint by detecting a loss of water signal, but NIR is the more sensitive technique and is additionally capable of detecting the endpoint of secondary drying [148, 153]. Diode array detector (DAD) measures the absorption of photons in UV–vis, typically in the range of 190–520 nm. In contrast to FTIR, in this high-energy range, delocalized π-electron systems are the main absorbers [154]. DAD is widely used in chromatography, but potential applications in earlier process stages exist [135]. Fluorescence measures the time-delayed emission of photons after excitation by a specific wavelength. The emission range is lower than the excitation wavelength, since the photons lose energy after absorption due to the non-radiative transitions [155]. Fluorescence was previously employed in monitoring upstream processes [156]. Integrity of IgG can also be monitored by fluorescence, e.g. in combination with circular dichroism measurement [137]. The limits of detection values given in Table 9.2 are for orientation purposes only since they vary with different measurement/integration times and product. However, the presence of too many side components can reduce the detection accuracy [157]. Therefore, detection accuracy enhancements through combination of multiple spectroscopic techniques have been proposed [158].

The above presented PAT approaches in combination with a digital twin can be used to achieve APC and in-line process optimization [13, 159–163]. PAT can compensate model inaccuracies by providing additional measurement data, and the digital twin is used for in-line process optimization and online process monitoring.

9.4.2 QbD-based PAT Control Strategy

In general, PAT is not limited to in-line analytics but is a consistent technology approach that is integrated into the QbD philosophy, demanded by regulatory authorities, and is becoming the standard in biopharmaceutical process development. In the QbD approach, a design space of operating parameters is defined to ensure specified QAs. This leads to multiparameter optimizations and a significant experimental effort. The modern approach for process development and quality assessment is shown in Figure 9.1.

These MPCs manipulate input variables to match the desired set points while maintaining process critical constraints. This is performed by utilizing optimization routines on process models, which predict the future process behavior for the next time frame [165, 166]. Common drawbacks of these models are that the model results tend to drift away from the real plant data over time, due to the aging, fouling, or blocking phenomena or summation of prediction errors in cyclic processes [167]. This is usually fixed by updating the internal model states, e.g. concentrations, with real plant data [168]. These real-time plant data have to be determined via potentially time-intensive and invasive offline analytics if no PAT tools are implemented, resulting in a gap between current process data and analytics [169, 170]. This gap risks not only a mismatch between current process state and model but also a general mismatch between current process and process analytics preventing data-driven process decisions, especially in continuous processes.

Starting with this overview, we will demonstrate that PAT is capable of filling the gap. The proposed control strategy is demonstrated exemplarily in simulation studies. Previous studies have focused on the implementation and distinct and quantitative validation of physicochemical process models to describe the unit operations shown in Figure 9.48 and enable APC. The successful operation of the proposed continuous process has been previously shown [11]. The missing link to achieve APC is a holistic PAT strategy [12]. Starting with this overview, we will demonstrate that PAT is capable of filling the gap. The proposed control strategy is demonstrated exemplarily in simulation studies.

9.4.3 Process Simulation Toward APC-Based Autonomous Operation

In the literature, concentration fluctuations or continuous mAb concentration decrease due to decreasing cell specific productivity over process time have been described [59]. Simulation studies are used to test whether PAT can enable a process control strategy that allows for compensation of titer fluctuations in the subsequent DSP. The scenario for the following simulations is that the mAb concentration and purity (i.e. high molecular weight (HMW) and low molecular weight (LMW) concentrations) are continuously measured in the outstream of each unit operation using the detector array presented above. The real-time measurement data is continuously forwarded to the following unit operation in the mAb manufacturing process. These information are fed into a process model that calculates the necessary process adjustment to reach either constant mAb concentration or constant volume flow.

Simulation of viable cell and product concentration in perfusion mode is shown in Figure 9.49. This increases over the first 72 hours. Then, the perfusion is started, and the product concentration decreases momentarily since more product is washed out than produced. Shortly after, the concentration increases again proportionally to the viable cell concentration until the steady state is reached after approximately 350 hours. Product concentration in the steady state reaches 1.8 g/l. As a basis for the following unit operations, a final steady-state concentration of 2 g/l is assumed.

The below images show how concentration fluctuations in USP are processed to a constant concentration in the product containing light phase. Polymer and salt solutions are mixed based on the phase equilibrium within the same tie line. Based on the lever-arm rule, either less or more light phase is produced; thereby concentration is kept constant. Operating on the same tie line is necessary to ensure constant yield and product phase properties (Figure 9.50).

In Figure 9.51, the concentration change in precipitation over process time is shown. Precipitation is a concentration-independent process [171] that can be also verified with these results. The ratio of PEG solution and light phase does not change with varying mAb concentration because the PEG content is calculated based on the volume coming from extraction and not to the concentration. In this way,

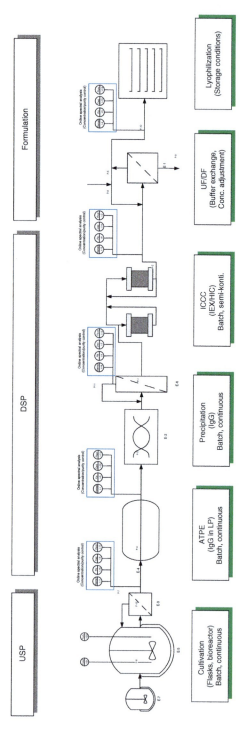

Figure 9.48 Proposed process flow sheet including spectroscopic PAT sensors. A PAT measurement array is placed at the outlet of each unit operation to provide real-time data on mAb, HMW, and LMW concentrations that are then used as input for the process model calculating the necessary process adjustments to ensure operation within a predefined design space.

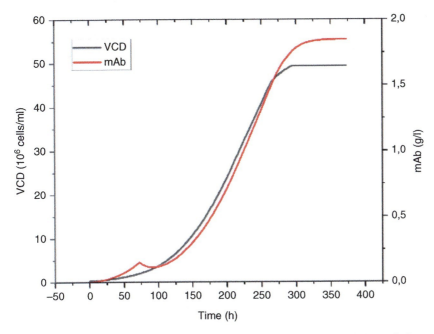

Figure 9.49 Simulation results of viable cell and product concentration in perfusion.

concentration variations do not affect precipitation. For higher titers, precipitation takes slightly longer (few seconds) as shown in Figure 9.51. However, the result is the same and leads to complete precipitation that is a function of mixing of light phase and PEG solution.

In contrast, dissolution is concentration dependent and able to react to decreased or increased feed concentration by adjusting the dissolution ratio. In this manner the precipitation unit can provide a constant concentration for chromatography, shown in Figure 9.51b. For different feed concentrations, the dilution ratio was adjusted, resulting in a constant output concentration of 2.6 g/l (±0.05). Changes in flow rate can be compensated with a higher stream of PEG solution to keep precipitation condition constant and ensure complete precipitation. Due to the fact that dead-end filtration eliminates the complete volume of supernatant around precipitates, flow rate fluctuation does not affect dissolution. Hence, concentration and flow rate changes cannot be compensated at once since a constant mAb concentration is accompanied by a fluctuation in flow rate. Compensation of purity was not possible in precipitation.

Simulation results from iCCC modeling are illustrated in Figure 9.52. In (a), the chromatogram of IEX is given. Between 400 and 1000 seconds, the majority of LMW1 and LMW2 are eluted. This is due to the loading of the strong-binding fraction of HIC and the weak-binding fraction of IEX. IgG is eluted from 2250 to 2500 seconds. Cutting points are marked with dashed lines and are set at 0.05 g/l that was detectable in chromatography using the PAT system described above.

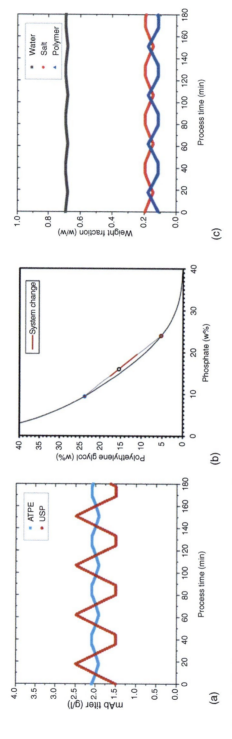

Figure 9.50 Simulation results for concentration fluctuations (a) and concentration compensation by system adjustment (b, c).

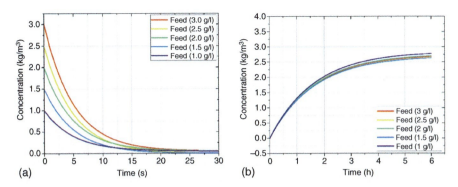

Figure 9.51 Simulation results for concentration fluctuations for (a) precipitation and (b) dissolution.

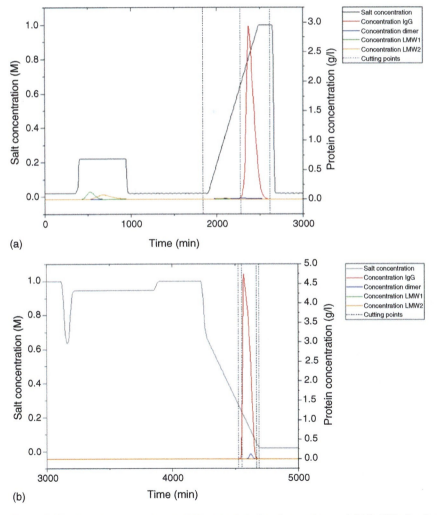

Figure 9.52 Simulation results of iCCC. (a) is IEX after five cycles, and (b) is HIC after five cycles.

This simulation shows that an in-line process control using a PAT system of DAD and/or fluorescence leads to a very high process yield, since cutting points in the chromatogram can be automatically detected, controlling the fractionation.

9.4.4 Applicability of Spectroscopic Methods in Continuous Biomanufacturing

The evaluation of the different detector and sensor data in USP and ATPE reveals that Raman was the most reliable technique not only for mAb but also for HMW and LMW prediction. It is also the most easily implementable spectroscopic technique as in-line probes and flow cells are widely available. Data acquisition is also sufficiently fast (few seconds) for USP and ATPE. FTIR data are also suitable for building a PLS regression, but less reliable when compared with Raman (Figures 9.53 and 9.54).

Due to the overall lesser observed variability in the spectra, it also appears to be less sensitive to changes in species concentration. DAD yielded very good correlation results; however, there are no in-line probes readily available, and the detection of

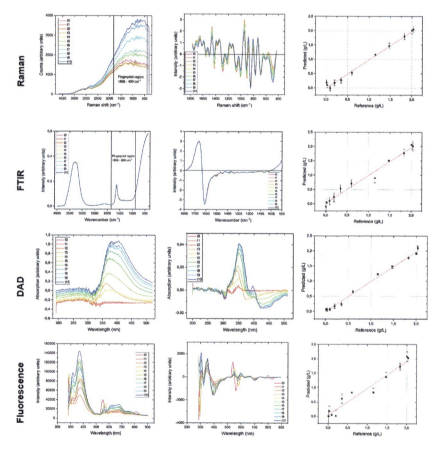

Figure 9.53 Spectral data overview in USP. From left to right: Raw data, processed data, and PLS regression.

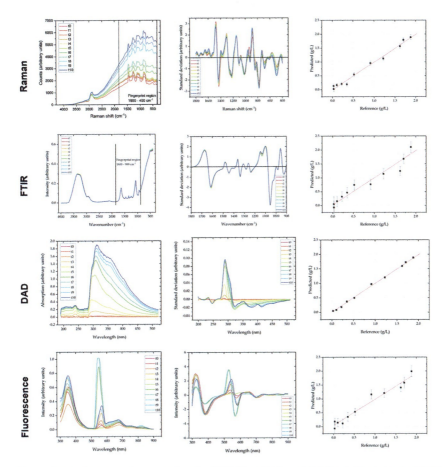

Figure 9.54 Spectral data overview in ATPE. From left to right: Raw data, processed data, and PLS regression.

absorption is by principle less species specific compared with Raman fingerprints and therefore more likely to give false-positive predictions. Fluorescence yielded the worst correlation results and is therefore less likely to be implemented as primary detector technique in USP and ATPE. Overall, for in-line mAb, HMW, and LMW analytics, Raman is the recommend spectroscopic technique.

Simulation results showed that based on the continuous measurement of mAb, HMW, and LMW concentrations by the PAT sensors, the unit operations in DSP could be controlled using process models that calculate necessary system adjustments to keep either concentration or volume flow constant.

For the precipitation unit, evaluation of Raman, FITR, DAD, and fluorescence yielded different results for precipitation and dissolution. During precipitation, experimental data show that the prediction of the target component and HMWs is poor. The results for HMWs are not surprising since most of this impurity is already removed in ATPE. For LMWs, excellent results can be obtained with Raman

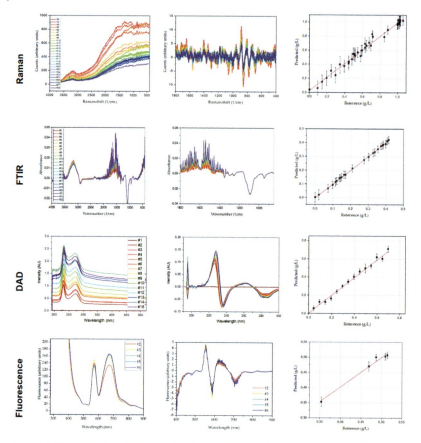

Figure 9.55 Spectral data overview in precipitation. From left to right: Raw data, processed data, and PLS regression.

(R^2 of 0.95), FTIR (R^2 of 0.94), and DAD (R^2 of 0.97). Fluorescence could not be tested due to the precipitates that are harmful for the flow-through cell of the detector. Unfortunately, no satisfactory results for the prediction of purity were found with any of the detectors during precipitation. For dissolution, all four detectors could reliably predict the concentration of the target component. Nevertheless, only poor correlations were found for HMWs and LMWs since most impurities are already separated before dissolution is performed. Only Raman was capable to predict the purity of HMWs (R^2 of 0.13), LMWs (R^2 of 0.22), and target component (R^2 of 0.85), whereby only the precision of the TC was convenient. Therefore, similar to USP and ATPE, Raman is the best-suited sensor for the precipitation unit (Figure 9.55).

According to this study, Raman is recommended as a detector for precipitation since it has shown persuasive results in detecting the target and side components (LMWs) in precipitation and dissolution. Additionally, purity of the target component in the dissolution could be correlated with Raman. As an orthogonal measurement strategy, a DAD is recommended due to its fast acquisition time

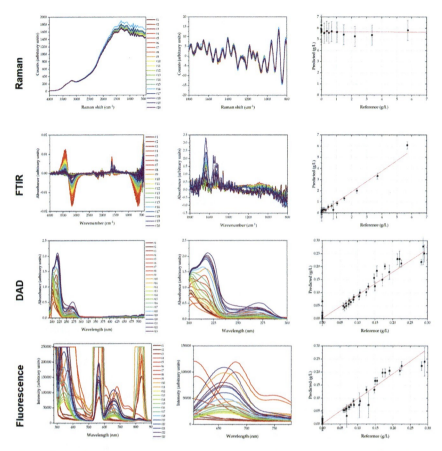

Figure 9.56 Spectral data overview in chromatography. From left to right: Raw data, processed data, and PLS regression.

and precision. Furthermore, a conductivity probe is recommended as PAT control strategy for detection of optimal conditions during precipitation.

In chromatography, APC using in-line measurements is the most promising way to establish continuous downstream manufacturing in the purification of biopharmaceuticals. In highly purified solutions of mAb, a way to detect impurities even at very low concentration is the combination of DAD and fluorescence. In this study, a concentration down to 0.05 g/l could be measured using this combination (Figure 9.56). Also, through this combination, an R^2 of 0.93 for the target component was achieved. Regression coefficients for HCP impurities were comparable, with 0.91 and 0.93, while regression coefficient for the dimer was 0.67, which probably resulted from the low concentrations observed in chromatography, with around 0.005–0.025 g/l, which is very close to the detection limit of the employed SEC chromatography. In DAD measurements, without simultaneously evaluating fluorescence, a high variation in zero concentration measurements for the product was observed. This would be a problem for in-line product detection, with resulting background noise interfering with peak detection. This problem is eliminated by using the combination of DAD and fluorescence. FTIR could be used in the present experimental setup; however, it has a disadvantage, such lower sensitivity. Therefore, detecting incoming peaks is more difficult. Later detection at higher concentrations would result in a lower process yield in preparative process. The lower sensitivity observed in chromatography likely results in the overlapping absorbance ranges of the changing buffer solution and the target component, which is discussed above. Raman was not feasible using the employed flow cell, which is due to its dead volume of 1 ml. In other works, Raman has been employed successfully for the breakthrough detection of IgG [142]. Other reasons might be the differing integration time or the salt gradient overlaying the elution.

9.4.5 Proposed Control Strategy Including PAT

Continuous, chromatography-reduced process is the proposed process as this offers the most economical production of mAbs, as shown in another study [12]. Robust process control is possible using either an APC based on a digital twin or in some cases traditional proportional–integral–derivative (PID) control as shown in this work. To enable process control, an in-line concentration measurement has to be implemented. Here, PLS-based spectroscopic methods enable a real-time, accessible process control in all used unit operations. The proposed process is illustrated in Figures 9.58–9.60. In Figure 9.57, an overview of the course of purity, yield, titer, and DNA concentration is given. Product titer is mainly increased by chromatography and adjusted by UF/DF. Purity is steadily increasing over the course of the process. In ATPE capture, most of the high-molecular-weight

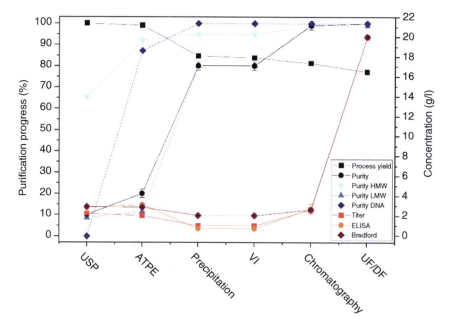

Figure 9.57 Purity, yield, and titer during the process.

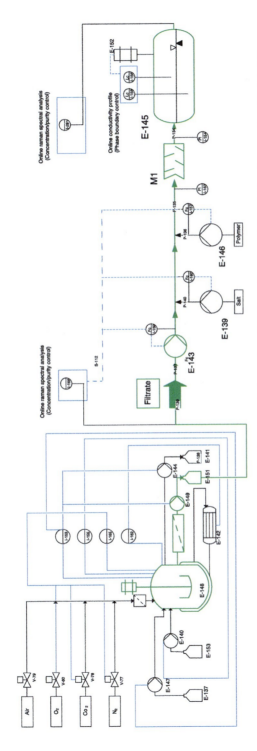

Figure 9.58 Proposed control strategy for USP and ATPE. A Raman probe is used as PAT at the filtrate outlet of USP and forwards mAb and side component concentrations to the ATPE, which adjusts the polymer and salt concentrations to produce light phase with a constant mAb concentration that is then forwarded to the precipitation unit.

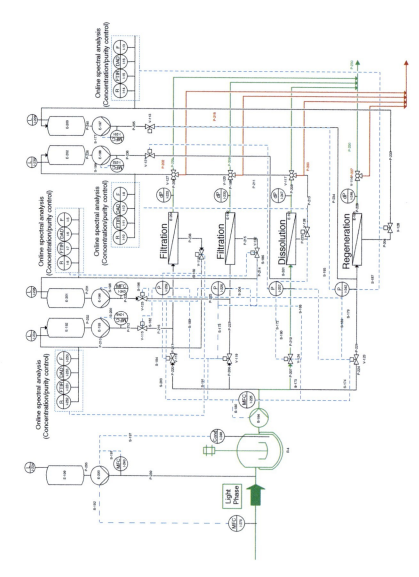

Figure 9.59 Process control strategy for precipitation. A PAT measurement array at the outlet of the ATPE forwards mAb and side component concentrations to the precipitation unit that adjusts filtration and dissolution times to obtain a constant mAb concentration in the dissolution solution.

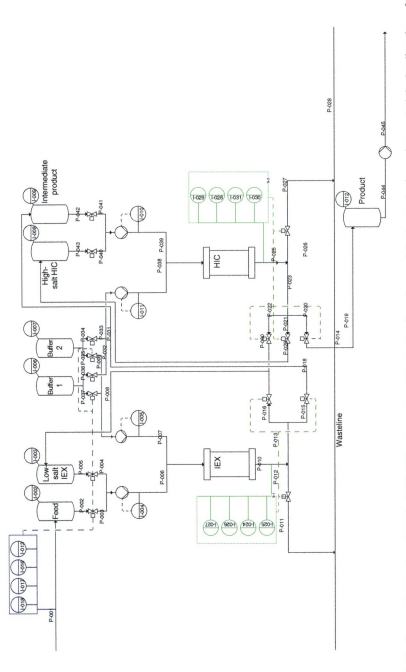

Figure 9.60 Proposed control strategy for iCCC. PAT measurement information from the precipitation outlet is used to control the loading time of the IEX. Fractionation in the IEX can also be monitored and controlled by the described PAT methods.

side components are eliminated. These mainly consist of multi-charged DNA molecules, which stay in the salt-dominated heavy phase [172]. Purification occurs in the precipitation as most low-molecular-weight host cell proteins stay in the supernatant and are eliminated through filtration of precipitates [10]. After chromatographic polishing (IEX and HIC), over 99% purity is achieved. HMW and LMW side components are below the detection limit of the applied analytic technologies.

As shown in the simulation studies, ATPE can sustain a constant mAb concentration, even when titer in USP fluctuates to ±50%. This is achieved by calculating the necessary polymer and salt concentration based on the phase equilibrium. As discussed before, Raman is the recommend spectroscopic technique to analyze mAb concentration in USP and ATPE and is therefore the primary detector at the inlet and product phase outlet stream.

Figure 9.59 shows the process control strategy for the precipitation unit. Precipitation is depicted in a red box and dissolution in a green box. The unit consists of four hollow fiber modules that are timed and enable continuous processing. Phase one is the filtration of precipitates, followed by a washing step in phase two. Redissolution takes place in phase three, and finally the module is regenerated by a rinsing process in phase four. After regeneration, the process restarts with the filtration of precipitates. Each module passes through all four phases with a time delay, which is controlled by pressure sensors to ensure the operation range for the hollow fiber membranes. In the dissolution recycling loop, in-line measurements and proteomics are installed to detect concentration and purity of the target component.

This measurement technique is used as switching criterion between valves in phase one to abort dissolution as soon as the target component has sufficiently redissolved into the buffer. Finally, the product is filtered through the membrane and transferred to the chromatography step, as well as values of product titer and purity, which are used as model input parameters for chromatography.

Figure 9.59 shows the flow sheet of the precipitation unit including online measurement trains and pressure sensors and mass flow controllers. The unit consists of four membrane modules that are controlled by the digital twin and the integrated in-line PAT. Each module passes through different phases with a time delay. In this way continuous processing is enabled. Phase one is the filtration of precipitates, followed by redissolution of the target component. Then, the module is regenerated by a rinsing process. After regeneration, the process restarts with the filtration of precipitates. The mass flow controller monitors the incoming flow rate of light phase after ATPE, which is used to calculate the needed precipitant flow rate. Pressure sensors measure the transmembrane pressure during filtration and ensure that the operation range is maintained. Furthermore, a critical pressure is used to switch valve positions and redirect the dispersion to the next filter. The first loaded module passes on to the next phase, the dissolution. Input concentration from ATPE and the predefined final mAb concentration are used to determine the buffer volume for redissolution of the target component. The in-line measurement train in the circuit controls the speed of the dissolution buffer pump and the valves to the product tank. As soon as the dissolution of the antibody has reached a stationary value, the dissolution is terminated, and the product solution is passed on to the next unit operation.

In Figure 9.60, a control strategy for iCCC is proposed. Data needed to set the loading time of IEX is either transferred by the previous unit operation or measured in-line in front of the feed tank (blue). Moreover, data used for release testing is obtained by measurement arrays located after the columns (green). From the obtained data, two different control strategies for the iCCC unit arise. Firstly, the process can be controlled using the real-time measured concentration as a switch criteria for the fractionation valves. Since the elution order of the components in known, and this order is not subject to change in the process, the product fraction can be cut easily when both light molecular weight side components were eluted. Cutting the product peak at the lowest concentration detectable (i.e. 0.05 g/l) resulted in a over 99% yield consistently. Secondly, iCCC process can be simulated with the help of a digital twin. Here, further optimizations of space–time yield or purity are possible since the separation can be further optimized. Using the data obtained by the PAT arrays (green), resin aging can be detected or even integrated continuously into the digital twin model by adjusting model parameters.

After chromatography, the mAb concentration is determined by the combination of DAD and fluorescence. This data determines the necessary concentration factor. Transmembrane pressure is adjusted to achieve the desired concentration. Conductivity measurements are used to determine if the UF/DF operates within the specification limits. Finally, transmembrane pressure for final buffer adjustment toward formulation is adjusted to achieve the desired concentration. Conductivity measurements are used to determine if the UF/DF operates within the specification limits. Lyophilization as the final formulation step before fill and finish has already been described [13].

9.4.6 Evaluation and Summary of PAT

For a whole mAb manufacturing process, this subsection presents the applicability of a combination of spectroscopic methods and has been evaluated to enable APC in CM by PAT.

In USP and the following direct ATPE, Raman, FTIR, fluorescence, and UV/vis spectroscopies have been successfully applied for titer and purity prediction. Raman spectroscopy was the most versatile and robust method and is recommended as primary PAT. In, similar results were obtained for titer determination. Prediction of purity was challenging for FTIR, fluorescence, and UV/vis spectroscopies but achievable for Raman spectroscopy. In chromatography the combination of UV/vis and fluorescence spectroscopies was able to overcome difficulties in titer and purity prediction induced by overlapping side component spectra, often already reported in the literature as well. In the final UF/DF, before lyophilization, UV/vis spectroscopy is applicable for titer concentration determination. However, for high concentration processes, it is important to ensure the DAD employed is able to operate at elevated concentrations. Raman spectroscopy is especially useful in early stages of the process, whereas more traditional detector technology concepts, such as DAD, can be used in the late stages. The combination of spectroscopic data improves the predictivity as shown for chromatography.

Continuous operation generates much smaller holdup volumes than batch processing. This causes much shorter start-up und shutdown times with smoother system responses. In addition, system response is much shorter and smoother, i.e. nearly constant, around the continuous operation point. Hence, detector signal acquisition times and corresponding sampling scan rates for continuous processing are much lower than batch operation. Changes due to the natural system variances are less steep, i.e. continuously near-constant system responses. In contrast, only typical gradient elution chromatography operation with fraction cut points at steep chromatogram concentration slopes changes the accuracy in time resolution. These changes in concentration occur within a few seconds, whereas the feasible number of measuring points is limited by the sampling rate that leads to a resolution of 10–50 points per peak. This results in an image of the concentration profile that is not sufficiently accurate for fractionation. Only typical breakthrough curves of flow-through operation mode in capture steps differ and are as well sufficiently describable with higher acquisition times and averaged scans.

Based on the developed spectroscopic predictions, dynamic process control of the unit operations was demonstrated for the total process in sophisticated simulation studies based on validated digital twins available for all unit operations. As such, a PAT development workflow for any holistic process development is proposed.

To our knowledge, this comprehensive demonstration of the combination of the applicability of different spectroscopic methods in the context of a holistic process development, including the complete total process simulation of an APC concept based on digital twins, has been shown for the first time, opening up innovative autonomous operation concepts for the future. The next steps will be the transfer toward different biologic types like antibody fragments, peptides, and virus-like particles (VLPs) available at the institute and piloting studies of the APC conception proposed.

9.5 Conclusion

Digital twins are of great aid in continuous biomanufacturing. All necessary methods and parts are available and have proven their feasibility. Many studies on single work steps like modeling or PAT tools have been published. Moreover, first studies combining the technologies already come up.

Appropriate models of all unit operations have been shown to be available, as well as an efficient time- and effort-reducing approach for model parameter determination.

All steps can be validated with regard to regulatory demands in documented data-driven predictive power.

One of the most utilized APC concepts are MPC. These manipulate input variables to match the desired set points while maintaining process critical constraints. This is performed by utilizing optimization routines on process models, which predict the future process behavior for the next time frame. Common drawbacks of these models are that the model results tend to drift away from the real plant data over time, due to aging, fouling, or blocking phenomena or summation of prediction errors in cyclic processes. This is usually fixed by updating the internal model states, e.g. concentrations, with real plant data. These real-time plant data have to be determined via potentially time-intensive and invasive offline analytics, if no PAT tools are implemented, resulting in a gap between current process data and analytics. This gap risks not only a mismatch between current process state and model but also a general mismatch between current process and process analytics preventing data-driven process decisions, especially in continuous processes.

If detector combinations within the PAT conception are applied in the process development, any experimental run could be easily directly evaluated with regard to further experimental planning – the actual limiting need for few weeks until offline analytical results are available, coping with the risk of many unworthy experimental efforts during that time of not evaluable experimental results.

Transfer to other component systems is straightforward, and a clear workflow is defined. It just needs to be done and shown.

Continuous biomanufacturing with autonomous operation based on digital twins on validated process models in combination with PAT in-line analytics for APC is key technology of the future to cope with molecular diversity challenges in process development and manufacturing operation – by time-to-market acceleration and manufacturing operation simplification and robustness.

Finally, further prototyping of demonstration projects is needed to convince regulatory authorities and management for data-driven decisions.

Acknowledgments

The authors would like to thank the ITVP team for Aspen Custom Modeler™, OpenFoam® and MATLAB® simulations, JMP® and Unscrambler® statistic evaluations, and COSMO-RS® thermodynamic calculations. Special thanks is given to Mourad Mouellef and Thomas Knebel and Professor Siemers and the Clausthal University of Technology (TUC) process automation group for their work on APC and PAT. Also, the authors want to thank the Bundesministerium für Wirtschaft und Energie (BMWi), especially Dr. M. Gahr (Projektträger FZ Jülich), for funding the scientific work.

References

1 Mats Lundgren (2021) Meeting the process development challenges of a diverse biologic pipeline: how to navigate the evolving biopharmaceutical landscape from a process development standpoint. White Paper, Cytiva See also: https://www.cytivalifesciences.com/en/us/solutions/bioprocessing/knowledgecenter/new-drug-process-development.

2 EMA (2017) Guideline on the requirements for the chemical and pharmaceutical quality documentation concerning investigational medicinal products in clinical trials. Rep. from the EMA-FDA QbD pilot program. EMA/213746/.
3 U.S. Food and Drug Administration (2004) FDA PAT – a framework for innovative pharmaceutical development, manufacturing, and quality assurance: guidance for industry.
4 Svrcek, W.Y., Young, B.R., and Mahoney, D.P. (2014). *A Real Time Approach to Process Control*. Chichester, West Sussex: John Wiley & Sons Inc.
5 Ferreira, A.P., Menezes, J.C., and Tobyn, M. (eds.) (2018). *Multivariate Analysis in the Pharmaceutical Industry*. London: Academic Press.
6 Zobel-Roos, S., Schmidt, A., Mestmäcker, F. et al. (2019). Accelerating biologics manufacturing by modeling or: is approval under the QbD and PAT approaches demanded by authorities acceptable without a digital-twin? *Processes* 7 (2): 94.
7 Zobel-Roos, S., Mouellef, M., Ditz, R., and Strube, J. (2019). Distinct and quantitative validation method for predictive process modelling in preparative chromatography of synthetic and bio-based feed mixtures following a quality-by-design (QbD) approach. *Processes* 7 (9): 580.
8 Schmidt, A. and Strube, J. (2019). Distinct and quantitative validation method for predictive process modeling with examples of liquid–liquid extraction processes of complex feed mixtures. *Processes* 7 (5): 298.
9 Roth, T., Uhlenbrock, L., and Strube, J. (2020). Distinct and quantitative validation for predictive process modelling in steam distillation of caraway fruits and lavender flower following a quality-by-design (QbD) approach. *Processes* 8 (5): 594.
10 Lohmann, L.J. and Strube, J. (2020). Accelerating biologics manufacturing by modeling: process integration of precipitation in mAb downstream processing. *Processes* 8 (1): 58.
11 Kornecki, M. and Strube, J. (2019). Accelerating biologics manufacturing by upstream process modelling. *Processes* 7 (3): 166.
12 Kornecki, M., Schmidt, A., Lohmann, L. et al. (2019). Accelerating biomanufacturing by modeling of continuous bioprocessing – piloting case study of monoclonal antibody manufacturing. *Processes* 7 (8): 495.
13 Klepzig, L.S., Juckers, A., Knerr, P. et al. (2020). Digital twin for lyophilization by process modeling in manufacturing of biologics. *Processes* 8 (10): 1325.
14 Huter, M.J., Jensch, C., and Strube, J. (2019). Model validation and process design of continuous single pass tangential flow filtration focusing on continuous bioprocessing for high protein concentrations. *Processes* 7 (11): 781.
15 Huter, M.J. and Strube, J. (2019). Model-based design and process optimization of continuous single pass tangential flow filtration focusing on continuous bioprocessing. *Processes* 7 (6): 317.
16 Grand View Research (2018) Digital Twin Market Size, Share & Trends Analysis Report By End-use (Automotive & Transport, Retail & Consumer Goods, Agriculture), By Region (Europe, North America, Asia Pacific), And Segment Forecasts, 2018–2025, Report ID: GVR-2-68038-494-9. 101.

17 Roush, D., Asthagiri, D., Babi, D.K. et al. (2020). Toward in silico CMC: an industrial collaborative approach to model-based process development. *Biotechnol. Bioeng.*

18 Woodcock, J. (2014). *Continuous Manufacturing as a Key Enabler*. MIT-CMAC International Symposium on Continuous Manufacturing of Pharmaceuticals, Cambridge, MA, See also: https://iscmp2014.mit.edu/sites/default/files/documents/ISCMP2014_Keynote_Slides.pdf.

19 Chanda, A., Daly, A.M., Foley, D.A. et al. (2015). Industry perspectives on process analytical technology: tools and applications in API development. *Org. Process Res. Dev.* 19 (1): 63–83.

20 Konstantinov, K.B. and Cooney, C.L. (2015). White paper on continuous bioprocessing May 20–21 2014 continuous manufacturing symposium. *J. Pharm. Sci.* 104 (3): 813–820.

21 Thiess, H., Zobel-Roos, S., Gronemeyer, P. et al. (2017). Engineering challenges of continuous biomanufacturing processes (CBP). In: *Continuous Biomanufacturing – Innovative Technologies and Methods* (ed. G. Subramanian), 69–106. Weinheim, Germany: Wiley-VCH Verlag GmbH & Co. KGaA.

22 Zobel-Roos, S., Thiess, H., Gronemeyer, P. et al. (2017). Continuous chromatography as a fully integrated process in continuous biomanufacturing. In: *Continuous Biomanufacturing – Innovative Technologies and Methods* (ed. G. Subramanian), 369–392. Weinheim, Germany: Wiley-VCH Verlag GmbH & Co. KGaA.

23 Gronemeyer, P., Thiess, H., Zobel-Roos, S. et al. (2017). Integration of upstream and downstream in continuous biomanufacturing. In: *Continuous Biomanufacturing – Innovative Technologies and Methods* (ed. G. Subramanian), 481–510. Weinheim, Germany: Wiley-VCH Verlag GmbH & Co. KGaA.

24 Jiang, M., Severson, K.A., Love, J.C. et al. (2017). Opportunities and challenges of real-time release testing in biopharmaceutical manufacturing. *Biotechnol. Bioeng.* 114 (11): 2445–2456.

25 Helling, C. and Strube, J. (2012). Quality-by-design with rigorous process modeling as platform technology of the future. *Chem. Ing. Tech.* 84 (8): 1334.

26 Helling, C. and Strube, J. (2012). Modeling and experimental model parameter determination with quality by design for bioprocesses. In: *Biopharmaceutical Production Technology* (ed. G. Subramanian), 409–443. Weinheim, Germany: Wiley-VCH Verlag GmbH & Co. KGaA.

27 Sixt, M., Uhlenbrock, L., and Strube, J. (2018). Toward a distinct and quantitative validation method for predictive process modelling – on the example of solid–liquid extraction processes of complex plant extracts. *Processes* 6 (6): 66.

28 Read, E.K., Park, J.T., Shah, R.B. et al. (2010). Process analytical technology (PAT) for biopharmaceutical products. Part I: Concepts and applications. *Biotechnol. Bioeng.* 105 (2): 276–284.

29 Read, E.K., Shah, R.B., Riley, B.S. et al. (2010). Process analytical technology (PAT) for biopharmaceutical products. Part II: Concepts and applications. *Biotechnol. Bioeng.* 105 (2): 285–295.

30 Yu, L.X., Amidon, G., Khan, M.A. et al. (2014). Understanding pharmaceutical quality by design. *AAPS J.* 16 (4): 771–783.

31 Simon, L.L., Pataki, H., Marosi, G. et al. (2015). Assessment of recent process analytical technology (PAT) trends: a multiauthor review. *Org. Process Res. Dev.* 19 (1): 3–62.

32 Ditz, R. (2012). Separation technologies 2030 – are 100 years of chromatography enough? *Chem. Ing. Tech.* 84 (6): 875–879.

33 Strube, J., Ditz, R., Kornecki, M. et al. (2018). Process intensification in biologics manufacturing. In: *Chemical Engineering and Processing: Process Intensification, Elsevier B.V.*

34 Kornecki, M., Mestmäcker, F., Zobel-Roos, S. et al. (2017). Host cell proteins in biologics manufacturing: the good, the bad, and the ugly. *Antibodies* 6 (3): 13.

35 Gronemeyer, P., Ditz, R., and Strube, J. (2016). DoE based integration approach of upstream and downstream processing regarding HCP and ATPE as harvest operation. *Biochem. Eng. J.* 113: 158–166.

36 Kornecki, M. and Strube, J. (2018). Process analytical technology for advanced process control in biologics manufacturing with the aid of macroscopic kinetic modeling. *Bioengineering* 5 (1): 25.

37 Meyer, U.A., Zanger, U.M., and Schwab, M. (2013). Omics and drug response. *Annu. Rev. Pharmacol. Toxicol.* 53: 475–502.

38 Schaub, J., Clemens, C., Kaufmann, H., and Schulz, T.W. (2012). Advancing biopharmaceutical process development by system-level data analysis and integration of omics data. In: *Genomics and Systems Biology of Mammalian Cell Culture*, 2012e (eds. W.S. Hu and A.-P. Zeng), 133–163. Berlin, Heidelberg: Springer.

39 Schaub, J., Clemens, C., Schorn, P. et al. (2010). CHO gene expression profiling in biopharmaceutical process analysis and design. *Biotechnol. Bioeng.* 105 (2): 431–438.

40 Hu, W.S. and Zeng, A.-P. (eds.) (2012). *Genomics and Systems Biology of Mammalian Cell Culture*, 2012e. Berlin, Heidelberg: Springer.

41 Klepzig, L. and Strube, J. (2018). Rigorous modeling of lyophilization for botanicals and biologics process integration. *Chem. Ing. Tech.* 90 (9): 1299.

42 Klepzig, L. (2018) *Rigorous modelling of lyophilisation for botanicals and biologics process integration*, Frankfurt am Main.

43 Klepzig, L. (2018) *Process Modelling in Combination with Experimental Model Parameter Determination, Sevilla. Pharmaceutical Freeze Drying Technology, 27–28 November in Seville, Spain, Organizer: PDA Europe.*

44 Sommerfeld, S. and Strube, J. (2005). Challenges in biotechnology production – generic processes and process optimization for monoclonal antibodies. *Chem. Eng. Process. Process Intensif., Elsevier B.V.* 44 (10): 1123–1137.

45 Strube, J., Sommerfeld, S., and Lohrmann, M. (2007). Processes development and optimization for biotechnology production – monoclonal antibodies. In: *Bioseparation and Bioprocessing: A Handbook*, 2e (ed. G. Subramanian), 65–99. Weinheim, New-York: Wiley-VCH.

46 Deibele, L. and Dohrn, R. (2006). *Miniplant-Technik in der Prozessindustrie*, 1e. Wiley-VCH: Weinheim.

47 Strube, J. (2012). Prädiktive Modellierung von Trennverfahren. *Chem. Ing. Tech.* 84 (6): 867.

48 Steude, H.E., Deibele, L., and Schröter, J. (1997). MINIPLANT -Technik - ausgewählte Aspekte der apparativen Gestaltung. *Chem. Ing. Tech.* 69 (5): 623–631.

49 Jochen Strube Website Institute for Separation and Process Technology – Clausthal University of Technology. https://www.itv.tu-clausthal.de/ (accessed 23 December 2007).

50 Uhlenbrock, L., Jensch, C., Tegtmeier, M., and Strube, J. (2020). Digital twin for extraction process design and operation. *Processes* 8 (7): 866.

51 Ben Yahia, B., Malphettes, L., and Heinzle, E. (2015). Macroscopic modeling of mammalian cell growth and metabolism. *Appl. Microbiol. Biotechnol.* 99 (17): 7009–7024.

52 Kornecki, M. and Strube, J. (2018). Process analytical technology mechanisms in biologics manufacturing. *Chem. Ing. Tech.* 90 (9): 1270.

53 Kornecki, M. (2018) *Process Analytical Technology Mechanisms in Biologics Manufacturing*, Frankfurt am Main, Wiley-VCH: Weinheim.

54 Kornecki, M. (2018) *Host Cell Proteins in Biologics Manufacturing: A Methodical and Systematic Integration of Upstream and Downstream Processing*, Frankfurt am Main, Wiley-VCH: Weinheim.

55 Thiess, H., Leuthold, M., Grummert, U., and Strube, J. (2017). Module design for ultrafiltration in biotechnology: hydraulic analysis and statistical modeling. *J. Membr. Sci.* 540: 440–453.

56 Huter, M. and Strube, J. (2018). Model-based optimization of SPTFF ultrafiltration for integration in continuous biopharmaceutical processing. *Chem. Ing. Tech.* 90 (9): 1251.

57 Huter, M. (2018) *Modeling of Continuous Ultrafiltration for Biopharmaceutical Processes*, Frankfurt am Main, Wiley-VCH: Weinheim.

58 Lucke, M., Koudous, I., Sixt, M. et al. (2018). Integrating crystallization with experimental model parameter determination and modeling into conceptual process design for the purification of complex feed mixtures. *Chem. Eng. Res. Des.* 133: 264–280.

59 Feidl, F., Vogg, S., Wolf, M. et al. (2020). Process-wide control and automation of an integrated continuous manufacturing platform for antibodies. *Biotechnol. Bioeng.* 117 (5): 1367–1380.

60 Otto, R., Santagostino, A., Schrader, U. Rapid growth in biopharma: challenges and opportunities. https://www.mckinsey.com.br/~/media/McKinsey/Industries/Healthcare%20Systems%20and%20Services/Our%20Insights/Rapid%20growth%20in%20biopharma/Rapid%20growth%20in%20biopharma%20Challenges%20and%20opportunities.pdf (accessed 23 December 2007).

61 Walther, J., Godawat, R., Hwang, C. et al. (2015). The business impact of an integrated continuous biomanufacturing platform for recombinant protein production. *J. Biotechnol.* 213: 3–12.

62 Subramanian, G. (ed.) (2018). *Continuous Biomanufacturing: Innovative Technologies and Methods*. Wiley-VCH: Weinheim.

63 Kamga, M.-H., Cattaneo, M., and Yoon, S. (2018). Integrated continuous biomanufacturing platform with ATF perfusion and one column chromatography operation for optimum resin utilization and productivity. *Prep. Biochem. Biotechnol.* 48 (5): 383–390.

64 Salvalaglio, M., Paloni, M., Guelat, B. et al. (2015). A two level hierarchical model of protein retention in ion exchange chromatography. *J. Chromatogr. A* 1411: 50–62.

65 Brooks, C.A. and Cramer, S.M. (1992). Steric mass-action ion exchange: displacement profiles and induced salt gradients. *AIChE J.* 38 (12): 1969–1978.

66 Mollerup, J.M. (2008). A review of the thermodynamics of protein association to ligands, protein adsorption, and adsorption isotherms. *Chem. Eng. Technol.* 31 (6): 864–874.

67 Velayudhan, A. and Horváth, C. (1994). Adsorption and ion-exchange isotherms in preparative chromatography. *J. Chromatogr. A* 663 (1): 1–10.

68 Guiochon, G. (2006). *Fundamentals of Preparative and Nonlinear Chromatography*, 2e. Amsterdam: Elsevier Academic Press.

69 Seidel-Morgenstern, A. (2004). Experimental determination of single solute and competitive adsorption isotherms. *J. Chromatogr. A* 1037 (1–2): 255–272.

70 Zobel-Roos, S. Entwicklung, Modellierung und Validierung von integrierten kontinuierlichen Gegenstrom-Chromatographie-Prozessen. Dissertation, Shaker Verlag GmbH and Technische Universität Clausthal.

71 Carta, G. and Rodrigues, A.E. (1993). Diffusion and convection in chromatographic processes using permeable supports with a bidisperse pore structure. *Chem. Eng. Sci.* 48 (23): 3927–3935.

72 Wilson, E.J. and Geankoplis, C.J. (1966). Liquid mass transfer at very low Reynolds numbers in packed beds. *Ind. Eng. Chem. Fundam.* 5 (1): 9–14.

73 DePhillips, P. and Lenhoff, A.M. (2000). Pore size distributions of cation-exchange adsorbents determined by inverse size-exclusion chromatography. *J. Chromatogr. A* 883 (1-2): 39–54.

74 Golshan-Shirazi, S. and Guiochon, G. (1988). Analytical solution for the ideal model of chromatography in the case of a Langmuir isotherm. *Anal. Chem.* 60 (21): 2364–2374.

75 Golshan-Shirazi, S. and Guiochon, G. (1989). Analytical solution for the ideal model of chromatography in the case of a pulse of a binary mixture with competitive Langmuir isotherm. *J. Phys. Chem.* 93 (10): 4143–4157.

76 Baur, D., Angarita, M., Müller-Späth, T., and Morbidelli, M. (2016). Optimal model-based design of the twin-column captureSMB process improves capacity utilization and productivity in protein A affinity capture. *Biotechnol. J.* 11 (1): 135–145.

77 Vajda, J. (2016) Spezifische Elektrolyteffekte in der Chromatographie von Biopharmazeutika. Dissertation, Universitäts- und Landesbibliothek Darmstadt.

78 Chen, J., Tetrault, J., Zhang, Y. et al. (2010). The distinctive separation attributes of mixed-mode resins and their application in monoclonal antibody downstream purification process. *J. Chromatogr. A* 1217 (2): 216–224.

79 Gomes, P.F., Loureiro, J.M., and Rodrigues, A.E. (2017). Adsorption of human serum albumin (HSA) on a mixed-mode adsorbent: equilibrium and kinetics. *Adsorption* 23 (4): 491–505.

80 Kaleas, K.A., Schmelzer, C.H., and Pizarro, S.A. (2010). Industrial case study: evaluation of a mixed-mode resin for selective capture of a human growth factor recombinantly expressed in *E. coli*. *J. Chromatogr. A* 1217 (2): 235–242.

81 Pezzini, J., Joucla, G., Gantier, R. et al. (2011). Antibody capture by mixed-mode chromatography: a comprehensive study from determination of optimal purification conditions to identification of contaminating host cell proteins. *J. Chromatogr. A* 1218 (45): 8197–8208.

82 Toueille, M., Uzel, A., Depoisier, J.-F., and Gantier, R. (2011). Designing new monoclonal antibody purification processes using mixed-mode chromatography sorbents. *J. Chromatogr. B* 879 (13–14): 836–843.

83 Zhang, K. and Liu, X. (2016). Mixed-mode chromatography in pharmaceutical and biopharmaceutical applications. *J. Pharm. Biomed. Anal.* 128: 73–88.

84 Vajda, J., Mueller, E., and Bahret, E. (2014). Dual salt mixtures in mixed mode chromatography with an immobilized tryptophan ligand influence the removal of aggregated monoclonal antibodies. *Biotechnol. J.* 9 (4): 555–565.

85 Nfor, B.K., Noverraz, M., Chilamkurthi, S. et al. (2010). High-throughput isotherm determination and thermodynamic modeling of protein adsorption on mixed mode adsorbents. *J. Chromatogr. A* 1217 (44): 6829–6850.

86 Gao, D., Lin, D.-Q., and Yao, S.-J. (2006). Protein adsorption kinetics of mixed-mode adsorbent with benzylamine as functional ligand. *Chem. Eng. Sci.* 61 (22): 7260–7268.

87 Gao, D., Lin, D.-Q., and Yao, S.-J. (2007). Mechanistic analysis on the effects of salt concentration and pH on protein adsorption onto a mixed-mode adsorbent with cation ligand. *J. Chromatogr. B* 859 (1): 16–23.

88 Saleh, D., Wang, G., Müller, B. et al. (2020). Straightforward method for calibration of mechanistic cation exchange chromatography models for industrial applications. *Biotechnol. Progr.* 36 (4): e2984.

89 Chilamkurthi, S., Sevillano, D.M., Albers, L.H.G. et al. (2014). Thermodynamic description of peptide adsorption on mixed-mode resins. *J. Chromatogr. A* 1341: 41–49.

90 Hunt, S., Larsen, T., and Todd, R.J. (2017). Modeling preparative cation exchange chromatography of monoclonal antibodies. In: *Preparative Chromatography for Separation of Proteins* (eds. A. Staby, A.S. Rathore and S. Ahuja), 399–427. Hoboken, NJ: John Wiley & Sons Inc.

91 Nič, M., Jirát, J., Košata, B. et al. (2009). *IUPAC Compendium of Chemical Terminology*. Research Triangle Park, NC: IUPAC.

92 Mathes, J. and Friess, W. (2011). Influence of pH and ionic strength on IgG adsorption to vials. *Eur. J. Pharm. Biopharm.* 78 (2): 239–247.

93 Gomes, P.F., Loureiro, J.M., and Rodrigues, A.E. (2018). Expanded bed adsorption of albumin and immunoglobulin G from human serum onto a cation exchanger mixed mode adsorbent. *Adsorption* 24 (3): 293–307.

94 Altenhöner, U., Meurer, M., Strube, J., and Schmidt-Traub, H. (1997). Parameter estimation for the simulation of liquid chromatography. *J. Chromatogr. A* 769 (1): 59–69.

95 Levenspiel, O. (1999). Chemical reaction engineering. *Ind. Eng. Chem. Res.* 38 (11): 4140–4143.
96 Guiochon, G., Felinger, A., and Shirazi, D.G.G. (2006). *Fundamentals of preparative and nonlinear chromatography*. Elsevier.
97 Josch, J.P., Both, S., and Strube, J. (2012). Characterization of feed properties for conceptual process design involving complex mixtures, such as natural extracts. *FNS* 03 (06): 836–850.
98 Josch, J.P. and Strube, J. (2012). Characterization of feed properties for conceptual process design involving complex mixtures. *Chem. Ing. Tech.*, 84 (6): 918–931.
99 Ndocko Ndocko, E., Ditz, R., Josch, J.-P., and Strube, J. (2011). New material design strategy for chromatographic separation steps in bio-recovery and downstream processing. *Chem. Ing. Tech.* 83 (1–2): 113–129.
100 Strube, J., Grote, F., Josch, J.P., and Ditz, R. (2011). Process development and design of downstream processes. *Chem. Ing. Tech.* 83 (7): 1044–1065.
101 Strube, J. (2000) *Technische Chromatographie: Auslegung, Optimierung, Betrieb und Wirtschaftlichkeit*, Zugl.: Dortmund, University, Habil.-Schr., 1999, Shaker, Aachen.
102 Young, M.E., Carroad, P.A., and Bell, R.L. (1980). Estimation of diffusion coefficients of proteins. *Biotechnol. Bioeng.* 22 (5): 947–955.
103 Pikal, M.J., Mascarenhas, W.J., Akay, H.U. et al. (2005). The nonsteady state modeling of freeze drying: in-process product temperature and moisture content mapping and pharmaceutical product quality applications. *Pharm. Dev. Tech.* 10 (1): 17–32.
104 Rambhatla, S. and Pikal, M.J. (2003). Heat and mass transfer scale-up issues during freeze-drying, I: atypical radiation and the edge vial effect. *AAPS PharmSciTech* 4 (2): E14.
105 Tang, X.C., Nail, S.L., and Pikal, M.J. (2006). Evaluation of manometric temperature measurement (MTM), a process analytical technology tool in freeze drying, part III: heat and mass transfer measurement. *AAPS PharmSciTech* 7 (4): 97.
106 Lewis, L.M., Johnson, R.E., Oldroyd, M.E. et al. (2010). Characterizing the freeze-drying behavior of model protein formulations. *AAPS PharmSciTech* 11 (4): 1580–1590.
107 Hammerschmidt, N., Tscheliessnig, A., Sommer, R. et al. (2014). Economics of recombinant antibody production processes at various scales: industry-standard compared to continuous precipitation. *Biotechnol. J.* 9 (6): 766–775.
108 Baur, D., Angarita, M., Muller-Spath, T. et al. (2016). Comparison of batch and continuous multi-column protein A capture processes by optimal design. *Biotechnol. J.* 11 (7): 920–931.
109 Papathanasiou, M.M., Avraamidou, S., Oberdieck, R. et al. (2016). Advanced control strategies for the multicolumn countercurrent solvent gradient purification process. *AIChE J.* 62 (7): 2341–2357.
110 Godawat, R., Konstantinov, K., Rohani, M., and Warikoo, V. (2015). End-to-end integrated fully continuous production of recombinant monoclonal antibodies. *J. Biotechnol.* 213: 13–19.

111 Jungbauer, A. (2013). Continuous downstream processing of biopharmaceuticals. *Trends Biotechnol.* 31 (8): 479–492.
112 Subramanian, G. (2017). *Continuous Biomanufacturing: Innovative Technologies and Methods*. Wiley-VCH.
113 Zobel-Roos, S., Stein, D., and Strube, J. (2018). Evaluation of continuous membrane chromatography concepts with an enhanced process simulation approach. *Antibodies* 7 (1): 13.
114 Zobel-Roos, S. (2018). *Entwicklung, Modellierung und Validierung von integrierten kontinuierlichen Gegenstrom-Chromatographie-Prozessen*, 1e. Herzogenrath: Shaker.
115 Zobel, S., Helling, C., and Strube, J. (2014). Integrated counter current chromatography (iCCC) – Von der SMB zum integrierten Prozess. *Chem. Ing. Tech.* 86 (9): 1504.
116 Hribar, G. and Gillespie, C. (2015). *Next Generation Biopharmaceutical Downstream Processing – Continuous Bioprocessing*. PDA meeting on continuous manufacturing, Berlin.
117 Müller-Späth, T. (2013). Productivity boost for biopurification: twin-column ultra-high resolution chromatography. *Gen. Eng. Biotech. News* 33: 34–35.
118 GE Healthcare *A flexible antibody purification process based on ReadyToProcessTM products*, Application note 28-9403-48 AB. www.gehealthcare.com (7 December 2015).
119 David Pollard (2015) *Merck, Advances towards automated continuous mAb processing, Power Point Presentation, 2015,* www.merck.com.
120 Munk, M. (2015). What is holding industry back from implementing continuous processing: can Asia adopt more quickly? *BioPharma Asia* 4 (6): 16–22.
121 Kurt Wagemann and Kathrin Rübberdt (2015). *Recommendation for a risk analysis for production processes with disposable bioreactors. Dechema, Gesellschaft fur Chemische Technik und Biotechnologie e.V.* https://dechema.de/dechema_media/SingleUse_RiskAnalysis_2015-p-20001335.pdf (24 February 2017).
122 Unger, C., Skottman, H., Blomberg, P. et al. (2008). Good manufacturing practice and clinical-grade human embryonic stem cell lines. *Hum. Mol. Genet.* 17 (R1): R48–R53.
123 Ottawa Hospital Research Institute, Cell Manufacturing. http://www.ohri.ca/bmc/CellManufacturing.aspx (accessed 23 December 2020).
124 Huter, M. and Strube, J. (2019). Model-based design and process optimization of continuous single pass tangential flow filtration focusing on continuous bioprocessing. *Processes* 317.
125 Zobel, S., Helling, C., Ditz, R., and Strube, J. (2014). Design and operation of continuous countercurrent chromatography in biotechnological production. *Ind. Eng. Chem. Res.* 53 (22): 9169–9185.
126 Jos Buijs, Willem Norde, and James W. Th. Lichtenbelt Changes in the Secondary Structure of Adsorbed IgG and F(ab')2 Studied by FTIR Spectroscopy.
127 ICH (2011) Endorsed Guide for ICH Q8/Q9/Q10 Implementation.
128 Wasalathanthri, D.P., Feroz, H., Puri, N. et al. (2020). Real-time monitoring of quality attributes by in-line Fourier transform infrared spectroscopic sensors at ultrafiltration and diafiltration of bioprocess. *Biotechnol. Bioeng.*

129 Buckley, K. and Ryder, A.G. (2017). Applications of Raman spectroscopy in biopharmaceutical manufacturing: a short review. *Appl. Spectrosc.* 71 (6): 1085–1116.

130 Santos, R.M., Kessler, J.-M., Salou, P. et al. (2018). Monitoring mAb cultivations with in-situ raman spectroscopy: The influence of spectral selectivity on calibration models and industrial use as reliable PAT tool. *Biotechnol. Progr.* 34 (3): 659–670.

131 Boulet-Audet, M., Kazarian, S.G., and Byrne, B. (2016). In-column ATR-FTIR spectroscopy to monitor affinity chromatography purification of monoclonal antibodies. *Sci. Rep.* 6: 30526.

132 Sauer, D.G., Melcher, M., Mosor, M. et al. (2019). Real-time monitoring and model-based prediction of purity and quantity during a chromatographic capture of fibroblast growth factor 2. *Biotechnol. Bioeng.* 116 (8): 1999–2009.

133 Sellick, C.A., Hansen, R., Jarvis, R.M. et al. (2010). Rapid monitoring of recombinant antibody production by mammalian cell cultures using fourier transform infrared spectroscopy and chemometrics. *Biotechnol. Bioeng.* 106 (3): 432–442.

134 Dziadosz, M., Lessig, R., and Bartels, H. (2012). HPLC-DAD protein kinase inhibitor analysis in human serum. *J. Chrom. B* 893-894: 77–81.

135 Zobel-Roos, S., Mouellef, M., Siemers, C., and Strube, J. (2017). Process analytical approach towards quality controlled process automation for the downstream of protein mixtures by inline concentration measurements based on ultraviolet/visible light (UV/VIS) spectral analysis. *Antibodies (Basel, Switzerland)* 6 (4).

136 Jasco Deutschland GmbH FP-2020 Fluorescence Detector. https://www.jasco.de/en/content/FP-2020/~tpl.index/.html.

137 Moore-Kelly, C., Welsh, J., Rodger, A. et al. (2019). Automated high-throughput capillary circular dichroism and intrinsic fluorescence spectroscopy for rapid determination of protein structure. *Anal. Chem.* 91 (21): 13794–13802.

138 Park, J., Nagapudi, K., Vergara, C. et al. (2013). Effect of pH and excipients on structure, dynamics, and long-term stability of a model IgG1 monoclonal antibody upon freeze-drying. *Pharm. Res.* 30 (4): 968–984.

139 Santos, R.M., Kaiser, P., Menezes, J.C., and Peinado, A. (2019). Improving reliability of Raman spectroscopy for mAb production by upstream processes during bioprocess development stages. *Talanta* 199: 396–406.

140 Popp, J. and Mayerhöfer, T. (2020). *Micro-Raman Spectroscopy.* De Gruyter.

141 Kosa, G., Shapaval, V., Kohler, A., and Zimmermann, B. (2017). FTIR spectroscopy as a unified method for simultaneous analysis of intra- and extracellular metabolites in high-throughput screening of microbial bioprocesses. *Microb. Cell Fact.* 16 (1): 195.

142 Feidl, F., Garbellini, S., Vogg, S. et al. (2019). A new flow cell and chemometric protocol for implementing in-line Raman spectroscopy in chromatography. *Biotechnol. Progr.* 35 (5): e2847.

143 Thakur, G., Thori, S., and Rathore, A.S. (2020). Implementing PAT for single-pass tangential flow ultrafiltration for continuous manufacturing of monoclonal antibodies. *J. Membr. Sci.* 613: 118492.

144 Faix, O. (ca. (2011). Fourier transform infrared spectroscopy. In: *Methods in Lignin Chemistry*, 1ste (ed. S.Y. Lin), 83–109. Berlin: Springer.

145 Barth, A. (2007). Infrared spectroscopy of proteins. *Biochim. Biophys. Acta* 1767 (9): 1073–1101.

146 Oelmeier, S.A., Ladd-Effio, C., and Hubbuch, J. (2013). Alternative separation steps for monoclonal antibody purification: combination of centrifugal partitioning chromatography and precipitation. *J. Chromatogr. A* 1319: 118–126.

147 Pei, Y., Wang, J., Wu, K. et al. (2009). Ionic liquid-based aqueous two-phase extraction of selected proteins. *Sep. Purif. Technol.* 64 (3): 288–295.

148 Barresi, A.A. and Fissore, D. (2011). In-line product quality control of pharmaceuticals in freeze-drying processes. In: *Modern Drying Technology* (eds. E. Tsotsas and A.S. Mujumdar), 91–154. Weinheim, Germany: Wiley-VCH Verlag GmbH & Co. KGaA.

149 Fissore, D., Pisano, R., and Barresi, A.A. (2018). Process analytical technology for monitoring pharmaceuticals freeze-drying – a comprehensive review. *Drying Technol.* 36 (15): 1839–1865.

150 Johnson, R.E., Teagarden, D.L., Lewis, L.M., and Gieseler, H. (2009). Analytical accessories for formulation and process development in freeze-drying. *Am. Pharm. Rev.* 12: 54–60.

151 Nail, S., Tchessalov, S., Shalaev, E. et al. (2017). Recommended best practices for process monitoring instrumentation in pharmaceutical freeze drying-2017. *AAPS PharmSciTech* 18 (7): 2379–2393.

152 Patel, S.M. and Pikal, M. (2009). Process analytical technologies (PAT) in freeze-drying of parenteral products. *Pharm. Dev. Technol.* 14 (6): 567–587.

153 Stefan Schneid (2010) Investigation of novel process analytical technology (PAT) tools for use in freeze-drying processes. Dissertation, Friedrich-Alexander-Universität Erlangen-Nürnberg. https://opus4.kobv.de/opus4-fau/files/1025/StefanSchneidDissertation.pdf.

154 Yadav, L.D.S. (2005). *Ultraviolet (UV) and Visible Spectroscopy Organic Spectroscopy*. Netherlands: Springer http://dx.doi.org/10.1007/978-1-4020-2575-4_2.

155 Lakowicz, J.R. (ed.) (2010). Introduction to Fluorescence. In: *Principles of fluorescence spectroscopy*, 4e, 1–26. New York, NY: Springer.

156 Claßen, J., Aupert, F., Reardon, K.F. et al. (2017). Spectroscopic sensors for in-line bioprocess monitoring in research and pharmaceutical industrial application. *Anal. Bioanal.Chem.* 409 (3): 651–666.

157 Rolinger, L., Rüdt, M., Diehm, J. et al. (2020). Multi-attribute PAT for UF/DF of proteins-monitoring concentration, particle sizes, and buffer exchange. *Anal. Bioanal.Chem.* 412 (9): 2123–2136.

158 Jungbauer, A. et al. (2017). Real-time monitoring of product purification. WO2017174580A1.

159 Barresi, A.A., Velardi, S.A., Pisano, R. et al. (2009). In-line control of the lyophilization process. A gentle PAT approach using software sensors. *Int. J. Refrig* 32 (5): 1003–1014.

160 Bosca, S., Barresi, A.A., Fissore, D. (2013) *Fast freeze-drying cycle design and optimization using a PAT based on the measurement of product temperature.*

161 Fissore, D. and Pisano, R. (2015). Computer-aided framework for the design of freeze-drying cycles: optimization of the operating conditions of the primary drying stage. *Processes* 3 (2): 406–421.

162 Pisano, R., Fissore, D., and Barresi, A.A. (2013). In-line and off-line optimization of freeze-drying cycles for pharmaceutical products. *Drying Technol.* 31 (8): 905–919.

163 Tang, X.C., Nail, S.L., and Pikal, M.J. (2005). Freeze-drying process design by manometric temperature measurement: design of a smart freeze-dryer. *Pharm. Res.* 22 (4): 685–700.

164 Dittmar, R. *Advanced Process Control: PID-Basisregelungen, Vermaschte Regelungsstrukturen, Softsensoren, Model Predictive Control.* De Gruyter.

165 Keerthi, S.S. and Gilbert, E.G. (1988). Optimal infinite-horizon feedback laws for a general class of constrained discrete-time systems: stability and moving-horizon approximations. *J. Optimiz. Theory App.* 57 (2): 265–293.

166 Song, I.-H., Amanullah, M., Erdem, G. et al. (2006). Experimental implementation of identification-based optimizing control of a simulated moving bed process. *J. Chromatogr. A* 1113 (1-2): 60–73.

167 Sommeregger, W., Sissolak, B., Kandra, K. et al. (2017). Quality by control: towards model predictive control of mammalian cell culture bioprocesses. *Biotechnol. J.* 12 (7).

168 Fanali, S. (2009). Editorial on "Simulated moving bed chromatography for the separation of enantiomers" by A. Rajendran, G. Paredes and M. Mazzotti. *J. Chromatogr. A* 1216 (4): 708.

169 Diederich, P., Hansen, S.K., Oelmeier, S.A. et al. (2011). A sub-two minutes method for monoclonal antibody-aggregate quantification using parallel interlaced size exclusion high performance liquid chromatography. *J. Chromatogr. A* 1218 (50): 9010–9018.

170 Rüdt, M., Andris, S., Schiemer, R., and Hubbuch, J. (2019). Factorization of preparative protein chromatograms with hard-constraint multivariate curve resolution and second-derivative pretreatment. *J. Chromatogr. A* 1585: 152–160.

171 Martins, D.L., Sencar, J., Hammerschmidt, N. et al. (2020). Truly continuous low pH viral inactivation for biopharmaceutical process integration. *Biotechnol. Bioeng.* 117 (5): 1406–1417.

172 Kruse, T., Kampmann, M., and Greller, G. (2020). Aqueous two-phase extraction of monoclonal antibodies from high cell density cell culture. *Chem. Ing. Tech.* 256 (23): 41.

10

Regulatory and Quality Considerations of Continuous Bioprocessing

Britta Manser[1] and Martin Glenz[2]

[1]*Pall International Sàrl, Avenue de Tivoli 3, 1700 Fribourg, Switzerland*
[2]*Pall GmbH, Philipp-Reis-Strasse 6, 63303 Dreieich, Germany*

10.1 Introduction

For over 30 years, biopharmaceutical manufacturers have successfully produced valuable therapies such as antibodies or recombinant proteins using batch or fed-batch manufacturing concepts. In the past decade, cell lines for expressing the proteins of interest have been advanced by many biopharmaceutical companies and have increased the productivity of *upstream processing* (USP) significantly. To match these increases, advances in *downstream processing* (DSP) have been developed, leading to an optimization of auxiliary materials such as chromatography adsorbents with higher binding capacity or high-flux membranes. However, these efforts downstream have not been significant enough to reach the productivity needed. In addition, the increasing number of biosimilars on the market and the availability of multiple drugs targeting the same indication have shifted the industry need toward smaller and more flexible multiproduct facilities [1].

To meet the needs for intensified and flexible processing, single-use technologies have become state of the art, and several manufacturers have explored the means of process intensification through continuous bioprocessing. Continuous manufacturing steps when applied to both USP and DSP – and when connected – have enabled the fully integrated manufacture of drug substances [1]. Continuous or integrated technologies bring the potential to react to manufacturing variability, provide flexibility, and at the same time increase productivity while reducing manufacturing costs [2, 3].

While only few companies have chosen to move to fully integrated manufacture today, high potential is seen in hybrid platform approaches. This combines continuous processing steps with standard batch operation or repetitive batch. Several technologies supporting continuous processes have matured to the stage where they can be applied for current good manufacturing practice (cGMP) manufacture. With manufacturers moving toward the application of integrated bioprocessing in cGMP environments, the regulatory roadmap needs consideration. Regulatory authorities

Process Control, Intensification, and Digitalisation in Continuous Biomanufacturing, First Edition.
Edited by Ganapathy Subramanian.
© 2022 WILEY-VCH GmbH. Published 2022 by WILEY-VCH GmbH.

are promoting and supporting the adoption of continuous manufacturing through, for example, the Food and Drug Administration's *Emerging Technology Team* (ETT) or the European Medical Agency's *Innovation Task Force* (ITF), providing early engagement for the quality assessment teams in submission reviews. Recently, a draft guideline for continuous manufacture was released. There is, however, no specific guideline outlining a framework for continuous biomanufacturing available today [4].

From a regulatory perspective, the continuous process is not inherently different from batch processing. Both processes rely on the same fundamental concepts and operations. As a result, many concepts for patient safety, quality assurance, or process consistency could be directly correlated with the knowledge from the equivalent batch process. This document describes regulatory considerations for integrated operations based on standard batch manufacturing and shows where differences – especially with regard to linking of unit operations – are to be considered.

10.2 Integrated Processing

Continuous bioprocessing has been applied to the USP in the past decades, and several commercial products from perfusion cell culture exist today [5]. In DSP, however, the continuous approach has only gained momentum in more recent years. While in the 2000s the focus of the industry has been on "fast to first in human," targeting high-throughput process development and single-use technologies to reduce capital and timelines for drug to market, the industry has turned to integrated and intensified bioprocessing since the 2010s. However, it has shifted toward flexible platforms for applying diverse operating modes including continuous bioprocessing or hybrid processing.

Continuous bioprocessing describes a cascade of interlinked unit operations allowing for a seamless manufacturing platform. The difference between a batch and an integrated process is that several unit operations of the platform run simultaneously. A subsequent unit operation is started before the previous step is complete. The basic unit operations thereby remain unchanged compared with batch manufacture: the protein is expressed in a fed-batch or perfusion bioreactor and purified through a similar series of chromatography and filtration operations.

Through step integration, unit operations tend to be smaller as flow rates decrease, which reduces the facility footprint and the risk of scale-up. The platform approach using mostly single-use equipment provides the agility and flexibility to respond to market dynamics rapidly and minimizes the risk of product shortage. The highly automated and controlled platform provides opportunities for advanced data management to improve product quality and release products faster [6].

As of 2020, several biopharmaceutical manufacturers have either evaluated fully continuous manufacturing platforms or the impact of hybrid processing for monoclonal antibodies [1, 7, 8]. For the purpose of this review, a generic continuous *monoclonal antibody* (mAb) production platform is used as a context to discuss the regulatory aspects of continuous manufacturing; see Figure 10.1. The concepts and

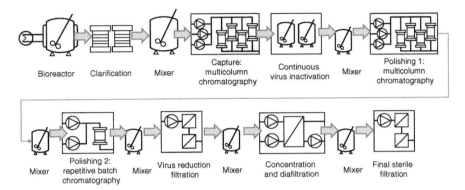

Figure 10.1 Process of a mAb platform using continuous technologies for downstream processing from capture to final sterile filtration of the drug substance.

approach presented for this fully integrated platform also apply to hybrid manufacturing processes that use continuous manufacturing technology, in one, or more unit operations.

10.3 Process Traceability

In integrated processing, drug product may not be as easily attributable to a lot as in standard batch processing. A lot in batch manufacture is typically associated to the upstream bioreactor. The integrated platform however can easily see continuous or continual introduction of raw material from multiple bioreactors, purified seamlessly in the same downstream platform. Also, material is not typically held in surge containers between unit operations to the same extent as in batch operation. Material cascades to the following unit operation before all quality attributes have been assessed. A deviation or nonconformity occurring in one step may therefore propagate to the following unit operations before any actions can be taken. As such it is essential to understand the rate at which material flows through the cascade of unit operations. To identify and trace material at all times in the dynamic process, establishing the connection between material and batch is a necessity [9].

10.3.1 Batch and Lot Definition

Batch definitions given by regulatory authorities do not reference the manufacturing principle and have been applied to both standard batch and continuous processes [4]. A batch is defined in *21 CFR 210.3* as a specific quantity of a drug or other material that is intended to have uniform character and quality, is within specified limits, and is produced according to a single manufacturing order during the same cycle of manufacture. A lot characterizes a batch or a specific portion of a batch [10]. *EMA ICH Q7* defines the batch as homogeneous material within specific limits and states that in the case of continuous production, a batch may correspond to a defined fraction of the production.

The regulatory framework allows the manufacturer to define a batch size based on fixed quantity or a fixed time interval [11]. In continuous processing, a batch definition could follow different rationales:

- The batch can be based on a fixed time or set amount of product irrespective of the raw material lots used in the manufacture of the product.
- The batch can be based on time constants of the process, such as the cyclic behavior of multicolumn chromatography or low-pH virus inactivation steps.

When working with time constants, it can be meaningful to harmonize the cycle time of several unit operations [9]. It must be considered that the cycle time can vary throughout the process in certain settings, for example, in dynamic column loading during chromatography steps. While column loading is generally based on a fixed volume with fixed time, dynamic column loading describes a chromatography column being loaded based on the protein concentration of the upstream feed or breakthrough of the column during the load step. With varying product titers from upstream operations and reduced binding capacity of the chromatography adsorbent over time, the load time and hence the cycle time of the chromatography process change.

Examples from the industry have used both time-based and volume-based strategies in biologics platforms using continuous USP or DSP: Alvotech (Iceland) presented on defining a batch based on time, with 15 days adding up to one batch [12]. A similar approach has been described by Sanofi Genzyme (USA) for a monoclonal antibody from a 31-day continuous process, where a batch is defined as one day of operation [13]. BiosanaPharma (Netherlands) apply a mass-based approach for a biosimilar antibody from continuous processing [14].

With the boundary conditions given by the regulators, it is the responsibility of manufacturers to provide a rationale for batch and lot definition. It is advisable to perform a failure mode and effects analysis to allow consideration of the process risks within the batch definition. This approach minimizes the business impact in the case of batch rejection. It should also be considered that a batch may be flexible in size but in any case needs to be defined prior to manufacturing [4, 15]. For integrated processing, defining procedures and establishing collection criteria for start-up and shutdown of every unit operation should also be considered [9].

10.3.2 Lot Traceability and Deviation Management

In integrated processing it is essential to understand how fast material propagates through the cascade of unit operations, allowing tracking batches and potential process upsets as they move through the platform. The FDA has requested for a scientific approach to characterize material flow, for example, by characterizing the *residence time distribution* (RTD) [4] that is a probability distribution that allows evaluating propagation velocity and residence time of product or impurities. By characterizing the RTD of each unit operation and the combination thereof in the integrated cascade, it is possible to identify product at any stage of the process train. This allows retrospective determination of batches or lots that have been affected by a process deviation or even indicate at what stage of the cascade to isolate or divert product.

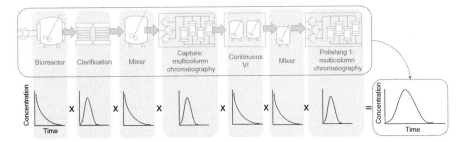

Figure 10.2 Conceptual design of a residence time distribution for part of a continuous processing platform for biologics. The residence time distribution is characterized for an individual unit operation and calculated for the overall process cascade to identify back mixing and axial dispersion in the system. VI, virus inactivation.

An RTD can be determined (i) experimentally through tracer experiments where a representative non-reactive tracer is introduced into the system and concentration changes are measured over time, (ii) through online process monitoring of specific products and their attributes, or (iii) by modeling of process steps [4, 16]. Pulse-response measurements or step changes are visible in characteristic peak shapes that determine back mixing in the unit operation and axial dispersion. While RTD has been implemented successfully in continuous manufacture for powder or granulation steps of small-molecule drugs [17], the more complex liquid handling steps in biotechnological processes and the higher susceptibility of substances to changing physical properties constitute additional challenges. It is important that the RTD experimental settings represent the full range of planned operating conditions and that it takes variables such as material attributes, flow rates, or equipment operation into consideration [4]. Experimental proof of concept of the RTD approach has been provided by Merck (US) and Sencar et al. who demonstrated how material can be virtually tracked in a continuous bioprocess using a model-based RTD approach [18].

For integrated bioprocessing, a traceability evaluation is conceptually shown in Figure 10.2. While the strategy indicates performing an analysis on every step of the process, it can be considered to implement a risk-based approach to reduce the extent of the RTD study. In this concept, the risk is based on the probability of a process upset and the detection time of a resulting deviation. If performed for unit operations and clusters thereof, high-risk segments are identified for RTD experiments. Such high-risk areas can include post-use filter integrity test failure, e.g. viral or sterile filtration steps, or column fouling and peak-cutting failure in multicolumn chromatography.

10.4 Process Consistency

Integrated processes are typically run over extended operating times of several days or weeks compared with batch processes where unit operations are completed within only a few hours. With the extended operation time and repetitive cycling of unit operations comes the need for long-term process consistency to assure that

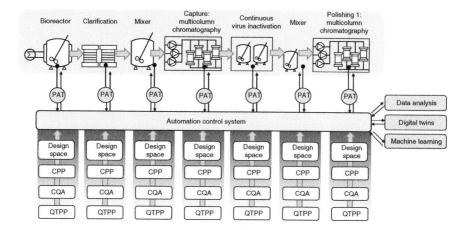

Figure 10.3 Automation Control System assuring that the process stays within the design space given from the *quality-by-design* (QbD) approach by assessing process information in real-time through *process analytical technologies* (PAT). Advanced data analysis tools (e.g. multivariate data analysis), digital twins, and machine learning can thereby improve the responses from the control system. QTPP, *quality target product profile*; CQA, *critical quality attribute*; CPP, *critical process parameter*.

the product continuously meets the quality specification. While it may seem that process consistency needs are a hurdle in continuous bioprocessing, technology advances in process control are playing a vital role in facilitating the implementation of integrated manufacture. The availability of automation technologies and possibilities unlocked by *process analytical technologies* (PAT) enable the higher degree of process control desired for integrated processes. Also, continuous bioprocessing naturally results in vastly more data from the many cycles run over the operating time. Here, advanced data interpretation such as multivariate data analysis (MVDA) can be used to provide statistically relevant process and product quality information as the process is running. Using these technologies can also facilitate the *quality-by-design* (QbD) approach.

Figure 10.3 gives a high-level overview of how principles from QbD and PAT, together with multivariate data analysis or even digital twins and machine learning, can assure the consistency of an integrated biomanufacturing platform.

10.4.1 Process Control

10.4.1.1 Automation

A continuous platform connects multiple systems that operate simultaneously and therefore requires a robust automation platform for real-time process monitoring and decision making. An automated supervisory control system can be used to control the platform, collect data, and make fast decisions during operation. Well-designed software for integrated and centralized process control can operate a process with minimum operator intervention and provides a high degree of process consistency. Harmonizing control systems across different scales from process development to manufacturing and data portability enables a seamless technology transfer.

The regulatory expectation toward such supervisory control systems is that they include both process control functionality and quality unit oversight [4]. The control software can control flow management by harmonizing product flow between individual unit operations. Flow control is propagated in the downstream direction by matching the inlet flow rate of a downstream unit operation to the outlet flow rate of the previous unit operation. Such an automation concept has been presented for an integrated mAb platform by the *Center for Process Innovation* (CPI) (UK). A weight-based flow control allows automated start-up and shutdown of all individual unit operations in the process sequence. Also, every unit operation comes with multiple inlet options (feed or buffer) and outlet choices (product or waste) that enable the system to condition or regenerate unit operations, execute buffer chases, or divert product during non-conforming sequences or interruptions [19]. Advanced automation concepts have been implemented by Merck (US) [16] or Bayer where in-line monitoring of pH and conductivity are coupled with feedback control for deviation management [20].

While the control system itself relies on relatively simple concepts of flow control and start/stop triggers, the advanced automation planning is the actual challenge because it is still new to the biopharmaceutical industry. For integrated processes, it aspires to engender more advanced control systems that include feedback from sensor readings within the PAT strategy and adapt operating parameters to the fluid quality throughout the production; see Figure 10.3. Alarms and control measures assure that the system operates within predefined limits according to its design space. Automatic responses in the case of process shifts can include automated pauses to temporarily interrupt sections of the process, automatically bringing intermediate product solutions to safe condition, for instance, by adjusting buffer conditions, or to divert product, as shown in Figure 10.4. This level of control not only allows to automatically react to process upsets or variation but also can improve product yield and intensify the platform. An example is given in dynamic loading of multicolumn chromatography depending on the feed titer and binding capacity of the chromatography column. With varying feed titer from USP or decreasing binding capacity as the sorbent ages, the load time in the chromatography is varied to assure maximum use of chromatography resin and harmonize product concentration in the elution.

Considering the complexity of an integrated continuous biomanufacturing platform, it may be a realistic first step to combine local control on individual systems, with remote control through the supervisory control system. Additional functionalities of the supervisory control system can include data collection or raw data management to combine all data in one repository for advanced analysis like MVDA or feed information into digital twins or machine learning.

10.4.1.2 Process Analytical Technologies (PAT)

The PAT approach is designed as a process control strategy based on real-time measurement of parameters that correlate with *critical quality attributes* (CQA).

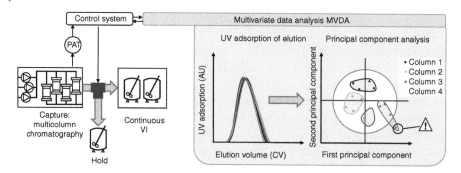

Figure 10.4 Multivariate data analysis (MVDA) applied to 4 cycles of a four-column multicolumn chromatography process to evaluate if UV adsorption values of elution peaks are within the design space. The example shows one elution of column 3 trending toward a potential out of specification. The control system can react in real time to direct product to the following virus inactivation (VI) step or divert the eluted product to a hold container.

PAT has been defined by the US FDA as a system to design, analyze, and control pharmaceutical manufacturing processes through the timely measurement of critical quality and performance attributes to ensure final product quality. This increases the understanding and control of a manufacturing process and holds the potential to efficiently detect process failures or variability with the option for active control. To help accelerate the adoption of PAT, FDA encourages manufacturers to engage with analytical technologies through the ETT.

In USP, online measurement of pH, dissolved gases, temperature, and foam level are parameters commonly monitored. They do not necessarily represent quality attributes but can be *critical process parameters* (CPPs). CQAs such as cell viability, cell density, biomass concentration, or chemical properties of media components have been assessed through capacitive methods or technologies based on *near-infrared* (NIR) *spectroscopy*, Raman spectroscopy, *infrared* (IR) *spectroscopy*, or fluorescence spectroscopy [21]. Raman spectroscopy has shown potential for biopharmaceutical processes particularly when used to determine glycoform patterns or structural forms of proteins [22, 23].

In DSP, standard online monitoring technology includes pH, conductivity, pressure, and temperature measurements. Those are also not necessarily quality attributes of the biologic product. Quality attributes such as concentration, purity, charge variants, and information on impurities including DNA, endotoxins, *host cell protein* (HCP), bioburden, and process-related impurities, can often not be detected directly. The current lack of sensors with a direct link to quality attributes has been described and recognized as a gap in biologics manufacture, especially for continuous processing [23–25].

Protein concentration, one of the main parameters, is monitored online through UV adsorption, multi-wavelength UV spectra, or index of refraction [26]. To assess product impurities or charge variants, *high-performance liquid chromatography* (HPLC), *ultra-performance liquid chromatography* (UPLC), or fluorescence methods have been applied through at-line systems. Examples of protein quantification in continuous DSP have been presented by Merck (USA), who implemented two

UPLC systems with a multicolumn chromatography system to determine product concentration in the eluate and breakthrough of protein during the load step [7]. In a different study, Merck has also demonstrated the use of UV adsorption differentials to dynamically control continuous downstream operations dependent on product titer [27]. CPI (UK) has presented an integrated biologics DSP platform that uses automated sampling for regular HPLC analysis at strategic points of the platform for potential future platform control [19].

The adoption of PAT has been perceived as complex with unclear deployment, especially due to the current gaps: certain procedures, such as bioburden, mycoplasma, and virus testing, require lengthy offline procedures that delay decisions in continuous processes. The development of new in-line methods is ongoing and is seen as a necessity for successful implementation of PAT or even *real-time release testing* (RTRT) in the future [28].

10.4.1.3 Data Analysis

Data analysis is one of the main challenges with integrated manufacturing platforms owing to the larger amount of data generated and the shorter response times for process control. While a batch process allows to hold product for hours or days to perform manual sampling and offline testing, an integrated platform relies on real-time process monitoring with automated in-line sampling and in-process testing. The increased data volume can be a challenge for analysis but provides more opportunities for advanced analysis methodologies. MVDA has shown to provide significant information on process consistency and is seen as a powerful tool for PAT and QbD [29].

To perform MVDA, data is analyzed by employing data reduction steps, for example, a *principal component analysis* (PCA) or partial least squares (PLS) regression. PCA developed in 1901 by Pearson [30] analyzes a data table in which observations are described by several interconnected quantitative variables. Their goal is to extract information from the table, present it as a series of new orthogonal variables called *main components*, and display the similarity pattern of the observations and variables as points on maps [31]. It is a powerful tool to gain in-depth information in the structure of the dataset and show similarities and differences of data points. The PLS on the other hand is most effective in assessing covariance between variables and outcomes of a process [29].

MVDA has been applied mostly to USP but has shown to also be an effective means for process control in downstream unit operations [29]. The example in Figure 10.4 shows how PCA can evaluate even the smallest column-to-column and cycle-to-cycle variations in a multicolumn chromatography process. The capture chromatography, in this example a protein A bind/elute step for mAb capture, continually elutes product from the four columns. The UV adsorption shows a good overlay from all columns and cycles, suggesting high process consistency. Only in the PCA do minor differences between columns and cycles become visible. A trend can be recognized through MVDA before it is visible through traditional methods and potentially even before it compromises product quality. Corrective actions can be taken before trends from upcoming column failure or process variations become problematic.

MVDA therewith provides the capability for real-time online and at-line monitoring and supports advanced automation approaches. It ensures consistent product quality by increasing the manufacturing robustness and process homogeneity. An integrated biopharmaceutical process provides numerous opportunities for MVDA, for example, in chromatography, for on-line column characterization and monitoring bed consistency. With the large amounts of data generated by the long-term operation or cyclic behavior of unit operations, MVDA can help manufacturers provide evidence of process control [32].

10.4.1.4 Real-Time Release Testing

RTRT has been described as the desirable end status of successfully implemented continuous bioprocessing. It is already a reality and being further developed for some small-molecule pharmaceuticals that employ continuous DSP, including Vertex Pharmaceuticals or Janssen where quantitative NIR is the basis for potential real-time product release [17, 33]. Both FDA and EMA encourage its implementation for integrated processes, and thanks to the advancement of PAT tools, a large amount of real-time process and quality data is captured throughout the continuous process that can support RTRT [4, 34].

With regard to RTRT for biological products, EMA states that elements such as product or process understanding, in-process parametric control, or attribute testing at an earlier process step can be used to justify RTRT instead of end product testing [34]. Thereby, it is referred to the possibilities of (i) a routine testing approach, (ii) a validation approach, or (iii) a combination thereof. While in routine testing an attribute is monitored at appropriate steps of the process to ensure acceptance levels in the final product, a validation approach is based on the evidence of successful product quality at a given purification scheme [34].

RTRT today is still a vision for biopharmaceuticals production, and it is expected that several gaps need to be closed before RTRT becomes a viable option. The main challenge is seen in quality testing: a RTRT requires real-time analysis of product quality or impurities, ideally in-line, where technical options for several critical quality attributes are missing today. An implementation of RTRT is only seen viable if all these elements can be brought together in the future.

10.4.2 Quality by Design

For well over a decade, FDA has advocated for QbD in pharmaceutical processes, and for new approvals, QbD approaches are requested by regulatory authorities [23]. QbD follows the rationale of quality being achieved by design and built in rather than confirmed through final testing. In continuous processing, a key concern has been the perception of higher efforts associated with process characterization compared with batch operation.

To address this concern, it has to be evaluated if CPPs for continuous operation could be identified and characterized using small-scale batch experiments [35]. This

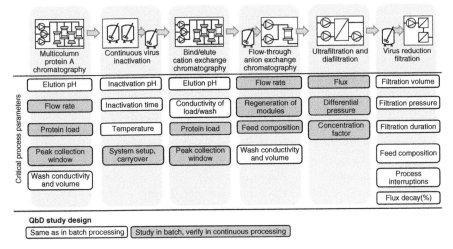

Figure 10.5 Critical process parameters (CPP) of typical downstream unit operations for mAb processing. Since the processing fundamentals for batch and continuous are similar, several of the CPPs can allow small-scale validation studies in batch. The remaining CPPs can be studied in batch, but a verification in continuous operation is meaningful.

model-based approach allows for significant savings in time and material. Since both batch and continuous processes use the same process steps that greatly rely on the same fundamental parameters, also the same QbD principles could be applied. Bridging small-scale validation data from batch to continuous requires a good process understanding and an assessment early on in the process to assure that the continuous manufacturing principle does not impact product aspects [4].

This section reviews the QbD approach for continuous downstream unit operations of an antibody process based on the critical and key process parameters identified in the batch process. The mAb process described in Figure 10.1 is used as a basis for the integrated process with reference data from the A-Mab case study used for describing the batch process in the assessment below [36]. The A-Mab case study gives an example of a QbD approach to a specific mAb production process and, even though not generically applicable to other antibodies, serves as example for this simplified assessment. The CPPs of the downstream steps and whether the QbD study could be performed entirely in batch or whether a verification in continuous processing is meaningful are summarized in Figure 10.5.

It is suggested that a *design of experiments* (DoE) for QbD can be performed in small-scale batch experiments and translated to continuous operation. A model-assisted translation from batch to continuous may also be applicable but is expected to require an extensive data foundation for justification.

10.4.2.1 Multicolumn Protein A Chromatography

Continuous capture is performed on multicolumn systems where a primary load column is overloaded and breakthrough protein gets captured on a secondary load column. This is to achieve higher binding capacities at short contact times. The non-load sequence that includes wash, elution, and regeneration steps remains unchanged. In batch chromatography, key process parameters have been identified as elution pH, protein load, process flow rates, wash volumes, conductivity, and peak collection window, as they all showed an impact on yield and HCP removal [36] or virus reduction.

A QbD strategy for continuous capture can include an identical control strategy for elution pH and wash conditions as in batch operation using adequate measures for buffer control and release. For protein load and load flow rate, operating ranges have been showed to be directly translatable from batch experiments: breakthrough data based on single-column experiments for evaluation of mass transfer kinetics and static binding capacity have been verified to be a solid basis for operating ranges in continuous operation [7, 8]. Impurity removal and yield have been described as equivalent when moving from single-column chromatography to multicolumn chromatography, despite the higher binding capacity and lower contact times [8, 37]. The data suggests performing DoE on single columns in batch mode, but to verify the results for protein load, load flow rate, and peak collection window in continuous operation.

10.4.2.2 Continuous Virus Inactivation

The virus inactivation concept assessed is based on repetitive batch inactivation in mixers [38]. Alternative approaches such as continuous flow systems using coiled tube reactors, for example, have raised questions with regard to adequate process control [24, 39]. For virus inactivation in batch operation, CPPs are linked to inactivation pH, incubation time, and temperature, as these parameters can affect inactivation efficiency and protein aggregation [36].

The repetitive batch mixing system utilizes two alternating mixers that repeatedly fill, inactivate, and empty neutralized material. The process sequence is identical to virus inactivation in batch, and it is expected that CPPs and design space are identical. As the inactivation is performed repetitively, the same mixers go through multiple rounds of inactivation. This can come with the risk of potential carryover of inactivated material that would then get inactivated more than once. In addition, holdup of product in dead volumes linked to the system setup may compromise inactivation efficiency. Adequate system design has shown to be an effective means to mitigate the risk of dead volume and carryover [40]. For continuous virus inactivation, the design space could be assessed through batch DoE, but the system design with its implication to carryovers would need additional review.

10.4.2.3 Bind/Elute Cation Exchange Chromatography

Continuous bind/elute *cation exchange chromatography* (CEX) is based on the same fundamentals as the capture step described in 10.4.2.1, and it is therefore expected that batch experiments can also be used to determine a design space for all CPPs that are seen in protein load, conductivity of load and wash buffer, elution pH, and peak collection window [36]. Studies have shown that small-scale batch experiments could well characterize parameters in continuous CEX [41].

10.4.2.4 Flow-Through Anion Exchange Chromatography

A continuous *anion exchange chromatography* (AEX) step is typically identical to the batch reference process but often operated at lower flow rates; columns or adsorbers are – depending on the design strategy – cycled through multiple operation and regeneration steps every day. CPPs of AEX in batch chromatography are seen in feed pH and conductivity, load capacity, and operating flow rate [36]. While load capacity and flow rate determination follow the same rationale in continuous operation, possible fluctuations in flow rate and pH or conductivity of the feed solution need to be considered for the design space. Also, the system design with its operating strategy should be included in the small-scale assessments early on.

10.4.2.5 Ultrafiltration and Diafiltration

In integrated downstream platforms, *single-pass tangential flow filtration* (SP-TFF) is used to adjust the product concentration and buffer composition. Batch processes typically use standard TFF systems where product is recirculated until it reaches the desired conditions. SP-TFF on the other hand concentrates or exchanges buffer in only one module passage.

CPPs such as flux, pressure differential, and concentration factor are expected to not differ from batch TFF process designs. The operating parameter design typically including *transmembrane pressure* (TMP), *cross-flow flux* (CFF), feed concentration, the amount of diafiltration volumes, and temperature is also crucial for robust processes. For small-scale design of an SP-TFF process, the extended operating times of the continuous process and the potentially higher throughput of liquid per membrane area should be included in the assessment as they can contribute to membrane blocking, fouling bioburden growth, product loss, or insufficient diafiltration.

10.4.2.6 Sterile Filtration

Sterile filtration is one of the most important unit operations in terms of patient safety: it prevents microbial contamination and growth in the drug substance or formulated drug product, and appropriate validation is therefore crucial.

FDA has issued various *question-based review* (QbR) documents [42–44] for different product types; all of the mentioned questions can be applied to batch but also to continuous processing. One main focus of attention should be the special characteristics of CBP in comparison with batch processing: longer filtration times and lower flow rates. Both would require an adjusted design space (DoE) and therefore modified sterile filtration assessment strategies. A brief overview about current proceedings can also be found in the white paper: "A Risk Based Approach to Validation Studies for Sterilizing Filtration and Single-Use Systems" [38].

10.4.2.7 Virus Reduction Filtration

The main difference between viral filtrations in batch and continuous processes is the mode of operation. While a batch process typically runs with constant pressure over a defined time and defined volume and/or flux decay over the filter, continuous viral filtration is operated under reduced, constant flow. Also, continuous virus reduction filtration is conducted mainly by applying low constant flow rates over an extended period of time, which shifts the operating space toward what is commonly regarded as higher-risk operating conditions for viral filtration. Pressure, duration, flux decay, volume throughput, and process interruptions have been identified as CPPs [36]. For continuous bioprocessing applications, low flow and extended filtration times must also be considered as part of the QbD approach for continuous viral filtration [45].

To establish boundaries of the filter design space, it is acceptable to use bacteriophages as an alternative for mammalian viruses as part of the QbD package to determine the worst-case test parameters for virus filter validation studies with mammalian viruses [46]. The multidimensional combination and interaction of input variables (e.g. material attributes) and process parameters have been demonstrated to provide assurance of quality. Prior knowledge will support the identification of the desired design space and can, for example, consist of end user DoE, core validation data, supplier performance data, published literature, platform knowledge, or established scientific principles [47]. The relevance and the justification for use of these data in risk assessments must be considered.

While it has previously been discussed to translate validation results from pressure-controlled batch filtrations to flow-controlled continuous filtrations, based on the common denominator of flow resistance, extensive mammalian virus data and proof of theoretical justification would be required to accept such an approach, according to the regulatory agencies. It is therefore expected that continuous viral filtration requires specific validation strategies with adequate virus-spiking strategies over the full filter lifetime. These are described in more detail in 10.5.2.3.

10.4.2.8 Connection of Unit Operations

In continuous manufacture, the connection of the individual unit operations needs to be considered for the design space. To harmonize fluid flow and fluid composition in between unit operations, a suitable surge container strategy is required. These containers can be placed in between unit operations to (i) balance flow variations of different unit operations, (ii) transform intermittent flow from previous unit operation to constant flow, or (iii) dampen or homogenize fluctuations in fluid composition.

When used to balance flow variations or intermittent product flows as typically observed in bind/elute chromatography or low-pH inactivation steps, surge containers can be sized large enough to hold one elution peak or low-pH inactivation volume. With a certain safety margin to prevent draining and pulling air, this container collects the intermittent effluent and allows following unit operations to continuously draw the product.

Some unit operations such as bind/elute chromatography are expected to naturally deliver liquid with fluctuating composition: an elution peak by default comes with a gradient in protein concentration, pH, and conductivity. It can be challenging to operate or validate the following unit operation with such feed variations, and a dampening or even homogenization of individual elution peaks can be required. It is important to note that this form of homogenization can require a solid rationale to assure the prevention of pooling of batches.

A certain degree of dampening can already be achieved in a single surge container. While this approach is simple in its hardware, single-use, and control requirements, relatively large surge containers are required to minimize concentration fluctuations significantly. In addition, sampling from a single container needs to be appropriately timed to be representative, and, if larger tanks are used, the RTD becomes wider. Alternatively, concentration fluctuations can be dealt with using two alternating surge containers: one container collects product from a unit operation, while the other already mixed product container feeds into the following unit operation. This strategy allows to proceed with homogeneous process conditions at a minimum surge volume with narrow RTD and representative sampling opportunities. It however requires a more complex setup or control; risk of carryover between sub-lots needs to be assessed.

A scientific rationale considering the process requirements, design spaces, and flow or composition harmonization requirements can be the basis of a surge container strategy within the QbD framework.

10.5 Patient Safety

Patient safety is fundamental for biopharmaceuticals and has been raised as a potential difficulty when applying continuous processing concepts. On the one hand, continuous processes require a stricter microbial contamination control in steps that are typically operated under low bioburden but not under sterile conditions. That is to assure that contamination cannot jeopardize the product over the extended operating time of the process. On the other hand, virus safety through chromatography and viral filtration is performed using a changed design space compared with standard batch processing that may require an adaptation of validation principles.

10.5.1 Contamination Control

For consistent operation of a continuous platform, bioburden control is essential and has been identified as a risk especially in downstream operations [24, 48]. While bioburden control through aseptic and closed systems in upstream bioreactors is well understood, downstream operations are typically only operated in a low-bioburden state. For continuous operation, downstream unit operations cover a longer duration than its equivalent batch process, and hence biological contamination could accumulate during a campaign and potentially propagate through unit operations. On the other hand, continuous operation shortens the contact time of the fluid in the unit operation and reduces the hold time in break tanks, which also reduces the contamination risk compared with batch operation. Considering that "lack of sterility assurance" is one of the main reasons for US recalls of sterile and nonsterile products, bioburden considerations for continuous processes are essential [49].

Mitigation strategies for continuous processing can rely on the same principles as bioburden control strategies for batch platforms. This includes measures for different stages of the process, namely, installation, operation, and detection.

Regarding installation, the use of γ-irradiated single-use assemblies and closed system design assures zero initial bioburden at the start of a campaign. Single-use assemblies are expected to be significantly smaller and compact than batch processing, owing to the increase in specific productivity through continuous operating principles [8]. Connections of single-use assemblies can be achieved through sterile connectors or welding to avoid opening the closed fluid path that would require sanitization before use. Certain parts of the platform, for example, protein A chromatography adsorbent, cannot tolerate harsh sanitization conditions. Bioburden reduction through γ irradiation or peracetic acid treatment have been described as options for bioburden control of protein A adsorbent in continuous processing [8]. To further mitigate bioburden risk upon installation, a system suitability and integrity test before use can provide evidence of fluid path integrity.

To assure low bioburden levels during operation, bioburden control filters can be implemented at strategic points of the platform. These could include filters between unit operations or in front of hold tanks, with the aim to segregate segments in the downstream platform and avoid propagation of microbial contaminants. A special attention should be given to the sterility barrier that separates sterile upstream from bioburden-controlled downstream. When integrating filtration steps, it has been discussed that microbial growth through membranes or membrane fouling could occur over the extended operating times [24]. It can therefore be considered to periodically exchange filters based on qualitative and quantitative bioburden risk assessments. Change-out strategies can also apply to certain parts of the single-use equipment such as sensors or tubing exposed to harsh conditions and ideally coincide with strategic points in operation such as transition between lots.

Microbial testing methods at strategic points in the platform and at specific points in time support early detection of contaminated material. A regular monitoring strategy cannot avoid a bioburden issue but allows a timely intervention to isolate the contamination and reduce the risk for the connected steps and the product. Current options and limitations of rapid microbial screening have been described in Section 10.4.1.2.

Continuous processing requires stricter bioburden control extending into the downstream operations of the platform. It is recommended to base the bioburden control on a risk assessment as suggested by the *Parenteral Drug Association* (PDA) and implement bioburden prevention and reduction strategies where necessary [48]. Platforms applying the different means of contamination control in continuous downstream steps have successfully been operated for several days [8].

10.5.2 Virus Safety

A challenge of continuous processing is meeting the regulatory requirement for assuring viral safety, as captured in ICH Q5A and ICH Q9 [50, 51]. Questions arise on how to implement unit operations for viral safety in continuous mode and how to validate in representative scale-down models [39]. Authorities require two orthogonal methods for virus reduction; very often the applied steps include chromatography, low-pH virus inactivation, and virus reduction filtration.

10.5.2.1 Virus Reduction in Chromatography

Viral clearance in multicolumn chromatography is not expected to be affected by the continuous flow operation or different system hardware, as viral clearance has been described as an inherent property of the resin and buffer conditions [39]. However, risk assessments have identified potential challenges for viral clearance caused by (i) higher load volumes and binding capacities, (ii) more frequent regeneration and high cycle numbers, and (iii) possible carryover of viruses through the cyclic operation [24, 35].

For assessing the impact of the higher load volumes and binding capacity, it is recommended to perform spiking studies in small scale [39]. While preliminary data suggests <1 *log reduction value* (LRV) difference between single-column batch loading and multicolumn continuous loading, adapted validation protocols for continuous loading may be required [39, 52]. A simple protocol has been presented by Just Biologics in which virus clearance is evaluated using virus-spiked load and flow-through material on a single bench-scale column [53]. A protocol for a modified spiking study to mimic continuous operation on a batch column could involve primary and secondary columns. In a first step, a primary column is loaded with virus-spiked feed solution. Effluent from the primary column is directed to a secondary load column to simulate the typical preloading in continuous chromatography. While the primary column is discarded, the secondary load column continues in the process by receiving virus-spiked feed and moving through all wash and elution steps. The eluted fraction is used to determine the *LRV* of the chromatography step. This scale-down model based on one or two columns is considered adequate for viral clearance studies of any multicolumn chromatography process, regardless of the number of columns in operation [35, 53].

Regarding the frequent cycling of chromatography resin in continuous chromatography, scientific data shows that resin age in protein A and ion exchange chromatography does not impact viral clearance [54, 55]. FDA therefore concluded that a clearance study would not necessarily need to be the same duration as the continuous operation itself and that end-of-processing performance could be predicted in batch operation under accelerated conditions [39]. Given that the resin is regenerated adequately, the cyclic behavior of continuous chromatography is not expected to lead to carryover of virus. This has been demonstrated over three cycles of multicolumn chromatography spiking studies with bacteriophages PR772 and PhiX 174 in a four-column protein A mAb process [52]. Virus reduction in continuous chromatography could therefore be validated with very similar protocols compared with batch chromatography.

10.5.2.2 Low-pH Virus Inactivation

Virus inactivation in continuous mode can be achieved with in-process mixing and hold. The strategy is based on at least two alternating mixers where one mixer collects the elution peaks continually eluted from the chromatography system while the other mixer incubates collected product. Once the product of one mixer is inactivated, the mixer is emptied and starts collecting fresh material. In the meantime, the second mixer that has been filled with product can start the inactivation cycle.

With in-process mixing and hold being based on a repetitive mini-batch operation that relies on the same fundamentals, the validation is expected to be possible in batch operation. Additional risks such as over-incubation through carryover or long-term equipment suitability may be addressed separately in the validation; see Section 10.4.2.2 [40].

10.5.2.3 Virus Reduction Filtration

To implement continuous viral filtration, it must be considered that the volumetric throughput and operating time of the filter is limited and that a replacement of the filter may be required. A risk-based approach can support the decision of the filter sizing and its operating time. Viral filtration in continuous operation has been successfully demonstrated for multiple weeks, but with increased product throughput, the impact of a post-use integrity test failure increases. One strategy to reduce this risk but also to address flow decay, pressure increase, or virus breakthrough due to clogging is to use an automated switch in/out system as described by FDA [39]. In such a system, devices are replaced by fresh modules before they reach the validated maximum volumetric throughput and/or flux decay.

To validate a virus filter switch in/out approach, process-specific factors such as the lower flow rates and pressures than batch operation and potential process interruptions need to be considered. Also, the same as in batch operation, product-specific factors need to be included in the setup. The validation method thereby needs to reflect the continuous process conditions.

Alternatively, the use of a virus filter that is not switched out is also an acceptable approach for virus safety. However, this requires additional validation considerations, including an understanding of virus stability and product aggregation over time, which may require traditional virus filter validation methods to be adapted.

For this latter approach, considerations for selecting a suitable validation strategy including initial spiking, spike replenishment, and in-line spiking have been described; see Figure 10.6 [45]. The initial spiking method follows the concept of applying one spike per study and represents the most simplistic validation strategy presented. It can however only be applied if the virus keeps stability throughout the study. In case the virus titer is lost over time, a spike replenishment strategy can be applied. Here, virus spike and process fluid are replenished in defined time intervals. This strategy requires additional sampling and adds complexity to determine the overall virus clearance. The third strategy presented is developed for settings with prefilters that can reduce the virus titer and may lead to product aggregation. In-line spiking requires careful consideration of flow dynamics to ensure adequate mixing of the virus, especially under low flow conditions [45].

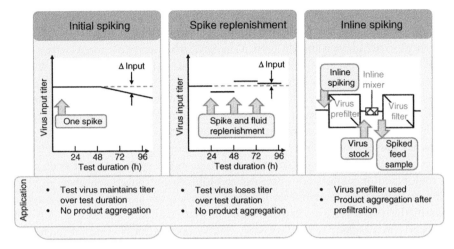

Figure 10.6 Three strategies for viral clearance in continuous viral filtration. Source: Modified from McAlister et al. [45].

10.6 Equipment Design

Fully closed systems are key for successful and bioburden-controlled implementation of continuous bioprocessing platforms. This can be achieved through adopting single-use systems that provide flexibility and facilitate sterile connections; however, they can also introduce risks linked to equipment robustness or leachable and extractable control.

With regard to equipment durability, an exchange of equipment might be necessary within the process: chromatography columns can decline in their separation performance, dead-end filters can create a high back pressure due to blocking or fouling, or parts of the single-use fluid pathway can reach the end of their validated lifetime. There is also a higher risk of sensor failures or of general outtakes in comparison with batch processes, owing to the longer processing times. Therefore, special effort should be put in to develop strategies to (i) minimize the probability of equipment failures through, appropriate sizing of filters, (ii) detect failures or abnormal equipment performance, and (iii) enable an easy and safe exchange of the affected components [4].

A simple concept to allow the exchange of components without process interruptions is the implementation of redundant equipment in a switch-in/switch-out setting. Preconfigured equipment assemblies can be used to replace exhausted equipment; they can be connected to the process stream by using sterile connectors and disconnectors or by simple tube welding. Such a concept is applicable to a variety of process steps including filtration devices, chromatography columns, and SP-TFF modules. With the appropriate auxiliary equipment, a parallel switch-in/switch-out assembly can enable online sanitization, conditioning, or integrity testing.

Due to the extended operating times, a sensor recalibration in the process may become necessary. Sensor readings can shift or drift over time: glass surfaces from UV sensors can nonspecifically bind proteins or chemical components from the product stream, or pH electrodes can move out of calibration [56–58]. In continuous processes with strict bioburden control, exchanging or removing a sensor for calibration is not an option. Instead, the means to detect the shift, verify readings, and apply corrective actions in closed systems is required. A verification of a sensor reading is, for example, possible through an independent monitoring system with a separate sensor for feedback control. This secondary sensor monitors and verifies proper operation of the system independently and detects potential drift early on [59]. Another method is to pass process buffers that have been checked with calibrated offline equipment, through the in-line sensors at certain time intervals, to verify the sensor readings and allow for one-point calibration if needed.

To perform a patient safety toxicology risk assessment, it is essential to characterize leachables and understand the possible interactions they may have with the drug product. According to FDA, qualification of equipment is expected to be performed on both separate unit operations and the integrated system to demonstrate equipment suitability for its intended purpose [4]. The qualification protocols should thereby reflect the operating conditions such as temperature, fluid components, and contact time. Industry guidance on risk-based extractable and leachable profiling is provided in the BioPhorum Operations Group (BPOG) protocol that uses a QbD approach to identify compounds [60] and the United States Pharmacopeia 1665. The QbD approach starting with well-characterized materials and prior knowledge can simplify equipment qualification and risk assessment of leachables.

It is expected that a continuous platform provides a better understanding of where leachables arise and get removed. Assessments of the impact of extended processing times of various continuous bioprocessing steps on the leachables profile suggest only little risk [61, 62]. The high throughput of liquid in continuous operation has even indicated to lower the risk of equipment-related leachables compared with batch operation [62]. To establish a leachable and clearance profile, simulation runs for continuous platforms could be performed as described by Song et al. [61].

10.7 Conclusion

With manufacturers moving toward implementation of continuous concepts under cGMP, regulatory implications gain importance. While regulators stress the need for flexible biomanufacturing platforms and actively support the implementation of continuous or integrated bioprocessing strategies, no formal regulatory guidance is available today. However, the continuous process is not inherently different from batch processing. Both processes rely on the same fundamental concepts and unit operations. The difference lies in how the concepts are applied in the process.

Continuous processes use the same chemistry as batch operations that leads to very similar CPPs and design space rationales developed thereof. Consequently, process characterization in small-scale batch DoE can be used as a profound basis for a continuous process. Also, parameters optimized through the intensification such as reduced contact times and increased binding capacity in bind/elute chromatography can be evaluated under batch conditions. Additional assessments are suggested to evaluate the impact of the cyclic nature of integrated process steps. This applies, for example, to chromatography steps where the risk of material carryover from one cycle to the next needs to be assessed.

Contamination control of integrated platforms is fundamental especially in multiday operations and can be achieved through the same strategies as applied in batch processing today: prevention and reduction of bioburden introduction through closed single-use systems and bioburden control filters. Virus safety is also achieved in the same manner as in batch processing – through chromatography, low-pH inactivation, and virus reduction filtration. Study results have shown similar removal capacities for continuous operating systems. Validation principles known from batch processing are a solid basis for continuous processes. Additional idea is to be put into adapting the validation protocols to fully reflect the continuous operating principle.

To establish a continuous bioprocessing platform that complies with the cGMP regulatory requirements, it is recommended to conduct a comprehensive risk assessment covering the entire continuous bioprocessing platform, inclusive of the interconnections and automation. Manufacturers who have been early adopters of continuous bioprocessing have successfully implemented means for traceability, consistency, and patient safety in continuous processes. With more manufacturers following suit, the industry is gaining extensive knowledge and deeper understanding of continuous processing and ways to meet regulatory expectations.

References

1 BiosanaPharma. BiosanaPharma gets approval to start phase I clinical trial for a biosimilar version of omalizumab. https://www.biosimilardevelopment.com/doc/biosanapharma-gets-approval-to-start-phase-i-clinical-trial-for-a-biosimilar-version-of-omalizumab-0001 (accessed 19 January 2021).

2 Walther, J., Godawat, R., Hwang, C. et al. (2015). The business impact of an integrated continuous biomanufacturing platform for recombinant protein production. *J. Biotechnol.* 213: 3–12.

3 Pollard, D., Brower, M., Abe, Y.L.A.G., and Sinclair, A. (2016). Standardized economic cost modeling for next-generation MAb production. *Bioprocess Int.* 14 (08): 14–23.

4 FDA (2018). *Quality Considerations for Continuous Manufacturing – Guidance for Industry*. Rockville, US: U.S. Department of Health and Human Services.

5 Pfister, D., Nicoud, L., and Morbidelli, M. (2018). *Continuous Biopharmaceutical Processes: Chromatography, Bioconjugation, and Protein Stability*. Cambridge: Cambridge University Press.

6 Lee, S.L., O'Connor, T., Yang, X. et al. (2015). Modernizing pharmaceutical manufacturing: from batch to continuous. *J. Pharm. Innov.* 10: 191–199.

7 Brower, M. (2016). Scale-up of continuous chromatography using Cadence BioSMB process system. In: *BioProcess International Conference.* Boston, US.

8 Ötes, O., Flato, H., Vazquez Ramirez, D. et al. (2018). Scale-up of continuous multicolumn chromatography for the protein A capture step: from bench to clinical manufacturing. *J. Biotechnol.* 281 (09): 168–174.

9 FDA (2017). *Advancement of Emerging Technology Applications for Pharmaceutical Innovation and Modernization. Guidance for Industry.* Rockville, US.

10 FDA. (2019). Code of Federal Regulations 21 CFR 210.3. Rockville, US.

11 European Medicines Agency. ICH Q7 good manufacturing practice for active pharmaceutical ingredients. London, UK, 2000.

12 Falconbridge, A. (2019). Implementation and development of continuous processing. In: *Driving Value Through Intensified Biprocessing.* UK: Oxford.

13 Godawat, R., Konstantinov, K., Rohani, M., and Warikoo, V. (2015). End-to-end integrated fully continuous production of recombinant monoclonal antibodies. *J. Biotechnol.* 213, 10 (11): 13–19.

14 Pennings, M. (2019). Viral clearance validation for a fully continuous manufacturing process for Phase 1 studies. In: *ICB IV Conference.* Cape Cod.

15 Hernan, D. (2017). Continuous manufacturing: challenges and opportunities. EMA perspective. In: *3rd FDA/PQRI Conference on Advancing Product Quality.* Rockville, US.

16 Pinto, N. (2020). *Automated Material Traceability in End-to-End Continuous Biomanufacturing for Batch disposition.* BPI.

17 Medendorp, J. (2019). Real time release testing and drug product continuous manufacturing. In: *ECA Continuous Manufacturing Conference.* Berlin, DE.

18 Sencar, J., Hammerschmidt, N., and Jungbauer, A. (2020). Modeling the residence time distribution of integrated continuous bioprocesses. *Biotechnol. J.* 5 (8): 1–12.

19 Taylor, L. (2019). Developing a fully automated and integrated continuous downstream process. In: *BioProcess International Conference.* Boston, US.

20 David, L., Schwan, P., Lobedann, M. et al. (2020). Side-by-side comparability of batch and continuous downstream for the purification of monoclonal antibodies. *Biotechnol. Bioeng.* 117 (4): 1024–1036.

21 Forno, G. and Ortí, E. (2017). Quality control and regulatory aspects for continuous biomanufacturing. In: *Continuous Biomanufacturing: Innovative Technologies and Methods,* 513–530. John Wiley & Sons.

22 Kuribayashi, R., Noritaka, H., Harazono, A., and Kawasaki, N. (2012). Rapid evaluation for heterogeneities in monoclonal antibodies by liquid chromatography/mass spectrometry with a column-switching system. *J. Pharm. Biomed. Anal.* 67: 1–9.

23 Fisher, A.C., Kamga, M.-H., Agarabi, C. et al. (2019). The current scientific and regulatory landscape in continuous biopharmaceutical manufacturing. *Trends Biotechnol.* 37 (3): 253–267.

24 Dechema (2020). *Continuous Bioprocessing in Up- and Downstream Processing: Technical State of the Art and Risk Analysis.* DE: Frankfurt am Main.

25 Rathore, A. and Kapoor, G. (2015). Application of process analytical technology for downstream purification of biotherapeutics. *J. Chem. Technol. Biotechnol.* 90: 228–236.

26 Botonjic-Sehic, E. and Brooke, H. (2019). Advanced technologies and computational modeling in continuous bioprocessing. In: *BioProcess International Conference*. Boston, US.

27 Chmielowski, R., Mathiasson, L., Blom, H. et al. (2017). Definition and dynamic control of a continuous chromatography process independent of cell culture titer and impurities. *J. Chromatogr. A* 1526: 58–69.

28 BioPhorum Operations Group (2017). *In-line Monitoring and Real-time Release*. London, UK: BioPhorum.

29 Rathore, A. and Singh, S. (2015). Use of multivariate data analysis in bioprocessing. *Biopharm. Int.* 28 (6): 26–31.

30 Pearson, K. (1901). On lines and planes of closest fit to systems of points in space. *Philos. Mag.* 2 (11): 559–572.

31 Abdi, H. and Williams, L.J. (2010). Principal component analysis. *WIREs Comput. Stat.* 2 (4): 433–459.

32 Bisschops, M. (2015). Capturing the value of continuous bioprocessing through MVDA. In: *Umetrics User Group Meeting*. Princeton.

33 Vannuffelen, I. (2019). The quality journey to continuous manufacturing. In: *ECA Continuous Manufacturing Conference*. Berlin, DE.

34 European Medicines Agency (2012). *Guideline on Real Time Release Testing*. London, UK: EMA Committee for Medicinal Products for Human Use.

35 M. Bisschops. (2020). Regulatory aspects of connected and/or continuous downstream processing. *Pall 2020*. https://www.pall.com/en/biotech/webinars/regulatory-aspects-continuous-downstream-processing.html (accessed 19 January 2021).

36 CMC Working Group (2009). *A-Mab: A Case Study in Bbioprocess Development Version 2.1*. International Society for Pharmaceutical Engineering ISPE.

37 Capito, F. (2019). Capturing the future: use of continuous chromatography for downstream processing of biologics-benefits & challenges. In: *ECA Continuous Manufacturing Conference*. Berlin, DE.

38 Pall Corporation. (2020). *A risk based approach to validation studies for sterilizing filtration and single-use systems*. https://www.pall.co.uk/content/dam/pall/biopharm/lit-library/non-gated/white-papers/USD3353-Risk-Based-Approach-Validation-Sterilizing-Filtration-Single-Use-WP-EN.pdf (accessed 19 January 2021).

39 Johnson, S., Brown, M., Lute, S., and Brorson, K. (2017). Adapting viral safety assurance strategies to continuous processing of biological products. *Biotechnol. Bioeng.* 114: 21–32.

40 Schofield, M. and Jones, K. (2018). *Assessing Viral Inactivation for Continuous Processing*. Cranbury, NJ: BioPharm International.

41 Utturkar, A., Gillette, K., Sun, C.-Y. et al. (2019). A direct approach for process development using single column experiments results in predictable streamlined multi-column chromatography bioprocesses. *Biotechnol. J.* 14 (4): 4.

42 FDA Center of Drug Evaluation and Research (2014). *Chemistry Review of Question-Based Review (QbR) Submissions*. Rockville, US: U.S. Food and Drug Administration.

43 FDA (2014). *Question-based Review (QbR) for Sterility Assurance of Aseptically Processed Products: Quality Overall Summary Outline*. Rockville, US: U.S. Food and Drug Administration.

44 FDA (2014). *Sterility Assurance Quality Summaries Not ANDA-Essential*. Rockville, US: U.S. Food and Drug Administration.

45 M. McAlister, N. Jackson and S. de Backer (2019) Virus filtration in continuous bioprocessing – considerations for filter design space and validation strategies. https://biotech.pall.com/en/posters-presentations/virus-filtration-continuous-filter-design-space-validation.html (accessed 19 January 2021).

46 Kleindienst, B., Kosiol, P., and Manzke, A. (2019). Continuous processing: challenges and opportunities of virus filtration. *Pharm. Technol.* 1: 38–40.

47 European Medicines Agency (2017). *Joint BWP/QWP Workshop with Stakeholders in Relation to Prior knowledge and its Use in Regulatory Applications*. London, UK: European Medicines Agency.

48 BioPhorum Operations Group (2019). *Continuous Downstream Processing for Biomanufacturing – An Industry Review*. London, UK: BioPhorum.

49 Parenteral Drug Association. (2015). Bioburden and biofilm management in pharmaceutical manufacturing operations. *Technical Rep. No. 69*.

50 ICH (1999). ICH guideline Q5 A: viral safety evaluation of biotechnology products derived from cell lines of human or animal origin. In: *International Conference for Harmonisation of Technical Requirements for Registration of Pharmaceuticals for Human Use*. Geneva, CH.

51 ICH (2015). ICH guideline Q9: quality risk management. In: *International Conference for Harmonisation of Technical Requirements for Registration of Pharmaceuticals for Human Use*. Geneva, CH.

52 Chiang, M.-J., Pagkaliwangan, M., Lute, S. et al. (2019). Validation and optimization of viral clearance in a downstream continuous chromatography setting. *Biotechnol. Bioeng.* 9 (116): 22292–22302.

53 Connell-Crowley, L. (2018). Considerations for virus clearance validation. In: *Continuous Manufacturing for the Modernization of Pharmaceutical Production: Proceedings of a Workshop*. Washington, DC: National Academies Press.

54 Miesegaes, G., Lute, S., Read, E., and Brorson, K. (2014). Viral clearance by flow-through mode ion exchange columns and membrane adsorbers. *Biotechnol. Progr.* 30: 124–131.

55 Lute, S. and Brorson, K. (2009). Bacteriophage and impurity carryover and total organic carbon release during extended protein A chromatography. *J. Chromatogr. A.* 1216: 3774–3783.

56 Undey, C., Low, D., Menezes, J.C., and Koch, M. (2011). *PAT Applied in Biopharmaceutical Process Development and Manufacturing: An Enabling Tool for Quality-by-Design*. CRC Press.

57 Eibl, R. and Eibl, D. (2019). *Single-Use Technology in Biopharmaceutical Manufacture*. John Wiley & Sons.

58 Teixeira, A.P., Oliveira, R., Alves, P.M., and Carrondo, M.J.T. Advances in online monitoring and control of mammalian cell culture: supporting the PAT initiative. *Biotechnol. Adv.* 27: 726–732.

59 Bull, K. (2006). A simple solution for sensor drift. *Pharm. Technol.* 30 (3): 2–3.

60 BioPhorum Operations Group (2017). *Best Practices Guide for Evaluating Leachables Risk in Biopharmaceutical Single-Use Systems*. Sheffield, UK: BioPhorum Operations Group.

61 B. Song, L. Liu and X. Gjoka (2018). Extractables and leachables in a continuous processing system. file:///C:/Users/BadertscherB/Desktop/20-07535-Extractables-Leachables-Continuous-Processing-System-A4-POS-EN.pdf (accessed 19 January 2021).

62 Hauck, A. (2019). The Prediction of extractables and leachables and the fate of leachables in intensified and/or continuous biopharmaceutical manufacturing. In: *BioTech 2019*. Wädenswil, CH.

Index

a

accelerated seamless antibody purification (ASAP)   183
acoustic waves   148–149
adsorption mechanism   158, 284, 285
adsorptive hybrid filters (AHF)   150
advanced process control (APC)   121, 266
advanced therapy medicinal products (ATMP)   95, 96
alternating tangential flow filtration (ATF)   46, 299
American type culture collection (ATCC)   107
anion exchange (AEX)   53
APC-based autonomous operation   323, 328
application programming interfaces (API)   114
aqueous two-phase extraction (ATPE)   270, 299, 312, 329
artificial intelligence (AI)   60–61, 75, 76, 79, 86–88, 114
asset performance management (APM)   103
axial dispersion coefficient   290

b

baculovirus expression vector system (BEVS)   105
batch and fed batch cultivation
  conventional approaches   7–8
  feeding and control strategies   8–9

batch, definitions   353–354
Bayesian optimization   244
big data   218, 219
bind elute cation exchange chromatography (CEX)   362
biobetter   179
biologics market   180
biomanufacturing   189, 190, 213
biomanufacturing 4.0   97
biomass concentration   23
biopharmaceutical industry   212
biopharma market dynamics   200, 201
bioprocess intensification   98, 100, 102
  advanced process control (APC)   121
  artificial intelligence   114
  automation   115–117
  autonomation   115–117
  bioprocess control   112–113
  bioprocess optimization   106–107
  bioprocess simplification   107–108
  bioreactor design   121–122
  cloud/edge computing   114
  commercialized systems   120–121
  continuous bioprocessing (CB)   108–109
  design of experiments   118
  digital biomanufacturing   110–112
  digital twins   113
  enterprise resource management (ERM)   103
  facilities   123–126
  genetic engineering   104–105
  high-throughput systems   119–120

bioprocess intensification (*contd*.)
  improved process and product development 118
  materials media optimization 109–110
  materials variability 110
  methods 120
  modeling 114–115
  new expression systems 105–106
  PAT 119
  process monitoring 117–118
  QbD 119
  single-use systems 122–123
  sustainability synergy 102–103
  synthetic biology 104–105
bioprocess monitoring 247, 248
bioreactor design 121–122
biosensors 117
biosimilar market 180
biosolve process model 197
black-box model 233, 234
body-feed filtration 150
budding market 201, 202
buffer handling 298
bulk drug substance (BDS) 160, 161

c

Cadence in-line concentrator (ILC) 216
Cadence in-line diafiltration (ILDF) 216
capillary electrophoresis mass spectroscopy (CE–MS) 118
capture chromatography 51
cation exchange (CEX) 53
cell clarification device 46
cell culture, process intensification 214
cell density of infection (CDI) 106
cell discard rate (CDR) 42
cell disruption 15
cell free fluid (CFF) 140
cell line development 186, 188
cell retention 46, 147
cell retention and harvest
  acoustic waves 148–149
  centrifuges 148
  floating filters 148
  hydrocyclones 148
  settlers 149
  tangential flow filtration (TFF) 147–148
cell separation and clarification 273–278, 306
cell-specific perfusion rate (CSPR) 42
centrifuges 148
chemostat cultures 11, 13
chemostats systems 11
chromatography 50, 53, 153–159, 282, 293, 314–318
  fiber-based cartridges 156
  flow through and tandem chromatography 157–158
  membrane absorbers 155–156
  mixed matrix membranes (MMM) 156–157
  mixed mode chromatography 158–159
  monoliths 155
  prepacked columns 154–155
citric acid cycle (CAC) 43
clean-in-place (CIP) 107
cloud/edge computing 114
codon optimization 104
coiled flow inverter (CFI) reactor 152
coiled flow inverter reactor (CFIR) 151
computational fluid dynamics (CFD) 44, 115, 192
computer-aided biology (CAB) 118
continuous bio-manufacturing (CBM) 9, 11
  capture and LLE 273–278
  vs. cell cultures 13–14
  cell separation and clarification 273–278
  chromatography 282–293
  E. coli
    plant usage 21–23
    subpopulations formation 27–29
  lyophilization 293–295
  mass-balancing and macroscopic effects 11–13
  membrane adsorption 282–293

in microbial   14–16
modelling and control strategies   19–20
PAT   320–338
precipitation/crystallization   282–283
SPTFF   278–281
subpopulation monitoring   16–18
UF/DF   278–281
USP fed-batch and perfusion   273–275
continuous bioprocessing (CB)   108–109, 352
continuous countercurrent tangential chromatography (CCTC)   164
continuous dead-end filtration   50
continuous downstream processing   220
continuous manufacturing   255, 257
continuous monitoring   84
continuous population balance equations (PBE)   253
continuous validation   84
continuous virus inactivation   362
contract manufacturing organizations (CMOs)   211
critical control points (CCP)   163
critical process parameters (CPP)   5, 80, 82, 119, 163, 220
critical quality attributes (CQA)   5, 40, 80, 82, 119, 162, 220
crossflow filtration   159
cross flow flux (CFF)   363
cryopreservant   143
Cryovault   162
cultivation mode   5, 6
current Good Manufacturing Practices (cGMP)   107

d

Darwin's principles   23
data analysis   358–359
decision-making process   78
depth filtration   149–151
  filter aids   150
  flocculation   150
  precipitation   150–151

design of experiments (DoE) approach   220
diafiltration processes   216
diethylaminoethyl (DEAE)   156
digital biomanufacturing (DB)   97, 110–112, 193
digital manufacturing   97, 222, 223
digital twin-based control   46
digital twins   113, 269, 295
distributed control systems (DCSs)   103
DMFCA model   45
DNA impurities   182
downstream process   180, 181
  continuous   184–185, 189
  development   188–189, 220–221
  process intensification in   214–216
  sizing   182–184
downstream process (DSP)   140, 358
downstream processing unit operations
  chromatography   153–159
  depth filtration   149–151
  drug substance freezing   161–162
  inline buffer blending and dilution   152–153
  inline virus inactivation   151–152
  tangential flow filtration   159–161
drug substance freezing
  in bags   161–162
  in bottles   161–162
  in containers   162
DSP sensors applications   164–165
dynamic mass balance model   45
dynamic soft sensor   250–252

e

E. coli   8, 9, 21, 23
  genomic integration   26
  genotypic diversification   23–25
  phenotypic diversification   25–26
elementary flux analysis (EFM)   45
end-to-end upstream   189
enterprise asset management (EAM)   103
enterprise resource management (ERM)   103

enzyme-linked immunosorbent assay (ELISA) 320
equipment design 369, 370
ethylene and vinyl alcohol (EVAL) 157
expanded bed adsorption (EBA) 182
extractionoemdash process modeling 276
extreme pathways (EPs) 45

f

fed-batch cell culture 235
fed batch cultivation 140
fiber-based cartridges 156
filter aids 150
first principle models (FPMs) 233
floating filters 148
flocculation 150
flow cytometry 16
flow through and tandem chromatography 157, 158
flow through anion exchange chromatography 363
fluid dynamics 277
fluidized bed centrifugation (FBC) 47, 148
Fourier-transformed-infrared-spectroscopy (FTIR) 321
fragment antigen binding (fABs) 21
fuzzy control systems 60

g

general rate model chromatography 282, 284
gene therapy 226, 227
genetic engineering 104–105
genomic integration 26
glycan analysis 194
glycosylation 270
good manufacturing practices (GMP) 88

h

HIC-SMA process model 287, 289
high cell density perfusion process 214
high-performance liquid chromatography (HPLC) 358

high pressure liquid chromatography (HPLC) 51
high-throughput bioprocess development (HTPD) techniques 120
high-throughput systems 119–120
host cell protein (HCP) 182, 358, 362
human machine interface (HMI) 112
hydrocyclones (HC) 148
hydrophobic charge induction chromatography 285
hydrophobic interaction chromatography (HIC) 53, 285, 299

i

iCCC method 314
independent primary test method (PTM) 85
industrial biotechnology 6
industrial internet of things (IIoT) 97, 114
Industry 1.0 212
Industry 2.0 212
Industry 3.0 212
Industry 4.0 97, 212, 213
infrared spectroscopy 320
inline buffer blending and dilution 152–153
In-line conditioning (ILC) 196, 197
inline DF (ILDF) 161
inline diafiltration 160–161
inline virus inactivation 151–152
In silico modeling 114
integrated counter current chromatography (iCCC) 299, 318, 325, 327
integrated process model (IPM) 19
internal rate of return (IRR) 125
Internet of Things (IoT) 77
ion exchange (IEX) 299
isoelectric point (pI) 151

j

just-in-time (JIT) system 81

k

Kalman filter 250, 251
kinetics 277

l

liquid chromatography–mass spectrometry (LC-MS) 110
liquid–liquid extraction (LLE)
  continuous biomanufacturing 273–278
  process integration 306
log reduction value (LRV) 367
lower molecular weight components (LMWC) 306
lyophilization 271, 293–295, 314, 319

m

machine learning (ML) based control 60–61
manufacturing cost of goods 197, 198
manufacturing execution systems (MES) 103
manufacturing vs. lab scales 252, 253
mass spectrometry (MS) 110
mass transfer coefficient 284
mass transfer kinetics 292
matrix assisted laser desorption/ionization (MALDI) 117
matrix-assisted laser desorption/ionization time of flight mass spectrometry (MALDI-TOF-MS) 44
membrane absorbers 155–156
membrane adsorption 282–283, 293, 314–318
metabolic flux analysis (MFA) 45
metabolite-based control 44, 45
metal-organic framework (MOF) 157
Michaelis-Menten kinetic 12
microbial biotechnology 3
mixed matrix membranes (MMM) 156–157
mixed mode chromatography (MMC) 158–159, 220, 221, 285
model parameter determination 289, 291
model predictive controllers (MPC) 121, 322

modified Henderson-Hasselbalch 287, 288
modified mixed mode 285
monoclonal antibodies (mAbs) 21, 40
  artificial intelligence control 60–61
  continuous chromatography control 50–53
  continuous dead-end filtration, control of 49–50
  continuous formulation 56–57
  continuous precipitation 54–56
  continuous upstream and downstream 46–48
  continuous viral inactivation 53–54
  high-level monitoring and control 59–60
  machine learning control 60–61
  process 353
  process digitalization 62–63
  process integration 48–49
  statistical process control 61–62
  surge tanks 57–59
  upstream mammalian bioreactor control 40–46
Monte Carlo simulations 19
multi-attribute method (MAM) 107
multicolumn protein A chromatography 361–362
multimodal mode chromatography (MMC) 158
multiple input multiple output (MIMO) 19
multitask elastic nets (MEN) 240
Multivariate Data Analysis (MVDA) 223, 357, 358, 360

n

N-1 bioreactor 214
Nelder-Mead estimation 287
net present value (NPV) 125
non classical IBs (ncIBs) 15
non-linear model predictive control (NMPC) 60
normal operating ranges (NOR) 5
N-1 perfusion 144–146

o

operations performance management (OPM)   103
optical character reader (OCR)   80
out-of-specification test results (OOS)   77
out-of-trend test results (OOT)   77
oxygen uptake rate (OUR)   43

p

patient safety, contamination control   365, 366
perfusion bioreactors   181
perfusion systems   10
periodic counter-current chromatography (PCC)   183
periodic forcing   25
phase equilibrium   277
phase-equilibrium isotherms   290
pH-shift   285
pH titration   153
physico-chemical models   271
PLS2 approach   240
plug flow reactors (PFR)   254
polycationic flocculation agent polydiallyldimethylammonium chloride (pDADMAC)   150
polyethylenglycol (PEG)   151
polyhexamethylene biguanide (PHMB)   156
polymerase chain reaction (PCR)   107
posttranslational modifications (PTMs)   21
precipitation/crystallization   282–283, 311, 313
principal component analysis (PCA)   223, 239
process analytical technology (PAT)   97, 119, 162–165, 222, 321
  in continuous biomanufacturing   320–338
  control strategy   332–335
  evaluation and summary   337–338
  process control   357–359
  process simulation   323–328
  QbD-based control strategy   322–324
  spectroscopic methods, applicability of   328–332
  tools   16, 18, 45

process analytical tools   186
process characterization studies (PCS)   5
process consistency   355
process control
  automation   356–357
  data analysis   358–359
  downstream processing   358
  PAT approach   357–359
  real time release testing   360
  upstream processing   358
process control system (PCS)   298
process design space   238
process development
  DoE   267
  QbD   266
  strategy   267
process dynamics   232, 234, 237
process economy   199, 200
process integration   295, 300
  capture and LLE   306
  cell separation and clarification   306
  chromatography   314–318
  CPD vs. experimental data   319–320
  lyophilization   314–319
  manufacturing platforms   216–217
  membrane adsorption   314–318
  precipitation/crystallization   311–313
  SPTFF   309–311
  UF/DF   309–311
  USP fed-batch and perfusion   301–306
process intensification
  in biomanufacturing   213
  in cell culture   214–216
  in continuous manufacturing   192–194
  current bioprocessing   140
  degree of   138
  development   141–142
  in downstream processing   214–216
  downstream processing unit operations   149–162
  in DSP   169
  fourth dimension   217
  general aspects of   140–141
  impact costs   169–170
  multi-product perfusion platform   190–191
  PAT and sensors   162–165

perfusion process 192
  in seed train development 192
  strategies 139
  and SU technology 180
  theory and practice 137–139
  time, influence on 170–171
  transition from traditional to intensified 165–169
  upstream processing unit operations 142–149
  in vitro requirements of 232
process optimization 242–247
process traceability 353
  batch and lot definitions 353–354
  deviation management 354–355
  lot traceability and deviation management 354–355
product quality attributes (PQA) 163
product quality prediction 238–257
product resistance 293–296
product temperature 295
pulse-response measurements 355
pure black box (PLS) model 237

q
quality assurance (QA) 80, 223–224
quality by design (QbD) 5, 42, 80, 97, 119, 360, 361
  bind elute cation exchange chromatography 362
  continuous virus inactivation 362
  flow through anion exchange chromatography 363
  multicolumn protein A chromatography 361–362
  PAT control strategy 322–324
  principles 220
  sterile filtration 363
  ultrafiltration and diafiltration 363
  unit operations, connection of 364–365
  virus reduction filtration 363–364
quality management, principle 81
quality risk management (QRM) 82
quality target product profile (QTPP) 82, 162
question based review (QbR) documents 363

r
radio frequency identification (RFID) transponders 163
rational design 104
raw material attributes (RMA) 5
real time release testing (RTRT) 162, 320, 360
recombinant protein formation 27
recombinant protein production (RPP) 13, 24
recombinase-mediated cassette exchange (RMCE) 104
red-but-not dead phenotype 16
residence time distribution (RTD) 354
retentostat cultivation 9
retentostat systems 11
ribozyme engineering 104
RNA interference (RNAi) 104
root mean squared error in prediction (RMSEP) 235, 237, 241, 242
run-to-run optimization 243, 244

s
Savitzky-Golay algorithm 249
scale down 233, 254, 255
scale up 233, 253, 254
seed-train intensification
  N-1 perfusion 144–145
  wave bag expansion 144
settlers 149
simulated moving bed (SMB) 183
single input single output (SISO) 61
single nucleotide polymorphism (SNP) 24
single pass diafiltration (SPDF) 160
single pass tangential flow filtration (SPTFF) 56, 159–160, 216, 270, 278–281, 363
  continuous biomanufacturing 278–281
  process integration 309–311
single-use manufacturing 194, 195
single-use systems 122–123
single use technology (SUT) 180
  cost management 195–196
  limitations of 198–199
size exclusion chromatography (SEC) 284, 320

SMART technologies 114
soft sensors 247, 248
 dynamic 250–252
 static 248–250
soft-sensors 118
SPC implementation 61
static soft sensor 248–250
statistical process control 61–62
statistical process control (SPC) charts 83
steam-in-place (SIP) 107
steric mass action (SMA) models 285
sterile filtration 363
stirred tank reactor (STR) 254
substrate-based control 44, 45
supervisory control and data acquisition (SCADA) 112
support vector regression, radial-basis kernel function (SVR-rbf) 240
surge tanks 57–59
sustainability synergy 102–103
SWOT-analysis 267
synergistic prediction 242, 243
synthetic biology 104–105

t

tangential flow filtration (TFF) 46, 147–148, 160, 216
 inline diafiltration 160–161
 single pass TFF 159–160
thermal conductivity 293
time-space yield should (TSY) 6, 21, 23
transcription activator-like effector nucleases (TALENs) 104
transcription induced mutation 24
transmembrane pressure (TMP) 363
turbidostat 301
Turing Test 76, 77

u

ultrafiltration/diafiltration (UF/DF) 56, 160, 270, 299, 363

continuous biomanufacturing 278–281
 process integration 309–311
ultra-performance liquid chromatography (UPLC) 358
upstream process (USP) 140, 180, 181
 continuous 184–185
 development 188–189
 fed-batch and perfusion 273–275, 301–306
 process control 358
 sensors applications 163
 sizing 181–182
 unit operations
  cell retention and harvest 145–149
  high density in bags 143–144
  large volume cell banking in bags 143–144
  seed-train intensification 144–145

v

viable cell densities (VCD) 142
virtual sensors 118
virus filtration 151
virus-like particles (VLP) 105, 106
virus reduction filtration 363–364, 368
virus safety
 in chromatography 367
 low pH virus inactivation 367–368
 virus reduction filtration 368

w

water for injection (WFI) 168, 197
water properties 295
wave bag expansion 144
wave bioreactor 141, 144, 197

y

yeasts 14, 105

z

zinc finger nucleases (ZFNs) 104

Printed and bound by CPI Group (UK) Ltd, Croydon, CR0 4YY